AF389204

Life Cycle Prediction and Maintenance of Buildings

Life Cycle Prediction and Maintenance of Buildings

Editors

Jorge de Brito
Ana Silva

MDPI • Basel • Beijing • Wuhan • Barcelona • Belgrade • Manchester • Tokyo • Cluj • Tianjin

Editors
Jorge de Brito
University of Lisbon
Portugal

Ana Silva
University of Lisbon
Portugal

Editorial Office
MDPI
St. Alban-Anlage 66
4052 Basel, Switzerland

This is a reprint of articles from the Special Issue published online in the open access journal *Buildings* (ISSN 2075-5309) (available at: https://www.mdpi.com/journal/buildings/special_issues/ Life_Cycle_Prediction_Maintenance_Buildings).

For citation purposes, cite each article independently as indicated on the article page online and as indicated below:

LastName, A.A.; LastName, B.B.; LastName, C.C. Article Title. *Journal Name* **Year**, *Article Number*, Page Range.

ISBN 978-3-03936-728-3 (Hbk)
ISBN 978-3-03936-729-0 (PDF)

Cover image courtesy of Ana Silva.

Contents

About the Editors . . . . **vii**

Jorge de Brito and Ana Silva
Life Cycle Prediction and Maintenance of Buildings
Reprinted from: *Buildings* **2020**, *10*, 112, doi:10.3390/buildings10060112 . . . . **1**

Mihail Vinokurov, Kaisa Grönman, Simo Hammo, Risto Soukka and Mika Luoranen
Integrating Energy Efficiency into the Municipal Procurement Process of
Buildings—Whose Responsibility?
Reprinted from: *Buildings* **2019**, *9*, 45, doi:10.3390/buildings9020045 . . . . **7**

Miguel Macedo, Jorge de Brito, Ana Silva and Carlos Oliveira Cruz
Design of an Insurance Policy Model Applied to Natural Stone Facade Claddings
Reprinted from: *Buildings* **2019**, *9*, 111, doi:10.3390/buildings9050111 . . . . **31**

Cláudia Carvalho, Jorge de Brito, Inês Flores-Colen and Clara Pereira
Pathology and Rehabilitation of Vinyl and Linoleum Floorings in Health Infrastructures:
Statistical Survey
Reprinted from: *Buildings* **2019**, *9*, 116, doi:10.3390/buildings9050116 . . . . **49**

Beata Nowogońska
Diagnoses in the Aging Process of Residential Buildings Constructed Using
Traditional Technology
Reprinted from: *Buildings* **2019**, *9*, 126, doi:10.3390/buildings9050126 . . . . **67**

Nicola Moretti and Fulvio Re Cecconi
A Cross-Domain Decision Support System to Optimize Building Maintenance
Reprinted from: *Buildings* **2019**, *9*, 161, doi:10.3390/buildings9070161 . . . . **83**

Vojtěch Biolek and Tomáš Hanák
LCC Estimation Model: A Construction Material Perspective
Reprinted from: *Buildings* **2019**, *9*, 182, doi:10.3390/buildings9080182 . . . . **109**

Jeanette Orlowsky, Franziska Braun and Melanie Groh
The Influence of 30 Years Outdoor Weathering on the Durability of Hydrophobic Agents
Applied on Obernkirchener Sandstones
Reprinted from: *Buildings* **2020**, *10*, 18, doi:10.3390/buildings10010018 . . . . **129**

Roberta Di Bari, Andrea Belleri, Alessandra Marini, Rafael Horn and Johannes Gantner
Probabilistic Life-Cycle Assessment of Service Life Extension on Renovated Buildings under
Seismic Hazard
Reprinted from: *Buildings* **2020**, *10*, 48, doi:10.3390/buildings10030048 . . . . **147**

Michael A. Lacasse, Abhishek Gaur and Travis V. Moore
Durability and Climate Change—Implications for Service Life Prediction and the
Maintainability of Buildings
Reprinted from: *Buildings* **2020**, *10*, 53, doi:10.3390/buildings10030053 . . . . **169**

Elaheh Jalilzadehazhari, Georgios Pardalis and Amir Vadiee
Profitability of Various Energy Supply Systems in Light of Their Different Energy Prices and
Climate Conditions
Reprinted from: *Buildings* **2020**, *10*, 100, doi:10.3390/buildings10060100 . . . . **189**

Pietro Croce, Paolo Formichi and Filippo Landi
Influence of Reinforcing Steel Corrosion on Life Cycle Reliability Assessment of Existing
R.C. Buildings
Reprinted from: *Buildings* **2020**, *10*, 99, doi:10.3390/buildings10060099 . . . . **207**

**Steinar Grynning, Klodian Gradeci, Jørn Emil Gaarder, Berit Time, Jardar Lohne
and Tore Kvande**
Climate Adaptation in Maintenance Operation and Management of Buildings
Reprinted from: *Buildings* **2020**, *10*, 107, doi:10.3390/buildings10060107 . . . . **227**

About the Editors

Jorge de Brito has worked in the area of service life prediction throughout the last 20 years and has supervised 2 Ph.D. and 10 MSc projects (both completed and ongoing) specifically on this theme. He has co-authored more than 40 papers in international journals (ISI database) and 12 papers in international conferences exclusively on this subject. He is also co-author of a previous book published by Springer. He is Full Professor at IST, University of Lisbon, Portugal, and has a 30+ years' experience in teaching concrete and construction technology related matters.

Ana Silva is co-author of more than 40 scientific publications in indexed journals related with the service life prediction of building components. According to the Scopus database, she is the author with the most articles on "service life prediction" in the world. She is the co-author of the book "Methodologies for Service Life Prediction of Buildings: With a Focus on Façade Claddings" by Springer International Publishing. She is the responsible investigator of the research project entitled "Buildings' Envelope SLP-Based Maintenance: Reducing the Risks and Costs for Owners (BEStMaintenance_LowerRisks)", which has been running since October 2018. She has supervised eight Master theses and one Ph.D. thesis on the implementation of an expert system based on fuzzy logic for estimating the functional life of a homogeneous set of buildings at University of Seville, and is currently supervising three Ph.D. thesis within the scope of service life prediction and the impact of climate change on the durability of buildings' envelope components. She is currently a postdoctoral researcher at IST, University of Lisbon, Portugal.

Editorial

Life Cycle Prediction and Maintenance of Buildings

Jorge de Brito and Ana Silva *

Department of Civil Engineering, Architecture and Georresources, Instituto Superior Técnico, Universidade de Lisboa, Researcher at CERIS, Av. Rovisco Pais, 1049-001 Lisbon, Portugal; jb@civil.ist.utl.pt
* Correspondence: ana.ferreira.silva@tecnico.ulisboa.pt

Received: 20 June 2020; Accepted: 22 June 2020; Published: 23 June 2020

The sustainability of the built environment can only be achieved through the maintenance planning of built facilities during their life cycle, considering social, economic, functional, technical, and ecological aspects. Stakeholders should be conscious of the existing tools and knowledge for the optimization of maintenance and rehabilitation actions, considering the degradation mechanisms and the risk of failure over time. Knowledge concerning the service life prediction of building elements is crucial for the definition, in a rational and technically informed way, of a set of maintenance strategies over the building's life cycle. Service life prediction methodologies provide a better understanding of the degradation phenomena of the elements under analysis, allowing relating the characteristics of these elements and their exposure, use, and maintenance conditions with their performance over time.

This Special Issue intends to provide an overview of the existing knowledge related to various aspects of "Life Cycle Prediction and Maintenance of Buildings". In this sense, 12 original research studies were published, with the relevant contribution of international experts from Canada, Czech Republic, Finland, Germany, Italy, Poland, Portugal, Norway, and Sweden. These outstanding contributions address the maintainability and serviceability of buildings and components, the maintenance and repair of buildings and components, the definition and optimization of maintenance and insurance policies, the financial analysis of various maintenance plans, and the whole life cycle costing and life cycle assessment.

Vinokurov et al. [1] performed a detailed and extensive literature review in order to clarify how municipal building departments can adopt life cycle cost-effective measures to promote energy efficiency and a high-quality indoor climate in buildings. This study is focused on the design phase of the building's procurement process, describing the relationship between indoor climate quality, energy use (and GHG emissions), and the life cycle economy from the perspective of design-related factors. A list of energy efficiency factors that need to be considered in the municipal building procurement process is defined in order to aid practitioners in the selection of a design solution that optimise the value for public money, contributing to a more transparent procurement and decision-making process.

In Macedo et al. [2], an innovative approach for tailoring insurance products is proposed in terms of the risk of failure of the building's components, as well as the financial charges related with the maintenance of these elements, channelling the risks to the market. For this purpose, in this study an insurance policy model applied to natural stone claddings is designed. Deterministic and stochastic service life prediction models were used, considering only the age of the elements or encompassing its different characteristics. This approach intends to identify the insurable risk of the degradation of these claddings, examining how these risks are managed through the insurance method and thus analysing different insurance premiums according to the expected claims and to the risk load. This study provides an interesting approach for the definition of realistic risk-based insurance policies, incorporating mitigation activities through knowledge related to the stochastic performance of the claddings over time. This type of insurance product, considering the risk of failure of the cladding, benefits not only the insurer but also the policy holder.

A statistical survey of the pathology and rehabilitation of linoleum and vinyl floorings is presented in Carvalho et al. [3]. In this study, 101 floorings were analysed in six healthcare facilities in the Lisbon area, Portugal. Healthcare facilities were chosen as case study due to the specificity of the maintenance activities in these buildings. An expert inspection and diagnosis system was created, identifying the most common types of anomalies, their probable causes, the most adequate in situ diagnosis methods, and the most useful repair techniques. Moreover, this information was converted into matrices that relate anomalies and causes, anomalies and diagnosis methods, anomalies and repair techniques, and anomalies with each other. This study identifies the main sensitive concerns regarding the maintenance of these claddings over its life cycle in order to minimise the susceptibility of these floorings to different degradation mechanisms.

Nowogońska [4] proposed an original diagnosis method to describe and predict the aging process of buildings and their components. This methodology intends to characterise the technical condition of the element analysed, predicting changes in the performance characteristics of buildings over their service life. For that purpose, a Prediction of Reliability according to Exponentials Distribution (PRED) approach is adopted, applying Predicted Service Life of a Component (PSLDC) danger curves. The forecasting model, designed to predict the changes in the technical condition of buildings, can be extremely useful in aiding decision-making regarding maintenance works during a building's life cycle. Knowledge related to the aging process of buildings over their service life and the diagnosis of their loss of performance, in terms of their technical condition as well as the reasons behind damage, can be used to define repair needs and to establish adequate maintenance policies.

A cross-domain Decision Support System (DSS) for maintenance optimization was proposed by Moretti and Re Cecconi [5]. In this study, the maintenance optimization is achieved through a wiser allocation of economic resources. For that purpose, four indexes are used: (i) a Facility Condition Index (FCI), (ii) an index measuring the service life of the assets, (iii) an index measuring the preference of the owner, and (iv) another measuring the criticality of each component in the asset. These four indexes are transformed into a Maintenance Priority Index (MPI), which can be used for maintenance budget provision. An average MPI for the whole building can be obtained based on the computation of the MPI of each asset within the building; however, the methodology proposed in this study does not allow comparing different elements among buildings within a portfolio. In this sense, the scalability of the methodology proposed needs to be further investigated. Nevertheless, the DSS proposed could be integrated into a Building Information Modelling (BIM) approach, allowing an effective asset and facility management. Furthermore, with the necessary adaptations, other parameters or metrics could be included in the DSS model in order to aid the prioritization of the maintenance interventions in buildings.

A methodology for building Life Cycle Cost (LCC) estimation, which supports investors in identifying the optimum material solution for their buildings on the level of functional parts, was established by Biolek and Hanák [6]. This methodology encompasses the investor requirements and relates them to a construction cost estimation database and to a facility management database. The methodology proposed is applied and tested for a case study, with a "façade composition" as functional part, with the sublevel "external thermal insulation composite system (ETICS) with thin plaster". The results obtained revealed that there is not a generally applicable optimum ETICS material solution, mainly because differing investors have different requirements and due to the unique circumstances of each building and its users. This study points out different future research directions, essentially: (i) the adoption of sustainable criteria in the selection of the best solution, combining LCC and LCA calculations; (ii) the incorporation of information attained from in-use buildings and BIM models to enable a more comprehensive LCC evaluation.

Orlowsky et al. [7] analysed the durability of 11 different water repellents applied on Obernkirchener Sandstones. The performance of the hydrophobic agents applied is analysed after the samples have been subjected to long-term weathering (30 years of outdoor weathering) in seven different locations in Germany. After 24 and 30 years of outdoor weathering, the treated stone surfaces revealed discolouration and staining. The authors measured the colour changes, identifying the presence of black crusts, the deposition of particles, and biogenic growth, which have caused the gradual darkening and significant changes in the sandstones' colour over time. After 30 years, all the agents show a decrease in performance, but some protective agents still provide an effective hydrophobic layer. Succinctly, the authors [7] concluded that: (i) the protective agents based on isobutyltrimethoxysilane show a clear loss of performance after 2 years of outdoor weathering; (ii) agents containing siloxane, the low-molecular methylethoxysiloxanes, show a good performance, which is similar to, partly better than, that of the oligomer methylethoxysiloxane; (iii) the agents with oligomer siloxane based on an isooctylmethoxy-structure have a higher performance loss than the agents containing low-molecular methylethoxysiloxanes; (iv) after 30 years of outdoor weathering, the effectiveness of the protective agents based on silicone resin is comparable to that of low-molecular siloxanes. Concerning the exposure conditions, the degradation of the treated stones is higher in southern Germany than in North Rhine-Westphalia, mainly due to a longer weathering time of 6 years as well as the rougher environment. On the other hand, in North Rhine-Westphalia, the prolonged exposure to temperatures under 0 °C and relative humidity above 80% leads in general to a higher degradation compared to Duisburg and Dortmund.

Di Bari et al. [8] proposed a methodology to consider the seismic hazard in the enhancement and extend of the buildings service life. For that purpose, a life-cycle-based decision support tool for building renovation measures was created and applied to a selected case study, as a "Proof-of-Concept". A probabilistic approach is proposed in this study in order to overcome the limitations of the "static" analyses; in this sense, the probabilistic methodology proposed allows considering dynamic effects and different sources of uncertainty. This probabilistic approach of life cycle assessment (LCA) and life cycle costs (LCC) analysis can reduce the risk of miscalculation due to uncertainties, while preventing misleading LCA-based decisions. This approach enhances the analyses through the addition of supplementary parameters related to environmental, economic contingencies, and external factors, leading to a more complex model but aiding the practitioners in making more conscious choices. The methodology proposed, performing both probabilistic LCA–LCC analyses, allows evaluating in a more accurate manner the relevance of a seismic retrofit, considering the performance of the construction under seismic actions and the risk of long-term losses due to the lack of a suitable anti-seismic structural system.

In Lacasse et al. [9], the impacts of climate change on the durability and maintainability of building envelope materials and elements are analysed. This study presents a literature review, related with the durability of building envelope components, considering the expected effects of climate change on the longevity and resilience of these components over time. For that purpose, the climate loads expected in the future under different climate change scenarios, were analysed. This study is especially focused on the climate change of Canada. The future climate loads were compared with the climatic effects arising from loads sustained under current historical climate conditions. This study [9] concludes that, in the next few decades, the general climate of Canada tends to become warmer, with some locations experiencing more intense and frequent rain events of longer duration, thus producing heightened wind-driven rain loads. The study provides theoretical specifications for the selection of products given climate change effects, aiding the maintainability and the selection of construction products to achieve climate resilient performance over the buildings' service life.

Jalilzadehazhari et al. [10] evaluated the profitability of a ground source heat pump, photovoltaic solar panels, and an integrated ground source heat pump with a photovoltaic system as three energy supply systems for a single-family house in Sweden. This study evaluates the profitability of the supply systems through the calculation of the payback period (PBP) and the internal rate of return (IRR) for these systems. The IRR and PBP are obtained by considering three different energy prices, three different interest rates, and two different lifespans. Moreover, the profitability of the supply systems was analysed for four Swedish climate zones. The authors [10] concluded that the ground source heat pump system was the most profitable energy supply system, providing a lower PBP and a higher IRR for all the climate zones analysed, when compared with the other energy supply systems. Furthermore, the results reveal that increasing the energy price improved the profitability of the supply systems in all climate zones. This study can be adapted and generalised to countries with similar climate conditions; nevertheless, the cost-effectiveness of the renewable energy resources varies according to the investment costs, the energy prices, and the evolution of energy policies.

The influence of reinforcing steel corrosion on life cycle reliability assessment of existing reinforced concrete structures is analysed in [11]. This study evaluates the influence of different degradation conditions and several reinforcing steel and concrete classes on the time-dependent reliability curves proposed. A special procedure to evaluate material properties and their statistical parameters based on cluster analysis was adopted for the implementation of the method proposed. Croce et al. [11] applied this method to thousands of historical test results dating back to the 1960s, concerning the concrete compressive strength and yield stress of steel rebars, to establish the resistance classes for both materials as well as for the estimation of the related statistical parameters. The application of the methodology is illustrated for significant case studies, consisting of reinforced concrete elements, part of residential, shopping and storage buildings, focusing on the effects of corrosion in steel rebars under different environmental conditions, resulting in no degradation to high degradation effects. The authors emphasize the relevance of the methodology proposed and the results obtained by comparing the time-dependent reliability curves with the target reliability levels currently adopted in the Eurocodes, performing a critical discussion about the results obtained.

Finally, Grynning et al. [12] adopted a multimethod research approach to evaluate the basic criteria, trends, applications, and developments related to climate adaptation in building maintenance and operation management (MOM) practices in Norway. The current status of the application and extent of climate adaptation practices in relation to MOM is analysed. For that purpose, three case studies involving different Norwegian building owner organizations were examined. The results of this study revealed a significant gap between theory and practice regarding the consideration of climate adaptation in MOM. This study reveals that the concept of climate adaptation is only addressed as a high-level strategic issue and that there is a need to incorporate the concept at lower organizational levels. The case studies analysed highlight the need for a generic and structured climate-adaptive MOM framework in order to support the incorporation of climate adaptation in current MOM practices at different scales and organizational levels. This study anticipates that the implementation of this flexible and transferable framework is expected to provide a basis for increasing further knowledge on climate adaptation. Further developments to the proposed model should include the introduction of more tangible and tailored tools and processes, including checklists or scoring systems accompanied by relevant climate adaptation factors and plans.

The editors would like to acknowledge the generosity of all the authors, who gently shared their scientific knowledge and expertise in different fields of knowledge related to various aspects of "Life Cycle Prediction and Maintenance of Buildings". Moreover, the editors would like to express their gratitude to the peer reviewers for their rigorous analysis of the different contributions, who have appreciably contributed to enrich the quality of this Special Issue and, last but not least, the managing editors of *Buildings*, who have continuously supported everyone involved in this Special Issue.

Author Contributions: All authors contributed to every part of the research described in this paper. All authors have read and agree to the published version of the manuscript.

Funding: This research was funded by FCT (Foundation for Science and Technology) through project PTDC/ECI-CON/29286/2017.

Acknowledgments: The authors gratefully acknowledge the administrative and technical support of the CERIS Research Institute, IST, University of Lisbon and the FCT (Foundation for Science and Technology).

Conflicts of Interest: The authors declare no conflict of interest.

References

1.   Vinokurov, M.; Grönman, K.; Hammo, S.; Soukka, R.; Luoranen, M. Integrating Energy Efficiency into the Municipal Procurement Process of Buildings—Whose Responsibility? *Buildings* **2019**, *9*, 45. [CrossRef]

2. Macedo, M.; de Brito, J.; Silva, A.; Oliveira Cruz, C. Design of an Insurance Policy Model Applied to Natural Stone Facade Claddings. *Buildings* **2019**, *9*, 111. [CrossRef]

3. Carvalho, C.; de Brito, J.; Flores-Colen, I.; Pereira, C. Pathology and Rehabilitation of Vinyl and Linoleum Floorings in Health Infrastructures: Statistical Survey. *Buildings* **2019**, *9*, 116. [CrossRef]

4. Nowogońska, B. Diagnoses in the Aging Process of Residential Buildings Constructed Using Traditional Technology. *Buildings* **2019**, *9*, 126. [CrossRef]

5. Moretti, N.; Re Cecconi, F. A Cross-Domain Decision Support System to Optimize Building Maintenance. *Buildings* **2019**, *9*, 161. [CrossRef]

6. Biolek, V.; Hanák, T. LCC Estimation Model: A Construction Material Perspective. *Buildings* **2019**, *9*, 182. [CrossRef]

7. Orlowsky, J.; Braun, F.; Groh, M. The Influence of 30 Years Outdoor Weathering on the Durability of Hydrophobic Agents Applied on Obernkirchener Sandstones. *Buildings* **2020**, *10*, 18. [CrossRef]

8. Di Bari, R.; Belleri, A.; Marini, A.; Horn, R.; Gantner, J. Probabilistic Life-Cycle Assessment of Service Life Extension on Renovated Buildings under Seismic Hazard. *Buildings* **2020**, *10*, 48. [CrossRef]

9. Lacasse, M.A.; Gaur, A.; Moore, T.V. Durability and Climate Change—Implications for Service Life Prediction and the Maintainability of Buildings. *Buildings* **2020**, *10*, 53. [CrossRef]

10. Jalilzadehazhari, E.; Pardalis, G.; Vadiee, A. Profitability of Various Energy Supply Systems in Light of Their Different Energy Prices and Climate Conditions. *Buildings* **2020**, *10*, 100. [CrossRef]

11. Croce, P.; Formichi, P.; Landi, F. Influence of Reinforcing Steel Corrosion on Life Cycle Reliability Assessment of Existing R.C. Buildings. *Building* **2020**, *10*, 99. [CrossRef]

12. Grynning, S.; Gradeci, K.; Gaarder, J.E.; Time, B.; Lohne, J.; Kvande, T. Climate Adaptation in Maintenance Operation and Management of Buildings. *Buildings* **2020**, *10*, 107. [CrossRef]

Article

Integrating Energy Efficiency into the Municipal Procurement Process of Buildings—Whose Responsibility?

Mihail Vinokurov *, Kaisa Grönman, Simo Hammo, Risto Soukka and Mika Luoranen

LUT School of Energy Systems, Lappeenranta University of Technology, FI-53851 Lappeenranta, Finland;
kaisa.gronman@lut.fi (K.G.); simo.hammo@lut.fi (S.H.); risto.soukka@lut.fi (R.S.); mika.luoranen@lut.fi (M.L.)
* Correspondence: mihail.vinokurov@lut.fi; Tel.: +35-850-449-1333

Received: 14 January 2019; Accepted: 11 February 2019; Published: 13 February 2019

Abstract: This study addresses the challenges in ensuring energy efficiency and high indoor climate quality with efficient use of public money in the municipal building procurement process. Energy efficient municipal building procurement provides a significant leverage when steering the built environment towards the low-carbon economy targets of the EU. Municipal building department professionals need more skills and knowledge to appropriately define the requirements and identify the energy efficient design options accounting for the building's changing operational environment. This study presents how to systematically integrate energy efficiency in the municipal procurement process of buildings by presenting the list of energy efficiency factors to be included into the procurement process. This list of factors clarifies how indoor climate quality, energy use, and the life cycle economy are related through technological solutions and how the optimal compromise solution can be determined. Furthermore, this list of factors explains the responsibilities in integrating energy efficiency within the municipal building procurement process. Applied in the design of the municipal building the list of factors contributes to more informed and transparent decision-making process.

Keywords: energy efficiency; indoor climate quality; life cycle economy; changing operational environment; municipal building procurement; climate targets

---

1. Introduction

Buildings are responsible for 40% of the energy consumption and 36% of greenhouse gas (GHG) emissions in the European Union (EU) [1]. Increasing the energy efficiency in buildings is recognized as an important policy objective in the EU for reaching the ambitious GHG emission reduction targets [2]. The Climate and Energy Framework of the EU for 2030 aims to reduce GHG emission reductions by 40% below the 1990 levels, to improve energy efficiency by 27%, and to increase the share of renewables by 27% by 2030 [3]. For 2050, the EU targets to reach a low-carbon economy with GHG emissions cut to 80% below 1990 levels [3].

The Energy Efficiency Directive and the Energy Performance of Buildings Directive of the EU have established a set of binding measures to help the EU reach the required emission reductions cost efficiently. The directives request the public sector to procure energy efficient buildings without compromising the indoor climate quality and to do so with the efficient use of public money [4,5]. Public procurements correspond to 14% of the overall gross domestic product (GDP) off the EU [2,6]. Buildings are a major part of public procurements. With major purchasing power, municipalities have the potential to provide significant leverage when seeking to steer the building market towards energy efficiency improvements [7].

The building department of the municipality is responsible for providing the users spaces with the required functionality while using public money efficiently in the long term [8–10]. In the case of

Finland, the municipal building department, the Department of Municipal Planning and Property Development, initiates and leads the building procurement project [9]. During the procurement process, the municipal building department uses local, national, and international energy efficiency strategies with project-specific requirements and targets while following the regulatory framework [11]. The national building codes of each EU member state provide the minimum energy performance requirements for building procurement projects [12,13]. In addition, each municipal building project has to fulfill the specific functional, economic, safety, cultural, and ecological requirements of the users [10,14]. The general requirements for a building are to provide heating, cooling, ventilation, and lighting so that safety and functionality requirements are met with efficient use of energy and costs [8,10,14].

Despite the regulatory pressure to transit towards energy efficient municipal construction, the procurement of energy efficient and low-carbon buildings remains low [15]. Insufficient understanding of how to integrate energy efficiency into building procurement is among the barriers to sustainable municipal construction [15,16]. Municipal building department officers need more skills and knowledge to appropriately define requirements, qualify suppliers, and identify energy efficient design options [10,15–17]. Unclear responsibility distribution and inefficient communication between the municipal building department and the project partners jeopardizes the success of the project [18]. The municipal building department has to take active leadership over the procurement project to successfully implement energy efficient procurement [17]. Failing to do this during the design-related decisions can partially shift the decision-making ability of the building department to the designers and contractors, which often compromises the quality of the procured building [10,19,20]. There is a demand for clear guidelines defining how to integrate energy efficiency in the municipal building procurement project and what is the responsibility distribution in the project.

Energy efficiency must be addressed through the optimization of design solutions based on the useful output (such as indoor climate quality and GHG emission reductions) that the alternative solution provides with the specific energy input [21–24]. The optimal energy efficiency level may not always reduce energy demand if this is justified by the improved indoor climate quality. To achieve the EU targets, municipal building departments and designers require more clarity on how to holistically assess and optimize the building solutions with the goals of indoor climate quality and the efficient use of both energy and public money [14,15,23,25]. Holistic and systematic frameworks to describe the relationship between indoor climate quality, energy use, and the life cycle economy are needed to identify the optimal compromise design solution [14,24,25].

The tendency to concentrate on capital costs as the main criteria when selecting building design options is another barrier to achieving cost efficient emission reductions in municipal construction [15]. Investment costs typically constitute a quarter of the life cycle costs, while the majority of costs occur during the utilization of the building [15]. Up to four fifths of the total life cycle cost is fixed during the design phase (see Figure 1). The selection of the design solutions has to be based on more advanced, life cycle economy considerations to bring up the viability of the energy efficient design solutions along with the more efficient use of public money. In order to improve the life cycle economy of the building, the design has to accommodate the technological, economic, climatic, and regulatory changes in the building's operational environment in the long term [26]. Ongoing changes in the building's operational environment include decreasing costs of renewable energy technologies, increasing electricity market price volatility, emerging peak power fees, new kinds of demand-response options, and potential GHG reduction-oriented economic steering [27]. The feasibility assessments have to favor design solutions that respond to both present and to future circumstances with a low risk of becoming prematurely obsolete due to the high utilization costs [26,27]. The additional economic elements that need to be included in economic assessments were identified by [27]. The economic assessment should also include the monetizable benefits, such as indoor climate quality, derived from each design option [28]. In municipal service buildings, indoor climate quality can be monetized in the life cycle economy through externalities related to increased productivity and reduced employee sick leaves [29].

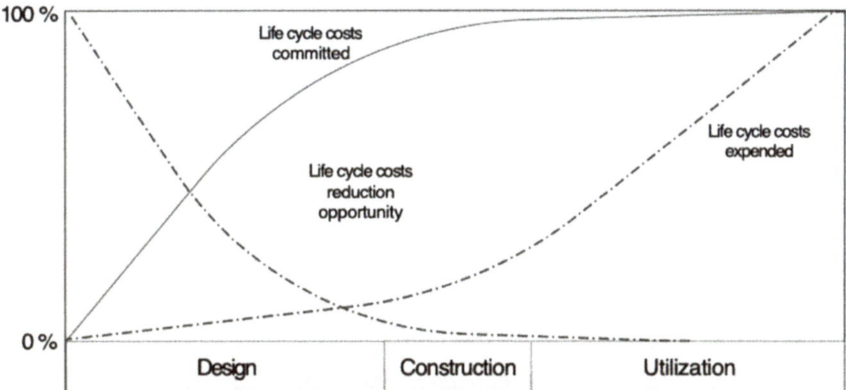

Figure 1. The general behavior of a commitment to life cycle costs [30–35].

This study presents how to systematically integrate energy efficiency into the municipal procurement process of buildings. The study clarifies the procurement process of energy efficient building by answering the following research questions:

- How should one describe energy efficiency factors in relation to the municipal building procurement process?
- What are the responsibilities of each actor at different stages of the municipal procurement process from an energy efficiency point of view?

The municipal building department is provided with a list of design factors that need to be considered when leading the procurement project towards energy efficiency and a high-quality indoor climate, which complies life cycle cost-effectively with international and national targets and regulations. Changes in the building's operational environment are given special emphasis in the list of energy efficiency factors. The list of factors points out which choices made in the design phase affect energy efficiency and clarifies the responsibilities in the building procurement process. Finnish municipal building procurement and building standards are used as an example of applying the list of factors, as these reflect the common EU regulations. The list of energy efficiency factors was developed alongside the actual procurement process of the local kindergarten by utilizing the experience of the municipal building department of the town of Lappeenranta, Finland.

2. Methodology

The step-by-step structure of the conducted research process and the methods applied in each step are presented in Figure 2.

As a starting point of the study, the procurement procedure of an energy efficient municipal building was identified systematically and was described with a review of the relevant literature. The phases of the procurement process were identified where the municipal building department can affect the realization of the energy efficiency. The relevant legislative frameworks providing the basic requirements for new building designs and the municipal procurement practices in EU and Finland were studied. The literature review also included the procurement-related documentation of the Finnish town Lappeenranta. Besides the literature review, the description of the procedure is based on the conversations with the municipal building procurement officers of Lappeenranta. The procedure was described by the phases of the procurement process, along with the phase-specific tasks, the outcomes, and the responsible players.

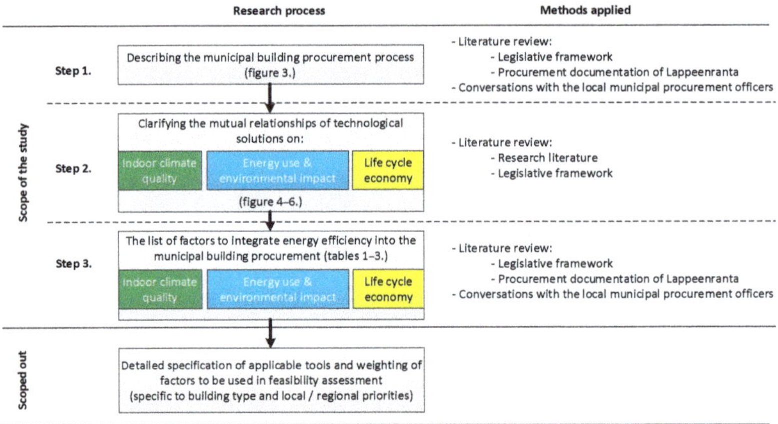

Figure 2. The research process and applied methods.

The relationships between indoor climate quality, energy use (and GHG emissions), and the life cycle economy were described with a review of the relevant research literature and the regulations. Based on the literature review, the design-related factors influencing these three aspects were clarified with figures to ensure the ability of the municipal building department to demand energy efficient solutions in the critical phases of the design. The findings were further structured into the list of factors that need to be considered in the feasibility assessment of alternative design solutions. The structure of the list of factors was developed based on discussions with the local procurement officers involved in the Myllymäki kindergarten procurement project. The paper does not present detailed specifications of tools and weighting of factors to be applied in the feasibility assessment, as these have to be chosen on project-specific basis. Versatility of municipal building types and different local and regional priorities make weighting of factors and tools to be applied difficult to predefine.

3. Results: Integrating Energy Efficiency into the Municipal Building Procurement Process

3.1. Describing Municipal Building Procurement Process—An Energy Efficiency Perspective

As the manager of the building procurement project, the municipal building department is the head responsible for the integration of energy efficiency into the building procurement process. The building department is involved in every step of the procurement process, and the building department's ability to affect the outcome is high. In general, procurement process follows the common framework by the EU and the national procurement legislation, while also some municipality-specific adaptation differences exist [36–40]. The municipal building procurement process is described in Figure 3.

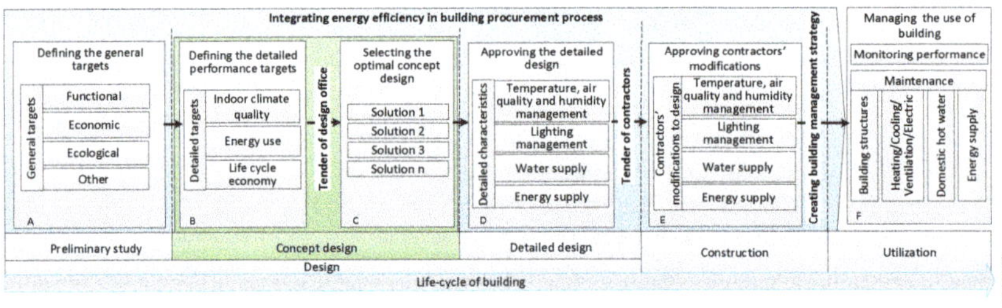

Figure 3. The municipal building procurement process from the municipal building departments' perspective, a Finnish example. A–F order of the process phases. Based on [2,5,7,10,14,36–46].

In the preliminary study, the municipal building department in co-operation with the actual users establishes the general performance targets for the building project (A in Figure 3) [2,7,10]. The performance targets can cover the functionality (i.e., indoor climate quality and purpose of use of spaces) as well as the economic and ecological performance targets [10,20,36]. The national building codes define the minimum performance requirements [36]. The specific requirements of the end user have to be accounted for while establishing the performance targets [10,36].

The concept design (highlighted with green in Figure 3) is the most decisive stage of the building's procurement process form the perspective of the life cycle performance [10]. At the beginning of the concept design, the municipal building department defines more detailed targets for the performance of the building (B in Figure 3) [10]. Detailed performance targets can include specific energy use performance class (in Finland classes A–G), indoor climate quality class (in Finland classes S1–S3), GHG emission reduction targets, and the life cycle economy [10]. Later, the building department carries out the tendering of designers who plan the building [7,10,20]. Energy efficiency targets have to be clearly present throughout the tendering [7,44]. During the tendering of the design work, the building department reviews the received tenders, evaluating the ability of the design offices to fulfill the energy use and other performance targets [7,41,44]. The chosen design team proposes alternative concept design solutions to achieve the performance targets. The municipal building department together with the lead designer iteratively defines the one optimal design solution for further development based on a feasibility assessment (C in Figure 3). The optimal compromise between different performance targets is sought [10,20]. This paper concentrates on clarifying the municipal building department can lead the concept design towards the optimal design solution.

In the detailed design, the lead designer co-ordinates the design process and ensures that all the design elements work together [14]. The design group lead by the lead designer consists of multiple engineers, including the structural designer, the HVAC (heating, ventilation, and air conditioning) designer, the electric designer, and the automation designer. [10,14]. Ensuring that the proposed design solution meets the performance targets and the approval of the detailed design is the responsibility of the municipal building department (D in Figure 3) [10,14,36].

The procurement process proceeds with the construction of the building, led by the main contractor [10,20]. Ensuring the required expertise of the contractors through tendering is an important task of the municipal building department [7,41]. The main contractor ensures that the sub-contractors are aware of the performance targets and that they comply with them [10]. The main contractor may propose some changes to the final selection of the technology [10]. In theory, the proposed alternatives should not compromise the original performance targets. On practice, cheaper solution alternatives with lower quality might be proposed if this benefits the contractor. Co-ordinated by the municipal building department, the lead designer approves the proposed final selections of the equipment ensuring the required quality (E in Figure 3) [10,14]. Here, the impact of the final equipment selection on the performance has to be examined. Also, the energy use, as well as the other performance targets, has to be considered by the building department when establishing a building management strategy and creating the maintenance manual at the end of the construction phase. [7,10,14]

The building management strategy directs the appropriate operations and oversees the building's maintenance while it supports efficient energy use and other performance targets [7,10,14]. The actual performance of the building in terms of providing the purposed service quality and the life cycle costs has to be monitored against of the performance targets during utilization (F in Figure 3).

3.2. Defining Energy Efficiency in the Concept Design of the Municipal Building Procurement

In order to lead the concept design work (the light green in Figure 3) towards energy efficiency, the municipal building department has to be aware of how the indoor climate quality, the energy use, and the life cycle economy are related through technological solutions and energy markets [15,16,47]. This chapter shows how the design-related factors influence these three performance aspects in operational environment of the building undergoing major changes. The same color codes are applied throughout the all figures of this paper to visualize the indoor climate quality performance (dark green), the energy use performance (blue), and the life cycle economy (yellow). A deeper understanding of the relationship between design solutions and performance allows including all the necessary factors in the feasibility assessment of the alternative design solutions [15,16,23,24]. Furthermore, it allows better-informed dialogue between the municipal building department and the lead designer, as the optimal compromise solution is being co-operatively developed.

3.2.1. The Factors Affecting Energy Use and the Environmental Impact

Figure 4 presents how technological solutions define the purchase/sales energy balance and the GHG emissions of the building. The purchase and sales balance for energy and fuel is the basis for the energy efficiency feasibility assessment of alternative design solutions [13,47]. The required energy balance parameters include the purchase energy demand, the peak power demand, and the energy sales to the grid [27]. Technological feasibility depends on the local operational environment of the building and its future changes [27]. Future risk free designs need to accommodate the changes in the climate, energy costs, fuel costs, as well as the availability of energy service models and the regulatory constraints over the long term [26,27].

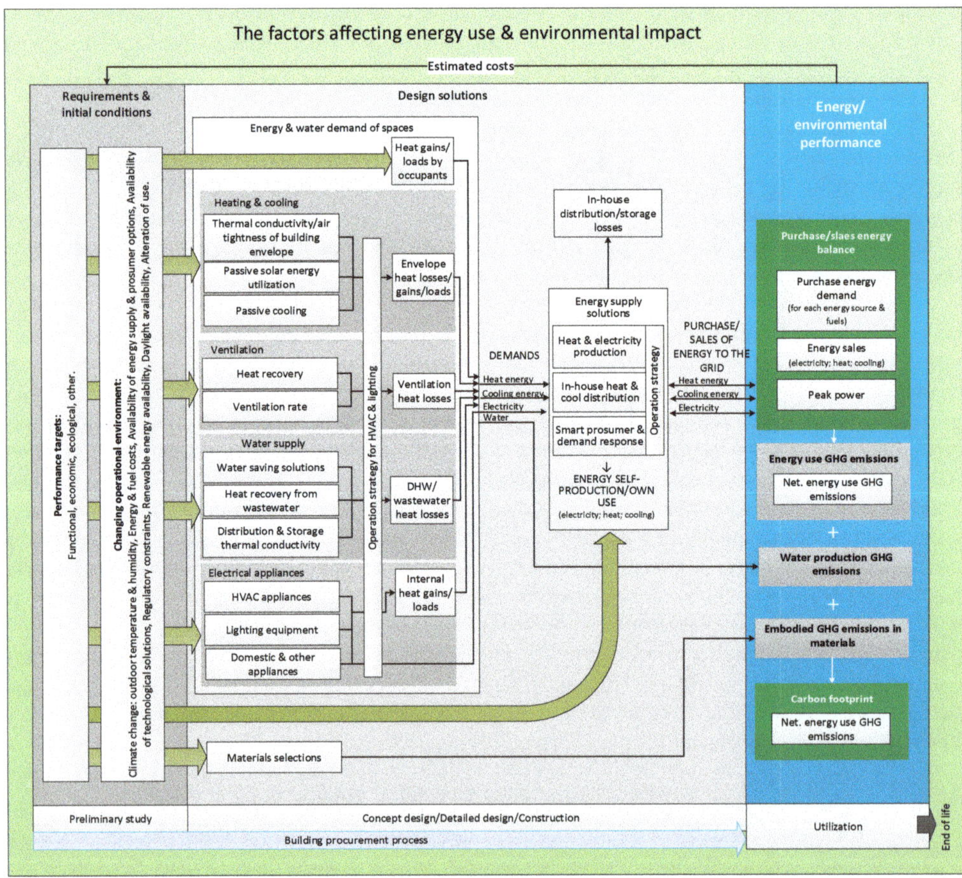

Figure 4. The identification of the factors affecting energy efficiency and the environmental impact of a building.

The heating energy and the cooling demands of the building are determined by the various heat losses, loads and gains [2,13,47]. Reducing heating and cooling demand mitigates the economic risks related to the unpredictable development of energy prices (generation charges, distribution network charges, and tax) [27]. Ventilation is a major heat loss component causing 30–60% of the energy demand in buildings in industrialized countries [48]. Up to 90% of the ventilation heat losses can be recovered with modern ventilation heat recovery systems, which reduce the heat demand [48]. Both heat recovery solutions, the ventilation-to-ventilation and ventilation-to-central water heating, can be applied. Energy efficient ground source heat pump-based pre-cooling, pre-heating, and after heating technologies can be applied to reduce the purchase heating energy demand [27]. The demand-based operation strategy and automation can further reduce the heat losses of ventilation, which would reduce the building's heating demand. In addition, heating and cooling demand can be reduced with the building envelope materials that have improved thermal insulation and air tightness [27,49,50]. The thermal inertia of buildings' structures can be utilized to cut heating and cooling energy demand peaks [27]. Domestic hot water (DHW) typically consumes between 10 and 20% of the heating demand for a typical house from the late twentieth century in industrialized countries. [51]. The heating demand of DHW can be reduced through water saving solutions and through minimizing the heat losses from the in-house domestic hot water distribution and the storage system [51,52]. The utilization of passive energy, such as passive solar energy, reduces the heating energy demand of the building [27]. Building envelope solutions, such as window sizing and orientation, as well as potential shadings and blinds, have to be optimized to reduce the purchase energy demand for heating and cooling [53].

Electrical appliances, including lighting, domestic appliances, and HVAC systems, in turn, create the demand for electrical energy [2]. High energy efficiency LED lighting solutions and high energy efficiency class (A+++) domestic appliances strongly reduce purchase electricity demand compared to incandescent bulbs [27]. Window sizing and orientation can be optimized for natural light utilization to reduce the demand for artificial electric lighting [54,55]. Buildings with a lower frame depth usually have better potential to utilize natural light. It is important to optimize illuminous intensity so that the visual performance required by the purpose of the building's use is met. An increase in lighting intensity beyond the optimal level will increase electricity demand without further improving the visual performance. Passive and active thermal radiation blocking solutions can be used to avoid increasing electricity demand for air conditioning [54]. The electrical energy consumption can be reduced by demand-based automation and by the operation strategy of the HVAC and the lighting equipment. Utilizing both system components with increased energy efficiency (frequency controlled fans, pumps etc.), and optimal dimensioning to increase the overall energy efficiency of the system, further reduces the electric energy demand [10,47,50].

The purchase/sales energy balance depends on the applied energy supply technology [13]. Energy supply solutions include centralized and decentralized heating, electricity, or cooling supply based on renewable or non-renewable sources. Minimizing purchase energy and fuel demand mitigates the economic risks related to energy costs [27]. The purchase energy demand can be reduced by efficient energy supply solutions, such as ground source heat pumps. Ground source technology can also be efficiently applied for cooling [55]. Since this technology provides heating throughout the year and scales down the peaks of electricity purchase demand, it mitigates the risks related with increasing peak power charges [27]. Self-produced renewable heating and electric energy reduce the purchase demand for more expensive grid energy [27,56]. The surplus of the self-produced heat and electricity can be sold into the local grid by using appropriate real-time metering and bi-directional connection, or it can be stored for future own use [57–61].

Efficient smart prosumer and demand response solutions can be introduced due to the emerged dynamic consumption/sales metering and IoT (Internet of Things) solutions. Smart metering allows the dynamic pricing of purchased and sales energy [58,62]. IoT solutions allow the control of devices' energy use based on real time online energy price data. The smart demand-side management technologies improve the flexibility of the building's energy use by shifting energy use to off-peak periods when electricity spot prices are lower [27,61,63]. Furthermore, reducing peak power demand mitigates economic risks related to increasing peak power fees [27]. Heat and electricity storages reduce peak power use and maximize the use of self-produced heat and electricity. The electric vehicles can be used as electricity storages through smart charging interfaces. [27,62–64]

The carbon footprint of the design solutions include GHG emissions related to energy use, water production, and emissions embodied in the materials [65]. Emissions of energy use can be estimated based on the purchase/sales energy balance of the building. Accountable emissions include production emissions of purchased energy and fuels used by the building as well as the abated emissions achieved by selling the excess self-produced energy to the grid [27]. Carbon footprint should include emissions embodied in the selected construction materials through production, transportation, installation, maintenance, replacement, demolition, and waste treatment [66]. The GHG emissions need to be estimated to account for the economic impacts of present and future emission-reduction oriented economic steering [2].

3.2.2. The Factors Affecting Indoor Climate Quality

Indoor climate quality has to be considered during the feasibility assessment of design solution alternatives when converting energy use into energy efficiency, as in Figure 5. In European countries, around 76% of the energy consumed in buildings goes to indoor climate comfort control—heating, ventilation, and air conditioning [67]. Energy efficiency can be considered by comparing the service quality that the alternative solution provides with the specific energy input [15,24,31].

Temperature comfort is commonly determined by the indoor temperature level, including temperature profile stability within the building and the actual sensation of the temperature [68,69]. Maintaining indoor temperature comfort in Finland is technologically challenging due to the difficult climatic conditions and the effects of climate change. Finland has high annual fluctuations of temperature, with cold winters and increasingly hot summers. In the case of Finland, the dimensioning of heating and cooling solutions has to be based on the outdoor temperature varying from below $-30\ {}^{\circ}C$ in the winter to above $+30\ {}^{\circ}C$ in the summer. Heating demand is high, and cooling demand is growing due to hot summer periods, which are forecasted to become longer and more frequent [70–72]. Heating and cooling demands are also affected by the building envelope design, including thermal insulation and openings [10,12,64]. Heating and cooling dimensioning have to account for internal heat loads, such as from occupants as well as from lighting and various electrical appliances [10,12]. The indoor temperature in occupied zones has to be comfortable and temperature comfort should not be disturbed by draught, heat radiation, uneven temperature profiles, temperature fluctuation, or surface temperatures [12]. The temperature profiles can be modified by applying different types of distribution systems for heating, cooling, and ventilation [64]. The actual temperature feeling, in turn, is highly affected by the perceptions of the individuals themselves. The presence of draught increases the sensation of coldness and decreases temperature comfort and should be avoided [12,72]. Sources of draught can be caused by structural air leakages or by supply air distribution issues, such as the velocity of the supply air being too high. The negative sensation of draught can be reduced by increasing the temperature of the air through heating. This, however, leads to an increase in heating energy demand. Furthermore, excess heating decreases the relative humidity of air, which can cause respiratory problems [69]. Structural solutions, the selection and positioning of heating radiators and air distribution devices, and demand-controlled automation all affect temperature comfort [64]. Also, the indoor air humidity level affects the temperature sensation.

Indoor air contains a certain amount of different impurities of both external and internal origin [2]. Among the internal sources of air impurities are those related to respiratory activity and substances released by specific building materials [69]. Ventilation has a major role in managing indoor air pollutant concentration by removing pollutant-containing air and replacing it with cleaner fresh air [12,69]. A ventilation operation strategy can include the timing of the ventilation rate in accordance with the actual needs and the control of the ventilation rate with CO_2 sensors. An efficient operation strategy increases indoor air quality while reducing energy demand. Ventilation equipment and construction materials rated with Finnish M1 emission classification do not emit hazardous substances or unpleasant odors into indoor air and are easily cleansable [73]. Increased ventilation rates two weeks before starting the building's use reduces the indoor air's concentration of impurities derived from the new construction materials [10]. Supply air filters of various efficiency levels are used to prevent impurities from spreading into the building from outdoor air, external loads [12,74]. Filtration is also used for recycled air. An increase in the filtration efficiency, however, commonly leads to higher energy costs due to the increased electricity consumption of the fan [75]. Thus, the filter selection depends on the targeted indoor climate quality and on the outdoor air pollution conditions at the location.

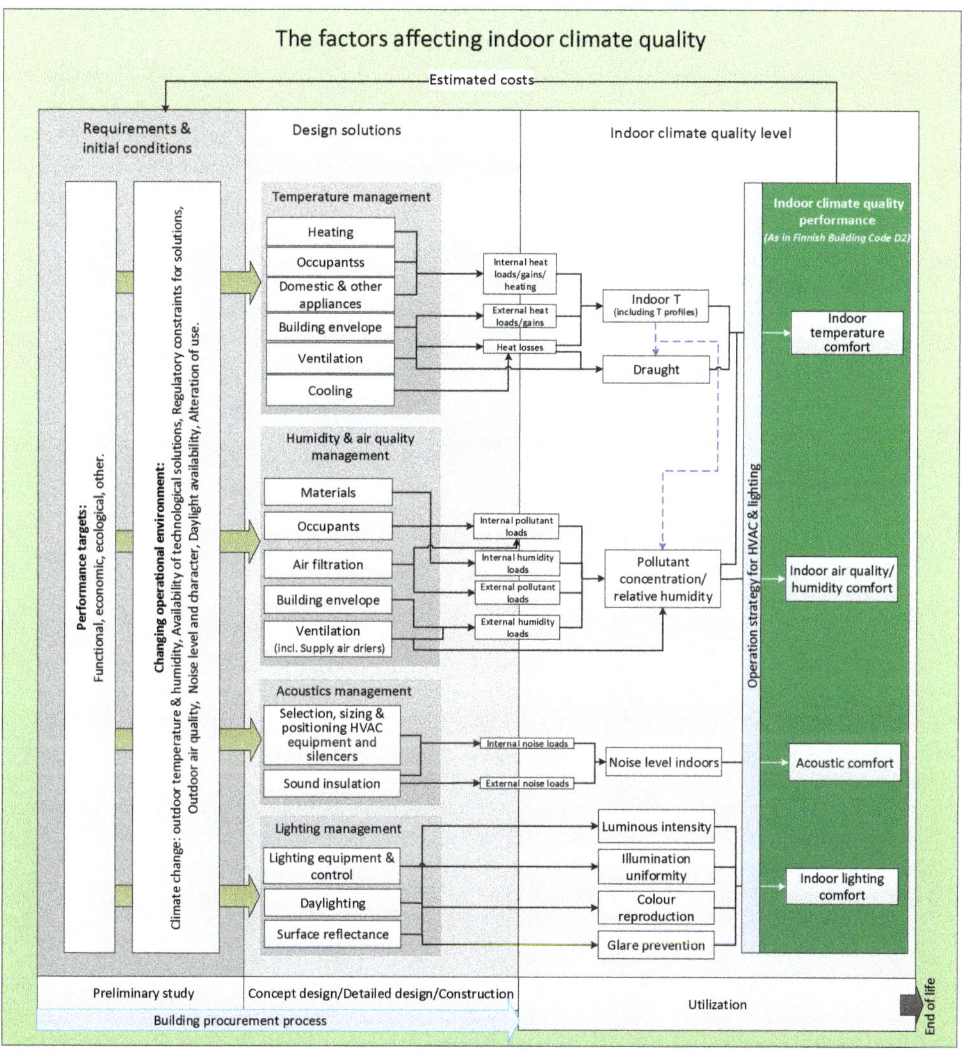

Figure 5. The identification of the factors affecting indoor climate quality in a building.

The relative humidity has to be maintained so that it does not cause moisture damage to the building's structures or mold growth, which could lead to respiratory symptoms among occupants [12,43]. The condensation of humidity inside the components of the ventilation system also has to be prevented. Low relative humidity, below 25%, might increase allergic reactions and respiratory inflammations [69]. Mold growth can occur with a relative humidity level above 70% [69]. The optimal relative humidity indoors should be between 25 and 45% [69]. The humidity loads in the building can be of an internal or external nature. In Finland, the external humidity loads are expected to grow in the future, as rainfall is becoming more frequent and the rising temperatures, caused by climate change, keep the water unfrozen for longer [71,76]. Ventilation is the core mean to remove the excess humidity from the indoor air, which keeps the relative humidity within the optimal level [10,43,47,64]. Supply air driers can be applied as part of the ventilation system to reduce external humidity loads. Reducing internal humidity loads decreases the need for excess ventilation, which lowers the building's purchase heating and electricity energy demand [27]. Post-construction structural humidity, released into indoor air, can be reduced by protecting the construction materials from elements during storage and construction [10]. Increased external humidity loads require the careful design of damp proofing and drainage. Building envelope materials with good hygroscopic performance reduce the risks related to the relative humidity due to the ability of these to balance short-term humidity fluctuations [27,77]. Appropriate ventilation pressure difference control is important to prevent outdoor humidity from infiltrating the envelope structures, which causes moisture damage [78].

Acoustic conditions are an important element of indoor climate quality. It is necessary to minimize the background noise of both an external and internal origin [79]. Typical noise sources are traffic, different human activities, and internal HVAC machinery [80]. Minimizing sound reverberation is also important, especially in spaces like classrooms [81]. The main solutions for acoustics management include utilizing sound insulation materials, positioning of sound-emitting devices from rooms with high acoustic requirements, and employing sound-absorption materials to reduce reverberation [80,81]. Silencers might be needed to be installed into the ventilation ducts to reduce the duct noise. Duct designs with fewer silencers have lower pressure losses, and the electricity demands of fan are reduced. If possible, positioning the building from significant sources of noise can also positively affect the indoor acoustic conditions.

Indoor lighting conditions are known to affect the productivity of people working in the space [82]. The perceptions of lighting are highly related to individual physiological and psychological characteristics [83]. The color characteristics of light in a space are determined by the spectral power distribution of the light source and by the reflectance properties of the surfaces in the room. The desired uniformity of the illumination depends on the type of activities in the space and the space itself. Lighting that is too non-uniform can cause distraction and discomfort [82]. Illumination uniformity can be affected by the lighting equipment, such as the type of bulb, the positioning and direction of the lighting equipment, and the reflectance of the different surfaces in the space [83]. Daylighting not only reduces the electrical energy demand of artificial lighting but also improves the color reproduction of indoor light, which increases the lighting comfort (Figures 4 and 5) [54,84]. Other factors that contribute to lighting quality include illuminance uniformity, luminance distribution, light color characteristics, and glare [64,85].

3.2.3. The Factors Affecting the Life Cycle Economy

Alternative solutions can be compared based on their ability to provide the quality of the indoor climate and GHG emissions reductions from the perspective of life cycle economy [2,10,36,47,86]. The life cycle economy assessment of alternative design solutions should include the costs elements presented in Figure 6.

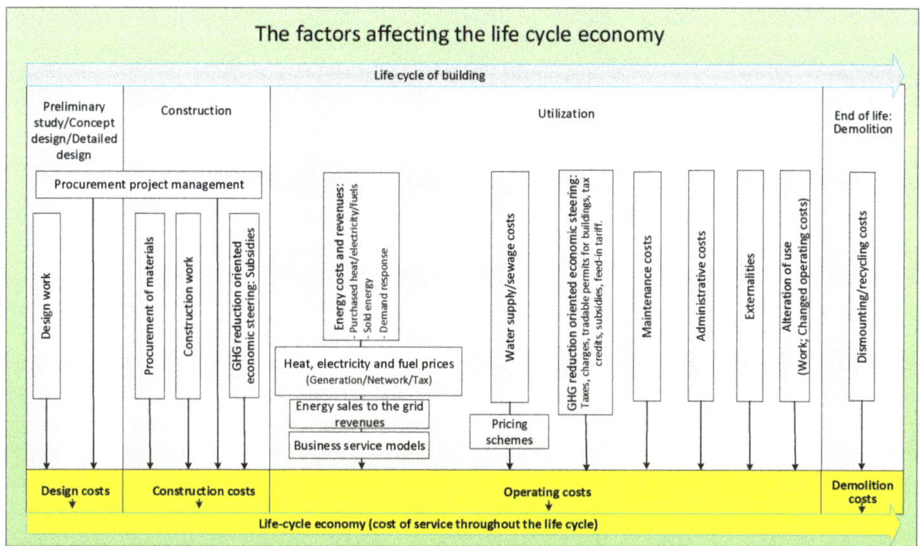

Figure 6. The identification of factors affecting the life cycle economy of a building. GHG: greenhouse gas.

The design phase is where the most important conceptual decisions that will have a crucial impact on the overall life cycle economy of the building are made [10,27,32]. The life cycle driven design process has to include the consideration of the economic opportunities and risks provided by the changing operational environment of the building [26–28]. Attention has to be given to the energy sector, as energy expenses dominate the operating costs. Climate change causes an increase in annual average temperatures, in annual temperatures' deviation, and in outdoor humidity [70–72]. These affect building's future costs of heating, cooling, and humidity management. The development of energy prices (generation charges, distribution network charges, and tax), the decreasing costs of decentralized renewable energy technologies (production and storages), emerging service business models and new potential GHGs reduction-oriented economic steering pose new economic risks and opportunities for the building's life cycle economy [26,27]. New economic steering mechanisms for curbing climate change might include taxes, charges, emission trading, or similar tradable permit schemes for real estate. Neglecting these changes might harm the life cycle economy of the building through high operating expenses and low resale value [26,27,87].

Considering opportunities and risks for the building's life cycle economy throughout the design phase is fundamental to defining the "future proofed" energy supply solution in the changing operational environment of the building [26,27]. The development of purchase energy (generation charges, distribution network charges, and tax) prices greatly affects the feasibility of the solutions aimed at energy conservation, energy use flexibility, and renewable energy self-production [26,27,88,89]. The future energy bills are affected by increasing maintenance fees, available energy pricing schemes, and increasing peak demand volatility of electrical energy spot prices and peak power charges [88,89]. The revenue from energy sales defines the profitability of renewable energy self-production technologies as well as the optimal sizing of production in relation to energy demand. The revenue from energy sales to the grid are affected by energy spot price development and by the available energy sales pricing schemes [90,91]. The development of spot prices, peak power fees, and business service models will affect the feasibility of demand response solutions steering energy purchases and sales towards profitable energy market prices and reducing peak power demand [26,27,92–96].

The municipal building department must be aware of the impact of potential low-carbon oriented economic steering on utilization costs. If introduced, carbon taxes, charges, subsidies, or tradable emission permit schemes for buildings will affect the profitability of energy supply investment in favor of renewable and low-carbon solutions [27,97].

The life cycle driven design has to pay specific attention to indoor climate quality, as it can be monetized in the life cycle economy of service building through the externalities related to increased productivity, reduced sick leaves, and reduced healthcare costs [28,29,87,98].

From the perspective of the life cycle economy, it is important that the potential future changes to the building do not compromise its energy efficiency. These changes might include the changing number of occupants, the alteration of use, such as from office to residential, and the retrofit of new and more advanced technologies [99]. An easily adaptable building requires less alteration work and has lower operating expenses [99,100]. The building frame design defines the adaptability of the building for alternative use and the magnitude of the related modification costs [99,101]. In general, deep building frames are more challenging to be converted for residential use, as all the apartments require windows [99]. Alternative purposes of use need to be accounted for during the frame and fenestration design to achieve the required lighting conditions during the alteration easier. Thermal and noise conductivity of the building envelope should be suitable for the building's potential alternative uses [99,101]. The sizing and positioning of the ventilation, heating, cooling, and water distribution system components with attention to potential future use would decrease the required modification work and the operating cost improving the building's life cycle economy [101]. Modification costs can also be reduced by leaving space and lead-through reservations for the future duct, pipe, and cable work required by altered use [27]. Technological development constantly produces new and more energy efficient building solutions to the market. At the same time, the updating of building codes narrows the applicability of the current basic solutions during capital renovations. Space, technological, and lead-through reservations would enable the more economical retrofit of novel advanced technologies in the future, as these become feasible or required. [27].

3.3. The Identification of the Optimal Compromise Design Solution during the Concept Design

The optimal compromise solution has to be identified by assessing the ability of the alternative solutions to provide sufficient indoor climate quality and to reduce GHG emissions with efficient use of invested public money. As the leader of the construction or renovation project, the municipal building department in co-operation with the lead designer can iteratively identify the optimal design solution following the order presented in Figure 7.

After considering the municipal building departments' targets, the alternative design solutions are proposed for the building department to choose from. The iterative feasibility assessment of the proposed solutions is conducted. The alternative solutions providing the same or different indoor climate quality class can be included into the energy efficiency assessment. The indoor climate quality performance should include temperature comfort, air quality, humidity comfort, acoustics, and lighting. To assess energy efficiency, the solution-specific purchase and sales energy balances are estimated. The estimates should include purchase energy demand and energy sales to the grid for heating, cooling, electricity, and fuels, as well as the peak power demand. At this point, the GHG emissions are also estimated based on the building's energy balance and the specific emission factors of the energy production technologies used. The design-specific life cycle economies can now be suggested based on the purchase/sales energy balance and the GHG emissions. The economic assessment includes future risks and opportunities related to the energy pricing, the economic GHG emission reduction-oriented economic steering and the alterations of the building's use. Externalities in the form of working productivity and sick leaves should also be considered. The municipal building department considers the ability of the solutions to provide indoor climate quality and GHG emission reductions from the perspective of the life cycle economy. Original performance targets and design solutions can be revised until the optimal service quality level (indoor climate and GHG emissions) for the public money invested is found. The optimal compromise solution highly depends on what weight the building department choses to give to indoor climate quality, energy use (and GHG emissions) and life cycle economy.

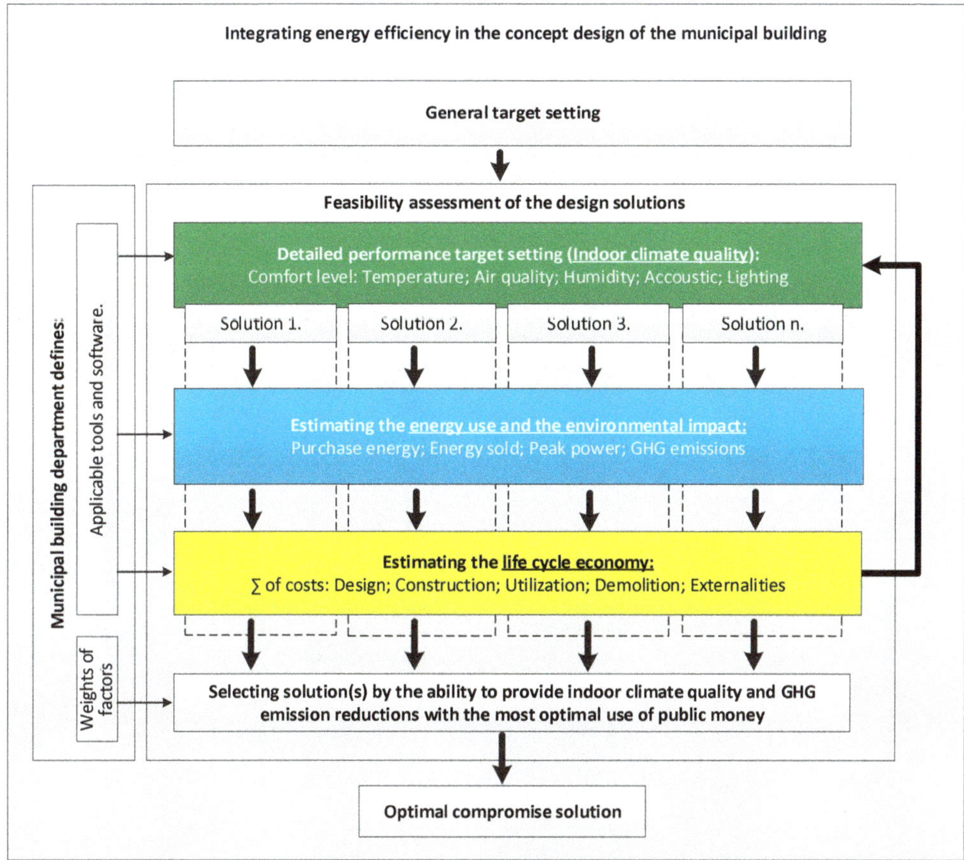

Figure 7. The identification of the optimal compromise design solution during the concept design.

3.4. The Energy Efficiency Responsibilities in the Concept Design of the Municipal Building

Figures 8–10 structure the findings of this study into a list of energy efficiency factors and clarify the responsibilities to integrate energy efficiency into the concept design of the municipal building procurement process. Same color codes are applied for the indoor climate quality performance, the energy use performance, and the life cycle economy as in Figures 2–7.

List of factors: Indoor climate quality (ICQ)

Indoor climate quality (ICQ) *Of the design solution*	Factors to be included in the municipal procurement process		Effect on energy use		Responsible player		Effects on the life cycle economy	
			Heating/cooling	Electricity	Municipal building department	Lead designer	Costs	Benefits
	Selection of the required indoor climate quality level	- Indoor climate quality class (in Finland: S1, S2, S3) - Material emission class (in Finland: M1, M2, M3) - Selection of the site based on local conditions (solar irradiance, air quality, humidity, shading, noise) - Other requirements	X	X	X	-	Design, construction, maintenance, & demolition costs of building (structural, HVAC, and lighting solutions). - Energy costs.	Safe & comfortable ICQ improves economy with increased working productivity and reduced sick leaves. - Reduced repair costs.
	Quality management of the procurement project	- Ensuring professional expertise of involved parties (tendering) - Selection of the design solutions - Reserving sufficient resources - Ensuring efficient communication - Quality control of design & construction work - Building management strategy & maintenance manual	X	X	X	-		
	↓ *Performance factors*							
	Temperature comfort, air quality, and air humidity comfort	**Solutions:** - Heating & cooling solutions and operation strategy - Building envelope (including doors, windows, etc.) - Management of pollutant concentration & relative humidity - Ventilation solutions and operation strategy *Quality assurance: Thermal & Indoor air quality condition simulation of the design solution* - *Temperature in different operating conditions* - *Stability of pressure ratios in different operating conditions* - *Stability of relative humidity & pollutant concentration* - *Effect of different supply/outlet/recycle air filtration levels.* - *Prevention of draught (air velocities in different operating conditions)*	↑ ↓	↑ ↓	-	X		

	Acoustic comfort	**Solutions:**
		- Sound insulation
		- Reverberation prevention
		- Selection, sizing, and positioning of the HVAC (incl. silencers) and other equipment with attention to the noise level
		Quality assurance: Acoustic condition simulation of the design solution
		- *Acoustics quality in different operation conditions*
		   - *Noise levels & noise characteristics*
		   - *Reverberation levels*
			-	↑ ↓	-	X		

	Lighting comfort	**Solutions:**
		- Lighting equipment, including:
		   - Artificial lighting
		   - Daylighting
		   - Operation strategy
		Quality assurance: Lighting condition simulation of the design solution
		- *Lighting quality in different operating conditions:*
		   - *Luminous intensity*
		   - *Uniformity of illumination*
		   - *Color reproduction quality*
		   - *Glare prevention*
		- *Utilization of natural light*
			↑ ↓	↑ ↓	-	X	-	-

Figure 8. The energy efficiency responsibilities in the design of the municipal building: Indoor climate quality. Based on [10–14,36,41,43,46,47,69,73,74,78–80,82–85].

List of factors: Energy use and environmental impact (heating, cooling and electricity)			Factors to be included in the municipal procurement process	Potential effect on indoor climate quality				Responsible player		Effects on the life cycle economy	
				T comfort	*Air quality*	*Lighting*	*Acoustics*	*Municipal building dep.*	*Lead designer*	*Costs*	*Benefits*
Energy use and environmental impact		**Selection of the energy performance requirements**	- Energy performance class (in Finland: A-G) - Renewable energy self-production and sales to the energy distribution network/grid - Other requirements	X	X	X	X	X	-	Design, construction, maintenance, & demolition costs (structural, HVAC, and lighting solutions). Energy costs (Through purchase/sales energy balance & peak power demand) Costs related to potential GHG emission reduction oriented economic steering.	Preparation for risks of future energy pricing: Reduction of demand for peak power and purchase energy with RE self-production, energy efficient solutions and demand response. Preparation for opportunities / risks related to low-carbon oriented economic steering: RE and low-carbon energy supply solutions & sa es of RE to the grid.
		Quality management of the procurement project	- Ensuring professional expertise of involved parties (tendering) - Selection of the design solution - Reserving sufficient resources - Ensuring efficient communication - Quality control of the design & construction work - Building management strategy & maintenance manual	X	X	X	X	X	-		
	↓ Performance factors										
	Heating & cooling demand	**Thermal conductivity of building envelope: (conduction & leakage air heat losses)**	- Basement (U-values) - Walls (U-values) - Roof (U-values) - Doors (U-values) - Windows (U-values, size) - Air tightness	*Temp. profile Draught*	*Relative humidity*	*Daylighting (window size)*	*Sound insulation*	-	X		
		Ventilation	- Ventilation demand - Ventilation heat recovery efficiency - Ventilation operation strategy	*Temp. profile Draught*	*Relative humidity/impurities*	*Equipment positioning*	*Noise emission*	-	X		
		Water supply	- Domestic hot water (DHW) heat losses (circulation & storage) - Water saving solutions - DHW temperature - Efficiency of wastewater heat recovery	*DHW heat losses*	*Impurities (Radon)*		*Noise emission*	-	X		
	Electrical energy demand	**Electric appliances**	-HVAC appliances (Including heating & cooling if electrical) - Lighting appliances (incl. passive) - Domestic and other electrical appliances - HVAC and lighting operation strategy	*Temp. profile Draught*	*Relative humidity/impurities*	*Lighting conditions*	*Noise emission*	-	X		
	Energy supply (Heating/Cooling/Electricity)	**Energy supply solutions**	- Energy production technologies (centralized/decentralized; renewable/non-renewable, incl. passive sources) -In-house heat & cool distribution system - Smart prosumer & demand response (Smart metering, bi-directional grid connection, peak power control, IoT energy use control, energy storage) → **Purchase/sales energy balance:** - Purchase electricity demand; - Energy to the grid; → **Peak power use** → **Carbon footprint of building, ∑ of:** - Net energy use GHG emissions: - Purchase energy GHG emissions; - Abated GHG emissions by selling self-produced renewable energy to the gird - **Water production GHG emissions** - **Embodied GHG emissions in materials** (Production, transportation, installation, maintenance, replacement, demolition, waste treatment)	*Temp. profile; Draught; Shading by PV panels*	*Relative humidity*	*Daylighting (PV roof coverage) Passive heating (window size)*	*Noise emission*	-	X		

Figure 9. The energy efficiency responsibilities in the design of the municipal building: Energy use and the environmental impact of a building. Based on [10–14,27,36,41,43,46,47,50,53–59,65,67,79].

List of factors: The life cycle economy				
Factors to be included in the municipal procurement process		**Responsible player** — *Municipal building dep.* / *Lead designer*		*Accountable requirements*
Selection of economy requirements		X / -		
Procurement project quality management	- Ensuring professional expertise of involved parties (tendering) - Selection of the design solution - Reserving sufficient resources - Ensuring efficient communication - Quality control of design & construction work - Building management strategy	X / -		
↓ *Performance factors (- Cost, + Income or saving)*				
Design costs	- Design work - Procurement project management cost	- / X		

Construction costs	- Construction work
	- Capital costs of the required:
	  - technology
	  - materials
	  - plot
	- Procurement project management cost
	+ GHG reduction oriented economic steering: Subsidies
		- / X		

Operating costs	+/- Energy costs (heating, cooling & electricity):
	  - prices for purchase energy (electricity & heat: generation, distribution network/grid & tax) and fuels: Development of available energy pricing schemes, effect of increasing peak demand volatility on electricity spot prices / capacity charges / peak power charges, increasing maintenance fees, and development of fuel market price.
	  - aging-related heat loss increase in thermal insulation
	  - aging-related drop in the efficiency of energy production/storage/distribution systems
	  + cost savings achieved due to the substitution of purchased energy with self-produced 'free' renewable energy: development of business service models, such as roof leasing for PV production.
	  + revenues/savings for electricity & heat sales to the grid and demand response: development of available pricing schemes, effect of increasing peak demand volatility on electricity spot prices.
	+/- GHG reduction oriented economic steering: taxes, charges, subsidies, feed-in tariff, tax credits, green certification, and tradable permit schemes for households, and other buildings.
	- Water supply/sewage costs
	- Maintenance costs: expected service life and repair/replacement costs of
	  - systems & components
	  - structures
	- Administrative costs:
	  - wages
	  - insurances
	  - taxes
	  - rates of interest on loans
	  - depreciation
	  + rental incomes
	  - other
	+/- Externalities
	+/- productivity
	  - sick leaves
	+/- Alteration of use
	  - Alteration work
	+/- Changed operating costs
		- / X		

Demolition costs	- Costs of demolition and demolition waste removal	- / X		

Left spanning labels: **Life cycle economy** · **Proposed design solution** *(Suggested materials, technology, dismountability)*

Right spanning labels: Meeting indoor climate quality requirements in all purposes of use and use profiles. · Energy performance in relation to use profiles.

Figure 10. The energy efficiency responsibilities in the design of the municipal building: The life cycle costs of a building. Based on [10,11,14,27,32,36,41,46,47,86,87,92,98].

This list of energy efficiency factors can be applied during the concept design of the building when the municipal building department together with the lead designer is seeking the design solution providing the optimal value for public money. The list of energy efficiency factors should be used when assessing the performance of the solutions to provide indoor climate quality and GHG emission reduction from the perspective of the life cycle economy in the changing operational environment of the building. Clarifying the energy efficiency factors and the responsibilities, the list contributes to more informed decision-making process.

The list of energy efficiency factors structure and clarify the factors affecting indoor climate quality, energy use (and environmental impact), and the life cycle economy, as identified in Figures 4–6. The connections between indoor climate quality, energy use, and the life cycle economy are presented at the factor level in order to promote energy efficiency through integrated building design solutions (Figures 8–10). The possible direction of the impact of indoor climate quality factors on heating energy, cooling, and electricity consumption are described with arrows in Figure 8. Figure 9 presents the potential effects of the energy use factors on indoor climate quality by means of a checklist and descriptions. If checked, the factors may have a positive or negative effect on the indoor climate quality. Figures 8 and 9 also explain how indoor climate quality and energy use performance factors affect the life cycle economy of the building. The costs and life cycle economic benefits achievable through the building design are explained with emphasis to the changing operational environment of the building. The life cycle cost elements required to be included in feasibility assessment of design solutions are presented in Figure 10.

It is for the municipal building department to decide how much weight each factor should be given and what are the most appropriate tools to be used in the feasibility assessment. Selection of tools and weighting of factors has to be done case-specifically based on the building type and the local and regional priorities. When applying the list of factors the building department demonstrates the applied energy efficiency factors, the chosen weights, and the responsibilities increasing transparency of the decision-making process.

The main responsible actors for the integration of the performance-specific factors into the concept design of the municipal building are defined in Figures 8–10. The importance of roles of the municipal building department and the lead designer in achieving energy efficiency performance are highlighted. The building department steers the project and is responsible to clearly establish the performance targets and ensure that all project partners have sufficient and timely information and resources to achieve the targets. The lead designer, by coordinating the consortium of sub-designers (structural, electric, ventilation, heating, automation, DHW), is responsible to deliver the design that complies with the energy efficiency performance targets.

A wide range of commercial and free software, tools, and methods can be applied by the designer to estimate the indoor climate, the energy use, and the life cycle economy of the design solutions. Energy simulation software can be utilized to assess the purchase/sales energy balance and the peak power demand. Life cycle cost modeling tools can be helpful with the complex life cycle economy estimations. The applied method should be chosen based on the case-specific modeling complexity and precision requirements.

4. Discussion and Conclusions

This paper clarifies how the municipal building department can lead the design towards energy efficiency and a high quality indoor climate, which is life cycle cost-effective and complies with the international and national targets and regulations. Specific emphasis is put on the concept design phase of the building's procurement process. The paper describes the relationship between indoor climate quality, energy use (and GHG emissions), and the life cycle economy from the perspective of design-related factors. The paper presents the list of energy efficiency factors that need to be considered in the municipal building procurement process. The list of factors can be applied during the concept design of the building when the municipal building department together with the lead designer is seeking a design solution providing the optimal value for public money. The responsibilities for integrating energy efficiency into the municipal building procurement process are clarified in the list of factors. The presented approach contributes to making the procurement and decision making process more transparent.

The public procurements have to provide the taxpayers the required quality service (indoor climate quality and GHG emission reductions) with the most optimal use of public money. The actual cheapest solutions, however, is not the one with the smallest investment costs. Selecting a design solution has to be based on solution's ability to provide indoor climate quality and GHG emission reductions from the perspective of the life cycle economy. The solution with the optimal compromise between indoor climate quality, the energy use performance, and the life cycle economy has to be defined. Procurement practices should not prevent the selection of the solutions with improved life cycle performance because, for example, of an undersized investment budget. Traditional procurement practices need to be revised in case they do not allow the selection of the optimal solution based on quality and cost.

New economic elements have to be included into the life cycle cost examination of alternative design solutions to achieve those with lower risks related to utilization costs and those with improved resale value preservation. In order to identify the sustainable "future proofed" solution, the feasibility assessments have to consider changes in the operational environment of the building. Considerable changes include the availability of service solutions, the future development of energy pricing, and the potential GHG reduction-oriented economic steering. This research contributes to rising the recognition of smart building solutions that improve energy efficiency. Thus, the research is in line with the smart readiness indicator initiatives of the European Commission.

Energy use during building's operation currently form the biggest share of the building's carbon footprint. However, the share of GHG emissions embodied in construction materials in building's carbon footprint is increasing as we move towards more energy efficient construction and lower emission energy production. In the future, material selection will play increasingly important role in low-carbon building design.

The responsibilities of the players involved in the procurement project have to be defined clearly by the municipal building department (client). No gaps in the chain of responsibility should exist. The building department steers the project and is responsible to clearly establish the energy efficiency targets and ensure that all project partners have sufficient and timely information and resources to achieve the targets. The overall control of design production is the responsibility of the lead designer. The lead designer is responsible to convey the performance targets and preferences of the client to consortium of sub-designers (structural, electric, ventilation, heating, automation, and DHW). The lead designer has to be aware about the progress of sub-designs and to coordinate the design work so that the final result complies with the energy efficiency targets. The building department choses the optimal concept design from the options proposed by the designers and approves the detailed design. Systematic and well-timed communication between the building department and the lead designer is important, especially during the concept design. The performance requirements have to be clearly communicated to the lead designer by the building department. To support the building department's decision-making the lead designer has to provide the information regarding the progress of design and its implications on the building's life cycle performance. Interpreting the results of energy efficiency assessment as presented in this approach require specific expertise in the field of construction technology, energy technology, indoor climate technology, automation and the economy. Tasks exceeding expertise of the municipal building department should be outsourced to a professional construction consultant experienced in the field, which can create new business opportunities for consulting companies and can provide new jobs. Outsourcing the tasks does not reduce the responsibility of the building department in the project. The municipal building department carries the final responsibility to provide the building with the functionality to offer the service required by the client, municipality, and taxpayers.

This research describes what factors are necessary to consider in energy efficiency assessment; however, it does not define in detail the actual tools or mathematical methods to be used in the assessments. The optimal method should be chosen based on the modeling complexity and precision required by the building type in question. Assessment will require making assumptions regarding the future development of the building's operational environment, including energy fees, transmission fees, peak power fees, energy taxation, GHG reduction-oriented economic steering, and climatic conditions. Since the future development of these issues is never certain, the assessment will always include some subjectivity. The outcome of the assessment will also strongly depend on the applied weighting of the factors. The freedom to decide the weight of criteria for optimization is left for the municipal building department. Factors affecting indoor climate quality, energy use (and GHG emission reductions), and the life cycle economy have to be weighted based on the building type and reflecting the local and regional priorities. The procedure can be applied universally during a municipal building procurement process; however, some country- and municipality-specific adaptations are required. Country-specific building standards and procurement regulations and practices have to be accounted for while applying the presented procedure. Besides, procurement procedures applied by different municipalities vary somewhat even within the same country. As the roles and responsibilities of the municipal building department and other project partners in the building procurement project vary in different countries and municipalities, they should be verified on a case-by-case basis. Case-specific energy efficiency and other targets, as well as the available resources, have to be taken into account. Also, attention is required on the effect of the local climatic conditions on energy use and indoor climate quality. For example, the significance of heating and cooling in the overall energy consumption of a building differs remarkably between southern and northern countries. The procedure can be adapted for use in private building procurement projects as well, if the differences in the procurement procedures, roles, and targets are accounted for. The presented procedure can be applied alone or alongside existing labeling systems supporting them, such as LEED or BREEAM.

The town of Lappeenranta integrated the presented approach into their building procurement practices initiating a dialogue regarding energy efficiency between the municipal building department and other stakeholders in building procurement. This dialogue provides a good basis for the further development of energy efficiency management procedures and the introduction of new innovative methodologies into everyday practice. The procedure should be updated continuously based on new evolving knowledge and technological development. In addition, the scope of the procedure can be expanded further to cover renovation projects.

Author Contributions: Conceptualization, M.V., K.G., M.L. and R.S.; Methodology, M.L.; Validation, M.V., S.H. and M.L.; Investigation, M.V.; Writing—Original Draft Preparation, M.V.; Supervision, M.L., R.S. and K.G.; Funding Acquisition, M.L.

Funding: This research was done as a part of the 'Innovatiivisuutta julkisiin investointeihin' (Innovative Public Investments) research project, which is funded by the European Regional Development Fund (ERDF), the town of Lappeenranta, and the Waste Treatment Company of South Karelia (EKJH). Grant number: A32168. We gratefully acknowledge this funding support.

Conflicts of Interest: The authors declare no conflict of interest.

References

1. EC (European Commission). Buildings. 2017. Available online: https://ec.europa.eu/clima/citizens/eu_en (accessed on 28 September 2017).
2. EC (European Commission). Green Public Procurement Criteria for Office Building Design, Construction and Management. 2016. Available online: http://ec.europa.eu/environment/gpp/pdf/report_gpp_office_buildings.pdf (accessed on 28 September 2017).
3. EC (European Commission). Energy. 2050 Energy Strategy. 2017. Available online: https://ec.europa.eu/energy/en/topics/energy-strategy/2050-energy-strategy (accessed on 31 January 2017).

4. EC (European Commission). Energy Efficiency Directive. 2018. Available online: https://ec.europa.eu/energy/en/topics/energy-efficiency/energy-efficiency-directive (accessed on 31 January 2017).

5. EUR-Lex. Directive of the European Parliament on the Energy Performance of Buildings. 2010/31/EU. 2018. Available online: https://eur-lex.europa.eu/legal-content/EN/TXT/HTML/?uri=CELEX:32010L0031&from=EN (accessed on 10 March 2018).

6. Parviainen, J. *Kuntien ja Maakuntien Ilmastotyön Tilanne 2015 [The State of the Climate Work in Municipalities 2015]*; Kuntaliitto: Helsinki, Finland, 2015; ISBN 978-952-293-335-5.

7. TEM (Ministry of Employment and the Economy). Energiatehokkuus Julkisissa Hankinnoissa (In English: Energy Efficiency in Public Procurements). 2017. Available online: https://www.motiva.fi/files/10919/Tyo-_ja_elinkeinoministerion_ohjeet_Energiatehokkuus_julkisissa_hankinnoissa.pdf (accessed on 1 March 2017).

8. Jätteenmäki, A. Opinion of the Committee on the Environment, Public Health and Food Safety for the Committee on Industry, Research and Energy on the proposal for a directive of the European Parliament and of the Council amending Directive 2010/31/EU on the energy performance of buildings. *Eur. Parliam.* **2017**. Available online: http://www.europarl.europa.eu/sides/getDoc.do?pubRef=-//EP//TEXT+REPORT+A8-2017-0314+0+DOC+XML+V0//FI (accessed on 2 August 2018).

9. Rava, Rakennuttajatjavalvojat ry (Association of builders and supervisors). Käsitteitä (In English: Terminology). 2017. Available online: https://rakennusvalvojat.fi/yleista/kasitteita/ (accessed on 29 September 2017).

10. RIL (Finnish Association of Civil Engineers). *RIL 249-2009 Matalaenergiarakentaminenasuinrakennukset (In English: Low-Energy Construction, Residential Buildings)*, 2nd ed.; Suomen Rakennusinsinöörien Liitto RIL ry; Saarijärven Offset Oy: Helsinki, Finland, 2009; ISBN 978-951-758-517-0.

11. RIL, (Finnish Association of Civil Engineers). Rakennuttajat. RIL Database 2017. Available online: http://www.ril.fi/fi/patevyydet/rakennuttajat-rap-ja-raps.html (accessed on 29 September 2017).

12. MotE (Ministry of the Environment). The Decree of the Ministry of the Environment (on Indoor Climate and Ventilation). 2017. Available online: https://www.finlex.fi/fi/laki/alkup/2017/20171009 (accessed on 25 July 2018).

13. MotE (Ministry of the Environment). The Decree of the Ministry of the Environment (on Energy Efficiency of Buildings). 2017. Available online: https://www.finlex.fi/fi/laki/alkup/2017/20171048 (accessed on 25 July 2018).

14. Rakennustieto (Building Information Group). Pääsuunnittelija ja suunnittelun johtaminen rakennushankkeessa (In English: Lead Designer and Leading the Design Process in the Construction Project). *Rakennustietosäätiö* **2010**. Available online: https://www.rakennustieto.fi/Downloads/RK/RK100202.pdf (accessed on 19 June 2017).

15. Sparks, D. Exploring Public Procurement as a Mechanism for Transitioning to Low-Carbon Buildings. Ph.D. Thesis, Queensland University of Technology, Brisbane, Australia, 2018.

16. Carlssona-Kanyama, A.; Carlsen, H.; Dreborg, K.-H. Barriers in municipal climate change adaptation: Results from case studies using backcasting. *Futures* **2013**, *49*, 9–21. [CrossRef]

17. Sparrevik, M.; Wangen, H.F.; Fet, A.M.; Boer, L.D. Green public procurement—A case study of an innovative building project in Norway. *J. Clean. Prod.* **2018**, *188*, 879–887. [CrossRef]

18. Rakennustieto (Building Information Group). Rakennusprojektin johtaminen (In English: Leading the Construction Project). *Rakennustietosäätiö, Finland*. Available online: https://www.rakennustieto.fi/Downloads/RK/RK060501.pdf (accessed on 6 February 2019).

19. Kim, J.; Goodwin, C.; Kim, S. Communication Turns Green Construction Planning into Reality. *J. Green Build.* **2017**, *12*, 168–186. [CrossRef]

20. Morledge, R.; Smith, A. *Building Procurement*, 2nd ed.; John Wiley & Sons: Hoboken, NJ, USA, 2013; ISBN 978-111-849-369-4.

21. Mikucioniene, R.; Martinaitis, V.; Keras, E. Evaluation of energy efficiency measures sustainability by decision tree method. *Energy Build.* **2014**, *76*, 64–71. [CrossRef]

22. Sorell, S. Reducing energy demand: A review of issues, challenges and approaches. *Renew. Sustain. Energy Rev.* **2015**, *47*, 74–82. [CrossRef]

23. Tang, Z. An Integrated Life Cycle-based Viable Procurement System to Enhance the Success Level of Sustainable Building. Ph.D. Thesis, The HKU Scholars Hub, Hong Kong, China, 2017.

24. Trianni, A.; Cagno, E.; De Donatis, A. A Framework to Characterize Energy Efficiency Measures. *Appl. Energy* **2014**, *118*, 207–220. [CrossRef]

25. Vidmar, K. Importance of Finding and Defining Energy Conservation Measures. *Strateg. Plan. Energy Environ.* **2010**, *30*, 45–63. [CrossRef]

26. Georgiadou, M.C. Future-Proofed Energy Design Approaches for Achieving Low-Energy Homes: Enhancing the Code for Sustainable Homes. *Buildings* **2014**, *4*, 488–519. [CrossRef]

27. Vinokurov, M.; Grönman, K.; Kosonen, A.; Luoranen, M.; Soukka, R. Updating the Path to a Carbon-Neutral Built Environment-What Should a Single Builder Do. *Buildings* **2018**, *8*, 112. [CrossRef]

28. Becchio, C.; Corgnati, S.P.; Orlietti, L.; Spigliantini, G. Proposal for a modified cost-optimal approach by introducing benefits evaluation. *Energy Procedia* **2015**, *82*, 445–451. [CrossRef]

29. Gvozdenović, K.; Zeiler, W.; Maassen, W.H. Towards nZEBs in 2020 in the Netherlands 2014. In Proceedings of the World Renewable Energy Congress 13 (WREC XIII), London, UK, 3–8 August 2014; ISBN 978-3-319-17777-9.

30. Barringer, H.P.; Weber, D.P.; Westside, M.H. Life Cycle Cost Tutorial. In Proceedings of the Fifth International Conference on Process Plant Reliability, Houston, TX, USA, 2–4 October 1996. [CrossRef]

31. Bogenstätter, U. Prediction and optimization of life-cycle costs in early design. *Build. Res. Inf.* **2000**, *28*, 376–386. [CrossRef]

32. Fuller, S. Life-cycle Cost Analysis (LCCA). Whole Building Design Guide. National Institute for Building Sciences 2010. Available online: http://www.wbdg.org/resources/lcca.php (accessed on 4 May 2015).

33. Sartori, I.; Hestnes, A.G. Energy use in the life cycle of conventional and low-energy buildings: A review article. *Energy Build.* **2007**, *39*, 249–257. [CrossRef]

34. Soukka, R. Applying the Principles of Life Cycle Assessment and Costing in Process Modeling to Examine Profit-Making Capability. Ph.D. Thesis, Acta Universitatit Lappeenrantaensis, Lappeenranta, Finland, 2007; p. 275.

35. Yates, W.; Beaman, D. Design Simulation Tool to Improve Product Reliability. In Proceedings of the Annual Reliability and Maintainability Symposium 1995, Washington, DC, USA, 16–19 January 1995.

36. EC (European Commission). *A Handbook on Green Public Procurement*, 3rd ed.; European Commission: Brussels, Belgium, 2016. Available online: http://ec.europa.eu/environment/gpp/pdf/Buying-Green-Handbook-3rd-Edition.pdf (accessed on 28 December 2017).

37. EUR-Lex. Directive of the European Parliament on Public Procurement. EUR-Lex Database 2014. 2014/24/EU. 2014. Available online: http://eur-lex.europa.eu/legal-content/EN/TXT/?uri=CELEX%3A32014L0024 (accessed on 10 March 2018).

38. Finlex. Directive on Changing the Law on Land Use and Construction (In Finnish: Valtioneuvoston asetus maankäyttö- ja rakennusasetuksen muuttamisesta). Finlex Database 2015. 215/2015. Available online: https://www.edilex.fi/data/rakentamismaaraykset/sk20150215.pdf (accessed on 9 November 2017).

39. Finlex. Law on Public Procurement (In Finnish: Laki julkisista hankinnoista ja käyttöoikeussopimuksista). Finlex Database 2016. 29.12.2016/1397. Available online: http://www.finlex.fi/fi/laki/ajantasa/2016/20161397?search[type]=pika&search[pika]=laki%20julkisista%20hankinnoista (accessed on 9 February 2017).

40. Finlex. Law on Land Use and Construction (In Finnish: Maankäyttö ja rakennuslaki). 5.2.1999/132. Finlex database 2017. Available online: http://www.finlex.fi/fi/laki/ajantasa/1999/19990132 (accessed on 9 November 2017).

41. Rakennustieto (Building Information Group). Suunnittelu- ja konsultointipalveluiden hankinta (In English: Procurement of planning and consulting servces). Rakennustietosäätiö 2016. Available online: http://docplayer.fi/41122549-Rts-16-60-1-0-suunnittelu-ja-konsultointipalveluiden-hankinta.html (accessed on 29 September 2017).

42. TEM (Ministry of Employment and the Economy). What Are Public Contracts? TEM Database 2016. Available online: http://www.tem.fi/en/consumers_and_the_market/public_procurement (accessed on 10 April 2016).

43. MotE (Ministry of the Environment). The Decree of the Ministry of the Environment (on Humidity Control of Buildings). 2017. Available online: https://www.finlex.fi/fi/laki/alkup/2017/20170782 (accessed on 20 July 2018).

44. CFCI (Confederation of Finnish Construction Industries). Hankintalainsäädännön soveltaminen (In English: Applying the Procurement Legislation). CFCI Database 2017. Available online: https://www.rakennusteollisuus.fi/globalassets/toimialat/talonrakennus/hyotytietoa-tyomaille/hankintamenettely_tiivistelma.pdf (accessed on 1 February 2017).

45. Alyami, S.; Rezgui, Y. Sustainable building assessment tool development approach. *Sustain. Cities Soc.* **2012**, *5*, 52–62. [CrossRef]

46. Peres Almeida, C.; Ferreira Ramos, A.; Mendes Silva, J. Sustainability assessment of building rehabilitation actions in old urban centres. *Sustain. Cities Soc.* **2018**, *36*, 378–385. [CrossRef]

47. Vinokurov, M.; Luoranen, M.; Hammo, S. Methodology for Optimization of Energy Efficiency, Indoor Climate and Economy Targets in Municipal Building Projects. In Proceedings of the Healthy Buildings 2015, Eindhoven, The Netherlands, 18–20 May 2015; ISBN 978-90-386-3889-8.

48. Dodoo, A.; Gustavsson, L.; Sathre, R. Primary energy implications of ventilation heat recovery in residential buildings. *Energy Build.* **2011**, *43*, 1566–1572. [CrossRef]

49. Al-Homoud, M. Performance characteristics and practical applications of common building thermal insulation materials. *Build. Environ.* **2005**, *40*, 353–366. [CrossRef]

50. Heikkilä, J. Energiatehokas suunnittelu (In English: Energy Efficient Design). Terveys ja talous 2012. Available online: http://www.terveysjatalous.fi/docs/Jari%20Heikkil%C3%A4%20Terveytt%C3%A4%20ja%20teknologiaa.pdf (accessed on 2 December 2016).

51. Meggers, F.; Leibundgut, H. The potential of wastewater heat and exergy: Decentralized high-temperature recovery with a heat pump. *Energy Build.* **2011**, *43*, 879–886. [CrossRef]

52. MotE (Ministry of the Environment). The Decree of the Ministry of the Environment (on Water and Sewerage Systems of Building). 2017. Available online: https://www.finlex.fi/fi/laki/alkup/2017/20171047 (accessed on 23 November 2018).

53. DoE (U.S. Department of Energy). Passive Solar Home Design. DoE 2015. Available online: http://energy.gov/energysaver/passive-solar-home-design (accessed on 2 March 2016).

54. Gregg, D. Daylighting. National Institute of Building Science 2017. Available online: http://www.wbdg.org/resources/daylighting.php (accessed on 4 February 2017).

55. Karunathilake, H.; Perera, P.; Ruparathna, R.; Hewage, K. Renewable energy integration into community energy systems: A case study of new urban residential development. *J. Clean. Prod.* **2016**, *173*, 1–16. [CrossRef]

56. VTT. Maalämmön ja–viilennyksen hyödyntäminen asuinkerrostalon lämmityksessä ja jäähdytyksessä (In English: Utilization of Ground Source Heat Pump in Heating and Cooling of the Residential Building). Available online: http://www.vtt.fi/inf/pdf/tiedotteet/2010/T2546.pdf (accessed on 5 March 2018).

57. Koroneos, C.; Tsarouhis, M. Exergy analysis and life cycle assessment of solar heating and cooling systems in the building environment. *J. Clean. Prod.* **2012**, *32*, 52–60. [CrossRef]

58. Li, D.H.W.; Yang, L.; Lam, J.C. Zero energy buildings and sustainable development implications—A review. *Energy* **2013**, *54*, 1–10. [CrossRef]

59. MotE (Ministry of the Environment). *Energiatodistusopas 2013 (In English: The Energy Certificate Guide)*; Minsitry of the Environment: Helsinki, Finland, 2013.

60. Jenkins, D.P.; Patidar, S.; Simpson, S.A. Quantifying Change in Buildings in a Future Climate and Their Effect on Energy Systems. *Buildings* **2015**, *5*, 985–1002. [CrossRef]

61. Crosbie, T.; Broderick, J.; Short, M.; Charlesworth, R.; Dawood, M. Demand Response Technology Readiness Levels for Energy Management in Blocks of Buildings. *Buildings* **2018**, *8*, 13. [CrossRef]

62. Karnouskos, S. Demand Side Management via prosumer interactions in a smart city energy marketplace. In Proceedings of the 2011 2nd IEEE PES International Conference and Exhibition on Innovative Smart Grid Technologies, Manchester, UK, 5–7 December 2011.

63. Wei, C.; Li, Y. Design of energy consumption monitoring and energy-saving management system of intelligent building based on the Internet of things. In Proceedings of the 2011 International Conference on Electronics, Communications and Control, Ningbo, China, 9–11 September 2011; pp. 3650–3652.

64. Akadiri, P.O.; Ezekiel, A.C.; Olomolaiye, P.O. Design of A Sustainable Building: A Conceptual Framework for Implementing Sustainability in the Building Sector. *Buildings* **2012**, *2*, 126–152. [CrossRef]

65. Bionova Oy. *Tiekartta Rakennuksen Elinkaaren Hiilijalanjäljen Huomioimiseksi Rakentamisen Ohjauksessa (In English: The Roadmap to Accounting Carbon Footprint in Steering of Construction)*; Ministry of the Environment: Helsinki, Finland, 2017.

66. Finnish Standards Association SFS. SFS-EN 15804, Sustainability of Construction Works. In *Core Rules for Product Category of Construction Products*; Finnish Standards Association: Helsinki, Finland, 2014.

67. Oldewurtel, F.; Parisio, A.; Jones, C.; Gyalistras, D.; Gwerder, M.; Stauch, V.; Lehmann, B.; Morari, M. Use of model predictive control and weather forecasts for energy efficient building climate control. *Energy Build.* **2012**, *45*, 15–17. [CrossRef]

68. Säteri, J. Sisäilmasto ja energiatalous (In English: Indoor climate and energy economy). Indoor Air Association of Finland 2014. Available online: http://www.korjausrakentaminen2014.fi/File/626/sisailmasto-ja-energiatalous-korjrak2014.pdf (accessed on 1 February 2016).

69. IAA (Indoor Air Association of Finland). Sisäilman tekijät (In English: Indoor Air Components). IAA 2008. Available online: http://www.sisailmayhdistys.fi/Terveelliset-tilat/Sisailmasto/Sisailman-tekijat (accessed on 10 November 2016).

70. Climate Guide. Projected Climate Change in Finland. Climate Guide 2018. Available online: https://ilmasto-opas.fi/en/ilmastonmuutos/suomen-muuttuva-ilmasto/-/artikkeli/74b167fc-384b-44ae-84aa-c585ec218b41/ennustettu-ilmastonmuutos-suomessa.html (accessed on 23 October 2018).

71. Jylhä, K.; Ruosteenoja, K.; Räisänen, J.; Fronzek, S. *Miten Väistämättömään Ilmastonmuutokseen Voidaan Varautua?—Yhteenveto Suomalaisesta Sopeutumistutkimuksesta eri Toimialoilla (In English: How to Adopt to Inevitable Climate Change?)*; Ministry of Agriculture and Forestry: Helsinki, Finland, 2012; pp. 16–23.

72. Ruosteenoja, K. Temperature and Rainfall Scenarios Based on the Global Climate Models. Sektoritutkimusohjelman ilmastoskenaariot (SETUKLIM) 1. part-Project. Finnish Meteorological Institute 2013. Available online: http://ilmatieteenlaitos.fi/c/document_library/get_file?uuid=c4c5bf12-655e-467a-9ee0-f06d8145aaa6&groupId=30106 (accessed on 23 October 2018).

73. IAA (Indoor Air Association of Finland). Sisäilmastoluokitus (In English: Indoor Climate Quality Classification). IAA 2018. Available online: http://www.sisailmayhdistys.fi/Sisailmayhdistys/Sisailmastoluokitus (accessed on 10 September 2018).

74. Burroughs, H.E.; Hansen, S.J. *Managing Indoor Air Quality*, 5th ed.; The Fairmont Press Inc.: Lilburn, GA, USA, 2011; ISBN 0-88173-661-9.

75. Kulmala, I.; Kalliohaka, T.; Kataja, J.; Leppälä, V. Reaaliaikainen tuloilmasuodattimen toimintakunnon mittausjärjestelmä (In English: Real Time Measurement System for Supply Air Filter Functionality). VTT 2015. Available online: http://www.vtt.fi/inf/julkaisut/muut/2015/OA-Reaaliaikainen-tuloilmasuodattimen.pdf (accessed on 29 January 2016).

76. Ruosteenoja, K.; Jylhä, K.; Kämäräinen, M. Climate projections for Finland under the RCP forcing scenarios. *Geophysica* **2016**, *51*, 17–50.

77. VTT. Puurakenteiden Kosteustekninen Toiminta (In English: Hygroscopic Performance of Wooden Stucturs). Available online: https://www.vtt.fi/inf/pdf/tiedotteet/1999/T1991.pdf (accessed on 2 March 2018).

78. OfRH (Organization for Respiratory Health in Finland (Hengitysliitto)). Kuiva ja kostea ilma (In English: Dry and Humid Air). Hengitysliitto 2016. Available online: http://www.hengitysliitto.fi/fi/sisailma/hiukkasmaiset-ja-kaasumaiset-epapuhtaudet/kuiva-ja-kostea-ilma (accessed on 5 June 2016).

79. MotE (Ministry of the Environment). The Decree of the Ministry of the Environment (on Building's Sound Environment). 2017. Available online: https://www.finlex.fi/fi/laki/alkup/2017/20170796 (accessed on 26 July 2018).

80. Paradis, R. Acoustic Comfort. Whole Building Design Guide. National Institute of Building Sciences 2012. Available online: http://www.wbdg.org/resources/acoustic.php (accessed on 19 December 2015).

81. Sala, E.; Viljanen, V. Improvement of Acoustic Conditions for Speech Communication in Classrooms. *Appl. Acoust.* **1995**, *45*, 81–91. [CrossRef]

82. Halonen, L.; Tetri, E.; Bhusal, P. *Guidebook for Energy Efficient Electric Lighting for Building*; Aalto University: Helsinki, Finland, 2010; ISBN 978-952-60-3229-0.

83. Boyce, P.R. *Human Factors in Lighting*, 2nd ed.; Taylor & Francis: London, UK; New York, NY, USA, 2003.

84. Sharp, F.; Lindsey, D.; Dols, J.; Coker, J. The use and environmental impact of daylighting. *J. Clean. Prod.* **2014**, *85*, 462–471. [CrossRef]

85. Veitch, J.A.; Newsham, G.R. Determinants of lighting quality I: State of the Science. *J. Illum. Eng. Soc.* **1998**, *27*, 92–106. [CrossRef]

86. Pal, S.K.; Takano, A.; Alanne, K.; Palonen, M.; Siren, K. A multi-objective life cycle approach for optimal building design: A case study in Finnish context. *J. Clean. Prod.* **2017**, *143*, 1021–1035. [CrossRef]

87. GBC (Green Building Council) Finland. Life-Cycle Cost—The Long-Term Cost Efficiency Indicator. GBC 2017. Available online: http://figbc.fi/en/building-performance-indicators/calculation-guide/life-cycle-cost-guide/ (accessed on 29 October 2017).

88. ACEEE (American Council for an Energy-Efficient Economy). Some Utilities Are Rushing to Raise Fixed Charges. That Would Be Bad for the Economy and Your Utility Bill. Available online: https://aceee.org/blog/2014/12/some-utilities-are-rushing-raise-fixe (accessed on 5 October 2018).

89. Brunekreeft, G.; Buchmann, M.; Meyer, R. New developments in electricity markets following large-scale integration of renewable energy. In *The Routledge Companion to Network Industries*; Routledge: Abington, UK, 2015.

90. EC (European Commission). Digitizing the Energy Sector: An Opportunity for Europe. European Commission 2018. Available online: https://ec.europa.eu/digital-single-market/en/blog/digitising-energy-sector-opportunity-europe (accessed on 16 March 2018).

91. FLEXe. The Final Report of the Research Project FLEXe (Flexible Energy System). Available online: http://flexefinalreport.fi/content/prosumers (accessed on 21 March 2018).

92. Hong, J.; Zhang, X.; Shen, Q.; Zhang, W.; Feng, Y. A multi-regional based hybrid method for assessing life cycle energy use of buildings: A case study. *J. Clean. Prod.* **2017**, *148*, 760–772. [CrossRef]

93. IEA (International Energy Agency). World Energy Investment. 2017. Available online: https://www.iea.org/publications/wei2017/ (accessed on 16 March 2018).

94. Pereira, R.; Figueiredo, J.; Melicio, R.; Mendes, V.M.F.; Martins, J.; Quadrado, J.C. Consumer Energy Management System with Integration of Smart Meters. *Energy Rep.* **2015**, *1*, 22–29. [CrossRef]

95. Simola, A.; Kosonen, A.; Ahonen, T.; Ahola, J.; Korhonen, M.; Hannula, T. Optimal dimensioning of a solar PV plant with measured electrical load curves in Finland. *Sol. Energy* **2018**, *170*, 113–123. [CrossRef]

96. Vimpari, J.; Junnila, S. Evaluating decentralized energy investments: Spatial value of on-site PV electricity. *Renew. Sustain. Energy Rev.* **2017**, *70*, 1217–1222. [CrossRef]

97. Kniefel, J. Life-cycle carbon and cost analysis of energy efficiency measures in new commercial buildings. *Energy Build.* **2010**, *42*, 333–340. [CrossRef]

98. Balaban, O.; Puppim de Oliveira, J. Sustainable buildings for healthier cities: Assessing the co-benefits of green buildings in Japan. *J. Clean. Prod.* **2016**. [CrossRef]

99. Gann, D.; Barlow, J. Flexibility in building use: The technical feasibility of converting redundant offices into flats. *Constr. Manag. Econ.* **1996**, *14*, 55–66. [CrossRef]

100. Jauhiainen, T.; Rahikka, A. Runkorakenteiden korjausrakentamisen suunnittelu (In English: Planning the Renovation of Building Frame). In *RIL 174-4. Korjausrakentaminen 4, Runkorakenteet*; Finnish Association of Civil Engineers: Helsinki, Finland, 1988.

101. Koskinen, J. Building's suitability for change of use. Master's Thesis, TUT DPub, Tampere University of Technology, Tampere, Finland, 2016.

 *buildings*

Article

Design of an Insurance Policy Model Applied to Natural Stone Facade Claddings

Miguel Macedo [1], Jorge de Brito [2], Ana Silva [2,*] and Carlos Oliveira Cruz [2]

[1] Department of Civil Engineering, Architecture and Georresources, Instituto Superior Técnico, Universidade de Lisboa, 1049-001 Lisbon, Portugal; miguel.macedo@ist.utl.pt
[2] CERIS, Department of Civil Engineering, Architecture and Georresources, Instituto Superior Técnico, Universidade de Lisboa, 1049-001 Lisbon, Portugal; jb@civil.ist.utl.pt (J.d.B.); oliveira.cruz@tecnico.ulisboa.pt (C.O.C.)
* Correspondence: ana.ferreira.silva@tecnico.ulisboa.pt; Tel.: +351-964186538

Received: 1 April 2019; Accepted: 30 April 2019; Published: 4 May 2019

Abstract: The insurance market deliberately excludes the buildings' envelope from their insurance policies, neglecting all the damage that can be caused by the degradation process or ageing of the materials. This stance is mainly due to the lack of knowledge in terms of risk and costs associated to the failure of these elements. Even though the building and its elements are the most valuable asset of any owner, most often homeowners do not adopt effective preventive measures to mitigate the deterioration and obsolescence of their assets. This study proposes an innovative methodology for the design of insurance policies for buildings' envelopes, applied to natural stone facade claddings. The insurance product is defined based on deterministic and stochastic service life prediction models, established through the past degradation history of 142 natural stone claddings analyzed in service conditions in Portugal. Single-parameter (only analyzing the cladding's age) and multiparameter (encompassing the relevant variables) models are applied in the calculation of the insurance premium. The expected claims are related with the performance of maintenance actions and established according to three degradation levels. The results obtained reveal that an increased knowledge about the insured cladding leads to a reduction of the risk margin and consequently, to a lower annual value of commercial premium paid by a household. This study proposes an innovative solution for tailoring the insurance products, in terms of the risk of failure of the buildings components, as well as the financial charges related with the maintenance of these elements, channeling the risks to the market.

Keywords: Insurance; mathematical models; service life prediction models; natural stone claddings; insurance premium; risk assessment

---

1. Introduction

An insurance policy is a classical management tool to mitigate the risks of a given event, redistributing the costs of unexpected losses. Generally, by using the law of large numbers, the insurance policy reallocates the cost of losses from the few members experiencing them to all the insured members who paid the insurance premiums [1]. The theory reveals that, from the perspective of risk aversion, individuals are willing to buy an insurance product if the premium is actuarially reasonable, i.e., the insured is willing to spend a small amount in the present to be protected against the risk of a large monetary loss in the future [2,3].

In the housing sector, the insurance market presents different solutions and products to protect the client from unexpected losses in their dwellings. Most of the insurance market is focused on the construction stage and a posteriori periods of guarantee, given that this is the most challenging stage in terms of risk exposure. Concerning the use stage, the most common housing insurance policies are fire and multihazard, which intend to cover the damages for both interior/filling and property damage when subjected to accidents of various natures according to the insured's concerns. These insurances are often mandatory in many countries, imposed by creditors under real estate mortgage.

In the case of failure of a building component, the first and most important question that arises is "who pays?". Blong [4] states that the insured always intends to know whether building damage is covered under normal household insurance policies. According to this author, the answer would appear to be 'yes, but not always'. A study performed by the Building Research Establishment (BRE) in England reveals that buildings' owners are responsible for ensure the safety of buildings and their components, being also responsible for all the damages inflicted by any accident to third parties. The insurance companies are not legally responsible for the financial losses caused by the failure of the construction elements, when it is proved that the owners do not perform adequate maintenance actions [5].

Common housing insurance policies are based on past statistical standards and are usually too simplistic and generalist. Rarely in the past have actuarial data been collected regarding other crucial aspects concerning the specificity of the different building components, such as characteristics of the materials applied, design and execution processes, building's environmental exposure, use and maintenance conditions, among other parameters. Although the existing actuarial models present an overall picture of losses, they are incapable of offering a view of the losses in terms of specific building components [6]. Therefore, the insurance sector is in transformation and is rapidly moving towards to a deeper segmentation and specification.

In this sense, new methodological approaches for the design of building components' insurance products must be developed. Currently, the insurance market for losses regarding the natural degradation of buildings and components is practically nil. In fact, there are many imponderables involved in defining the vulnerability of the building component to the degradation agents and mechanisms. Moreover, the financial costs associated with the deterioration of the built park are difficult to measure. Nevertheless, building components, and in particular external claddings, can be seen as an insurable asset at risk of failure. For instance, internationally, there are some unfortunate news, with huge human losses, concerning the insurers facing increased risk of claims on covering high-rises with combustible cladding solutions (e.g., Grenfell Tower fire in London, UK).

Therefore, this study intends to propose an innovative approach for the design of an insurance policy model, applied to natural stone claddings. This study intends to identify the insurable risk of degradation of these claddings, examining how these risks are managed through the insurance method, thus analyzing different insurance premiums according to the expected claims and to the risk load. In the proposed model, the insurance premiums are established according to the knowledge concerning the loss of performance of natural stone claddings over time and according to their characteristics. For that purpose, deterministic and stochastic service life prediction models are used, which allow evaluating the likelihood of failure of the claddings. The service life prediction models used are defined through the knowledge of the degradation history of 142 natural stone claddings analyzed in service conditions, in Portugal.

The methodology proposed in this study allows establishing a system in which the insurance rates are scientifically and effectively determined, depending on more detailed risk categories, established based on technical knowledge related with the degradation of the building component under analysis. Moreover, it allows expanding and improving the scope of compensation of the insurance policies within the coverage of buildings' claddings, adopting service life prediction methods to improve the risk assessment, thus reducing the risk of failure of the elements. This methodology, allied to the understanding of the decision processes of the property owner and insurer, allows providing some guidance for developing a set of strategies to reduce losses from future natural degradation of buildings' facades.

The availability of realistic risk-based priced insurance policies can boost responsible behavior by owners, encouraging them to adopt cost-effective measures to reduce their potential losses from future degradation of buildings' components. An insurance for the buildings' facades can be an effective tool in assisting the restoration of damaged property and in the preservation of the built park, and can be seen as a means to fulfil governmental standards regarding the maintenance and conservation of buildings and components.

2. Background

The process of an insurance subscription encompasses five main stages [7]: (i) the first stage is the application, in which the customer contacts an insurance company to inquire about the types of coverage and related-costs; the customer fills out an application, which is sent to the insurance company underwriter; (ii) the second stage, when the insurance company performs a risk assessment, based on the amount of coverage requested, the policyholder's insurance history, and other objective actuarial factors, and decides to accept or reject the application and the associated premium; (iii) in the third stage, the establishment of the policy contract, which includes a declaration page, an insurance agreement, modifications, and endorsements, is sent to the policyholder; (iv) the next stage corresponds to the claims; most policyholders do not suffer losses that lead them to submit claims (which does not occur in most cases) during the term of their insurance policy. In the event of a loss, the policyholder calls the company claims department to file a claim. The insurer investigates the loss, verifying the coverage for a particular risk, determining whether a policy is in force and, if so, estimates the related-costs; and (v) finally, the last stage is the renewal; most policies run for 12 months and are renewed at the end of that period. However, the policyholder can choose to cancel the policy during this period or move to another insurer at the end of the policy period.

Therefore, the definition of any insurance policy requires accurate knowledge regarding the: (i) insured, which is the party receiving a benefit on the occurrence of a specified event; (ii) the period of insurance, during which the policy covers a specific risk; and (iii) the indemnity, which is the total amount payable by the insurance company on each accepted claim.

A premium is the payment made by the policyholder for a partial or total coverage of a certain risk. In a simple actuarial perspective, the fair premium for a generic risk, for a specific insured, is estimated by the expected value of the claim (cost and probability of occurrence), when that risk occurs. However, in actuarial practice, a solidarity principle is usually applied, and the premium is estimated by evaluating the total damage of a specific risk in all the individuals belonging to the same risk class and equally dividing the costs among them [3].

There are no standardized principles or procedures to be used by insurers in the determination of insurance premiums, as this constitutes a typical managerial and commercial strategy of each insurance company. As mentioned by Ewald [8], insurance institutions are not repetitions of a single formula applied to different objects; instead, each insurance company presents different purposes, customers and legal basis. Since there is no legislation on the definition of premiums, each insurer is free to set its own prices in any of the insurance modalities according to its cost structure and the customer's claims history.

The premium charged by the insurer is calculated based on the coverage, the exemption imposed, and other criteria considered as relevant for the insurer. In practice, insurers must account not only for the characteristics of the risks they have to insure, but also for other factors such as premiums charged by competing companies [9]. The premiums are estimated mainly based on the expected costs of claims and the legal and administrative costs [10]. Operating expenses, although variable across firms, are relatively predictable. Claim costs, on the other hand, are not.

3. Materials and Methods

Currently, insurers use the claims' history data to develop an estimate of the amount of damages to be paid in the future. The accuracy of the estimations depends on the type of risk, the number and the characteristics of policyholders, as well as the number and value of accepted claims. The definition of the premium rate of a building is usually carried out by observing the risk inherent to its characteristics and based on an analysis of the historical data concerning the losses for similar types of buildings in the same risk situation. However, each building and component is a unique prototype subjected to unrepeatable conditions [5]. Therefore, the definition of insurance policies to cover the risk of failure of external claddings must consider the specificities of the insured object.

The methodology adopted in this study intends to define an expected rate of return for the investor, based on the age and estimated service life of the insured object, to determine the rates of degradation to be applied and to calculate the cost of the insurance policy, discounting the indemnity and adding the various administrative costs. In this study, the definition of an insurance premium for natural stone claddings encompasses three main steps:

- Risk assessment, in which service life prediction models are used to estimate the risk of failure of stone claddings over time and according to their characteristics. These models allow identifying the claddings' expected condition in each instant, as well as the instant after which they present any given probability of reaching the end of their service life;
- The definition of the claims; in this case, the claims are related with the need of performing different maintenance actions according to the cladding's degradation during the period of insurance;
- The estimation of the annual premium; the insurer reserves a monetary amount (provision) for the eventuality of the occurrence of the worst-case scenario to its customer portfolio and the respective indemnity payment. This amount depends on three factors: (a) number of subscribers; (b) cost of replacing the cladding (€/m^2); (c) area of the facade in inadequate conditions (m^2).

The insurer should validate the conditions of underwriting described by the policyholder, through an in situ inspection, carried out by an expert in this field. The inspection must be aided by an inspection and diagnostic data sheet, to systematize the data collected during the inspection. This file includes the characterization of the building and of the natural stone cladding, its maintenance history, and the surrounding environmental condition; moreover, the degradation condition of the cladding must be accurately described, through the identification of the anomalies observed, as well as their degradation level, the location of the anomalies in the stone plates, the determination of the probable causes, and the extent of the anomalies, as a percentage of cladding area. The information concerning the degradation condition of the cladding is essential to validate the insurance premium.

3.1. Proposal of Deterministic and Stochastic Actuarial Models

3.1.1. Overall Approach

One of the main challenges of actuarial science is to predict the costs of a future event (usually relying on past statistics) before the losses occur [1]. Inaccurate estimations could be dangerous, in two ways: if the insurance premium is underestimated, in the event of an accident, the losses may be too high to be borne by the insurer, but in the opposite sense, the lack of knowledge about the risk and costs of loss may lead to the premium being over evaluated, leading the potential consumers to choose another insurance company, or simply not subscribe at all.

In this sense, the service life prediction models applied in this study allow defining deterministic and stochastic actuarial models based on the analysis of the actual degradation condition of 142 stone claddings (located in Portugal) analyzed in situ and subjected to real exposure conditions and real degradation agents and mechanisms. The actuarial models proposed in this study adopt four different service life prediction models, of increasing complexity: (i) single-parameter deterministic model; (ii) multiparameter deterministic model; (iii) single-parameter stochastic model; and (iv) multiparameter stochastic model.

These models are based on the evaluation of the physical degradation condition of the stone claddings. For that purpose, a numerical index, called severity of degradation (Sw), is used [11], which evaluates the extent, condition, and cost of repair of the anomalies observed in a natural stone cladding (Equation (1)).

$$S_w = \frac{\Sigma(A_n \times k_n \times k_{a,n})}{A \times k},$$ (1)

where S_w represents the severity of degradation of the building component, expressed as a percentage, k_n is the multiplying factor of anomaly "n", as a function of their degradation level, within the range K = {0, 1, 2, 3, 4} (Table 1), $k_{a,n}$ is the weighting factor corresponding to the relative weight of the anomaly detected (Table 2), A_n is the area of cladding affected by an anomaly "n", in m^2, A is the facade area, in m^2, and k is the multiplying factor corresponding to the highest degradation level of a coating of area A.

Table 1. Degradation levels for natural stone claddings (data sourced from Silva et al. [12]; Mousavi et al. [13]).

Degradation Level	Anomaly		% Area of Cladding Affected
Level 0 ($S_w \leq 1\%$)		No visible degradation	-
Level 1 Good ($1\% < S_w \leq 8\%$)	Aesthetic degradation anomalies	Surface dirt Moisture stains Localized stains Colour change	>10% ≤15%
		Flatness deficiencies	≤10%
	Loss-of-integrity anomalies	Material degradation (*) ≤ 10% plate thickness Cracking width ≤ 0.2 mm	≤20%
Level 2 Slight degradation ($8\% < S_w \leq 20\%$)	Aesthetic degradation anomalies	Moisture stains Localized stains Colour change Biological growth Parasitic vegetation Efflorescence	>15% ≤30%
		Flatness deficiencies	> 10% e ≤ 50%
	Joints anomalies	Joints material degradation Loss of material—open joint	≤30% ≤10%
	Fastening to the substrate anomalies	Scaling of stone near the edges Partial loss of stone material	≤20%
	Loss-of-integrity anomalies	Material degradation (*) ≤ 10% plate thickness	>20%
		Material degradation (*) > 10% and ≤ 30% plate thickness	≤20%
		Cracking width ≤ 0.2 mm	>20%
		Cracking width > 0.2 mm and ≤ 3 mm	≤20%
		Fracture	≤5%
Level 3 Moderate degradation ($20\% < S_w \leq 45\%$)	Aesthetic degradation anomalies	Biological growth Parasitic vegetation Efflorescence	>30%
		Flatness deficiencies	>50%
	Joints anomalies	Joints material degradation Loss of material—open joint	>30% >10%
	Fastening to the substrate anomalies	Scaling of stone near the edges Partial loss of stone material	>20%
		Detachment	≤10%
	Loss-of-integrity anomalies	Material degradation (*) > 10% and ≤ 30% plate thickness	>20%
		Material degradation (*) > 30% plate thickness	≤20%
		Cracking width > 0.2 mm and ≤ 3 mm	>20%
		Cracking width ≥ 3 mm	≤20%
		Fracture	> 5% e ≤ 10%
Level 4 Generalized degradation ($S_w \geq 45\%$)	Fastening to the substrate anomalies	Detachment	>10%
	Loss-of-integrity anomalies	Material degradation (*) > 30% plate thickness	>20%
		Cracking width > 3 mm	
		Fracture	>10%

(*)—Material degradation is meant to be every anomaly that involves loss of volume of the stone material.

In the single-parameter deterministic model, the evolution of the degradation condition of stone claddings over time (the only parameter analyzed) is defined based on the analysis of a degradation curve. Figure 1 shows an illustrative example of the degradation curve obtained for natural stone claddings, with three different Sw thresholds, in which a third-degree polynomial curve is adjusted to the scatter of points corresponding to the severity of degradation (obtained through a fieldwork survey) and the ages of a set of 142 case studies analyzed.

Table 2. Weighting coefficients corresponding to the relative weight of the anomaly detected in natural stone claddings (data sourced from Silva et al. [12]).

Anomaly		Repair Operation (Cost in €/m^2)	Ratio between Repair Cost and Replacement Cost [*]	Weighting Coefficient $k_{a,n}$
Visual or surface degradation		Cleaning (11.7 €/m^2)	13%	0.13
Joints	Degradation of filling material	Joint repair (23.4 €/m^2)	25%	0.25
	Loss of filling material	Replacement of the joint material in cladding directly adhered to the substrate involves some risks, and may damage the natural stone	100%	1.0
Bond to substrate		Replacement of stone plates always costs at least as much as executing a new cladding, and may cost more because of having to remove the degraded original cladding	120%	1.2
Loss of integrity		Repairing loss-of-integrity anomalies may involve only a surface repair (epoxy resins or equivalent) or the replacement of the stone plate	100%	1.0

(*)—The cost of building a vertical granite cladding facade with a cementitious adhesive is around 93.10 €/m^2.

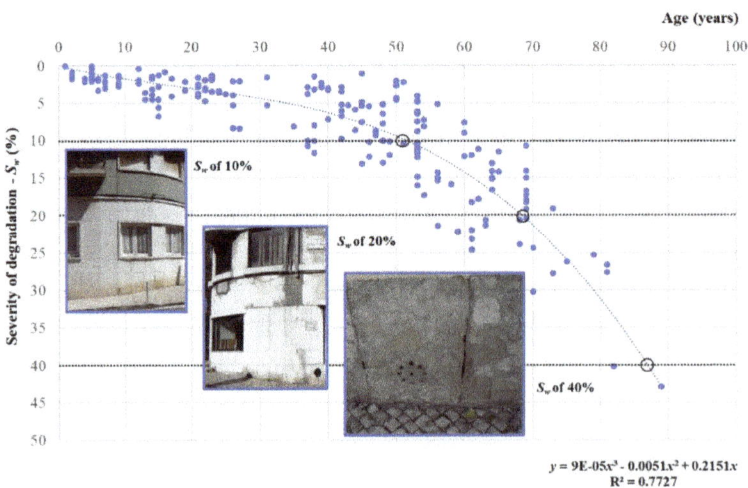

Figure 1. Degradation's evolution using the severity of degradation level for the 142 stone claddings inspected (data sourced from Silva et al. [12]).

The three S_w thresholds considered in this study correspond to different levels of claims for the insurance company. The first level, corresponding to a S_w of 10%, reveals a stone cladding with visual anomalies and slight cracking in the cladding, thus requiring a simple cleaning action. The second level, corresponding to a S_w of 20%, reveals a stone cladding with more serious defects, requiring a major intervention and partial replacement of the cladding. The third level, corresponding to a S_w of 40%, requires the replacement of the stone cladding.

According to the ISO 15686 [14], the service life of a building component can be defined as the period of time during which the element meets or exceeds the performance requirements. In this sense, the service life of a natural stone cladding can be predicted by determining the instant after which the cladding presents a degradation condition that it is no longer acceptable for the owners. The limit that establishes the end of service life (or the maximum acceptable degradation level) is not easy to define, and depends on the building context, social and legal requirements, as well as the funds available for maintenance actions [5]. Using the degradation curve, it is possible to obtain the estimated age of the natural stone claddings through the intersection of the degradation curve with a given degradation level. Table 3 shows the estimated age of the claddings obtained with the first model for three different S_w values. In this model, the of estimated values are independent of the cladding's characteristics, since the model only encompasses the age of the building as explanatory variable of the stone claddings' degradation phenomenon.

Table 3. Estimated age of natural stone claddings for three different S_w values (single-parameter deterministic model).

Degradation Level S_w (%)	Estimated Age (years)
10	52
20	68
40	88

This model presents accurate results but could be too simplistic to portray a complex phenomenon such as the degradation of natural stone claddings. In fact, the degradation of these claddings occurs due to a synergetic and simultaneous action of degradation agents and mechanisms, and the specific characteristics of the claddings must be considered, to obtain more reliable information. Therefore, this methodology can be used to disaggregate a generic vulnerability curve into several curves representing the vulnerability of specific building classes.

In this sense, a multiparameter deterministic model is also proposed, which allows encompassing different factors (e.g., environmental, design and type of materials, among others) in the determination of the estimated service life of stone claddings. Silva et al. [5,15] proposed different linear and nonlinear multiple regression models to estimate the severity of degradation of stone claddings. Among these, the generic exponential model, as shown in Equation (2), is the one with the best overall performance to describe the degradation of stone claddings; therefore, it is applied in this study.

$$S_w = 7.478e^{0.035*I-1.501*M-1.756*H-1.777*A-1.062*TP}, \tag{2}$$

where S_w represents the claddings' severity of degradation, I their age, M distance from the sea, A area of the cladding, H exposure to damp, and TP type of stone. To apply the model shown in Equation (2), the variables M, A, H, and TP should be replaced by their numerical quantification, presented in Table 4 [15].

The deterministic models provide accurate results and allow obtaining an average value of the instant after which it is necessary to intervene (the instant in which the insured makes the claim). However, more than an average estimated service life, the insurance company should know the probability of a stone cladding reaching the end of its service life, i.e., the probability of failure of the cladding over time and according to its characteristics. This information is crucial to convert those probabilities into costs, to predict unexpected situations during the claddings' service life, whatever the origin of the unforeseen events [16].

Table 4. Numerical quantification of the variables of the multiparameter deterministic model.

Independent Variables	Variables' Quantification		
Distance from the sea	≤5 km: 0.96		>5 km: 1.03
Dimensions of stone plates	Medium dimensions: 1.04		Large dimensions: 0.94
Exposure to damp	Low: 1.03		High: 0.91
Type of stone	Limestone: 1.04	Marble: 0.96	Granite: 1.39

In this sense, two stochastic models are proposed, both based on a logistic regression analysis. In the first model, a single-parameter stochastic model, it is possible to estimate the probability of the stone claddings reaching the end of their service life over time. Figure 2 shows the cumulative distribution functions concerning the probability of 142 stone claddings reaching the end of their service life, considering three S_w levels. For a S_w of 40%, a discrete cumulative distribution function is obtained, due to the small number of case studies with this degradation level. With this information, it is possible to define, for different risk thresholds, the probability of a given cladding reaching the specific degradation values (Table 5).

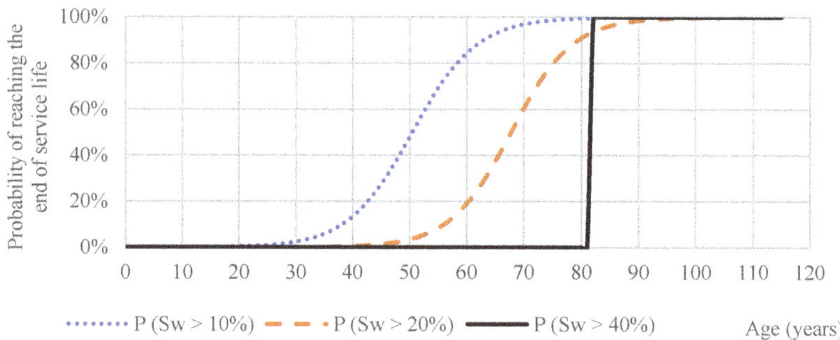

Figure 2. Cumulative distribution functions concerning the probability of 142 stone claddings reaching the end of their service life, considering three risk thresholds.

Table 5. Estimated claddings age obtained by the single-parameter stochastic model for different S_w values and risk margins.

		Risk Margin			
		5%	10%	20%	50%
	10	34	39	43	51
S_w (%)	20	53	57	61	68
	40	74	80	85	94

In the multiparameter stochastic model, it is possible to include all the relevant variables to the explanation of the severity of degradation of stone claddings, allowing estimating the probability of the cladding reaching the end of its service life according to its characteristics. In the case of natural stone claddings, the variables considered as statistically relevant are the distance from the sea, size of the stone plates, exposure to damp and type of stone. The multiparameter stochastic model adopts a similar approach of the single-parameter model, but the determination of the estimated service life requires the application to a specific case study, since it requires the quantification of the explanatory variables included in the model.

3.1.2. Models' Assumptions Concerning the Coverage of the Insurance Policy

The main coverage of the proposed insurance policy is related with the performance of three maintenance actions, when a specific degradation level is reached. The degradation levels that imply the performance of a specific action are intrinsically related with the owners' levels of demand, and the cost that the owners are willing to pay to reduce the likelihood or magnitude of failure of the claddings [17]. In this study, the claims occur for three different S_w thresholds, including the following maintenance actions: (i) for S_w of 10% (slight degradation), a cleaning action is performed; (ii) for S_w of 20% (moderate degradation), a major intervention is performed; and iii) for S_w of 40% (generalised degradation), a replacement of the cladding is required.

The insurance company must perform periodic inspections, to estimate the S_w value. Based on the results obtained, the insurer decides to claim maintenance actions and the related budget, or not. The different levels of maintenance include the following actions:

i.   Cleaning: includes scaffolding installation, cleaning with water jet and brushing [18], repair of cracking with a width ≤ 1 mm and replacement of 20% of joints;

ii.   Major intervention: includes all the previous actions but with the replacement of 30% of joints. Additionally, it includes the cleaning of the remainder 70% joints and the replacement of 20% of the stone plates;

iii. Replacement: includes scaffolding installation and the complete replacement of the cladding, with the application of the new cladding and the transportation to a landfill of the old cladding.

The costs of these actions are presented in Table 6 and are based on the most recent values of each operation found in the literature and from specialized companies.

Table 6. Fixed costs of the maintenance actions (data sourced from CYPE [19] and Mousavi et al. [13]).

S_w Index	Details of the Maintenance Task	Costs (Year 0) (€/m²)
10%	Scaffolding	3.69
	Cracking	8.28
	Cleaning	19.03
	Joint's replacement (20%)	2.05
	Total	33.05
20%	Scaffolding	3.69
	Cracking	8.28
	Cleaning	19.03
	Joint's replacement (30%) + Joint's repair (70%)	6.67
	Cladding's replacement (20%)	11.23
	Total	48.90
40%	Scaffolding	3.69
	Cladding's replacement (100%)	56.14
	Total	59.83

The insurance policy proposed presents some limitations regarding the claims:

i.   During the duration of the policy, the claim of each maintenance action occurs only once, i.e., after a cleaning action, the insurer will only intervene when the next degradation level is reached, thus performing a major intervention;

ii.   It is considered, as a simplification, that the maintenance actions have no effect on the cladding's degradation condition (in terms of the S_w value), i.e., the age in which the next maintenance action occurs is the same before and after the maintenance action;

iii.   It is assumed there are no periodic maintenance actions by the insured;

iv.   In the case of a condominium, the insurance is equally shared by the individual unit owners that pay the same premium;

v.   An annual policy with renewal option is adopted. The premium is fixed during the period of subscription, independently of the rate's volatility, which is beneficial for the insured, since he/she knows the total cost of the insurance policy, without surprises, hassles, or extra calculations.

3.1.3. Determination of the Annual Premium for the Insurance Policy

This study establishes a calculation method for an insurance policy to cover the natural degradation of buildings' components. According to the methodology proposed, the total present value of the insurance policy is given by Equation (3).

$$PV = \frac{C_{t,nom(C)}}{(1+r_{nom})^{t_C}} + \frac{C_{t,nom(MI)}}{(1+r_{nom})^{t_{MI}}} + \frac{C_{t,nom(R)}}{(1+r_{nom})^{t_R}}, \tag{3}$$

$C_{t,\,nom(C)}$, $C_{t,\,nom(MI)}$, and $C_{t,nom(R)}$ correspond to the nominal costs of the maintenance actions of the different degradation thresholds, t_C, t_{MI}, and t_R, indicating the year in which those costs occur and r_{nom} represents the nominal discount rate, which includes the global inflation risk, opportunity costs, and other costs. The $C_{t,nom}$ terms include the effect of inflation, which affects the nominal discount rate, as explained by Fisher's theory [20], as shown in Equation (4).

$$1 + r_{nominal} = (1 + r_{real}) * (1 + i) \Leftrightarrow r_{real} = \frac{1 + r_{nominal}}{1 + i} - 1, \tag{4}$$

The use of nominal or real values must be consistent, when the calculations are made with cash flows and discount rates. To work in nominal terms, the trends in equipment, labor, and material costs must be contemplated. In this study, an inflation rate (i_s) of 3% and a global inflation rate (i_g) of 2% are adopted, in accordance with the Portuguese context. To simplify the presentation of this study results, real values are used. Equation (3) can be used to determine the expected costs and premiums, as shown in Equations (5) and (7), respectively.

$$PV, cost = \frac{C_{t,real(10)}}{\left(1+r_{real,cost}\right)^{t_{10}}} + \frac{C_{t,real(20)}}{\left(1+r_{real,cost}\right)^{t_{20}}} + \frac{C_{t,real(40)}}{\left(1+r_{real,cost}\right)^{t_{40}}}, \tag{5}$$

$C_{t,real(10)}$, $C_{t,real(20)}$, and $C_{t,real(40)}$ are the real costs of the maintenance actions of the different degradation thresholds, which are the same for all the proposed models; t_{10}, t_{20} and t_{40} are the years in which those costs occur; $r_{real,cost}$ is the real discount rate, equal for all the models and obtained through Equation (4). Equation (6) presents the discount rate applied to the cost, only considering rate i_s.

$$r_{real,cost} = \frac{1 + r_{nominal}}{1 + i_s} - 1 = \frac{1 + 0.06}{1 + 0.03} - 1 = 0.029 = 2.9\%, \tag{6}$$

Equation (7) presents the present value of the premium, in which $C_{t,premium}$ is the annual premium in €/m², whose definition is the main objective of applying the proposed models.

$$PV, premium = \frac{C_{t,\,premium}}{\left(1+r_{real,premium}\right)^{1}} + \frac{C_{t,\,premium}}{\left(1+r_{real,premium}\right)^{2}} + \ldots + \frac{C_{t,\,premium}}{\left(1+r_{real,premium}\right)^{t_R}}, \tag{7}$$

The $r_{real,premium}$ is the real discount rate, equal for all the models, considering not only rate i_s but also rate i_g, as shown in Equation (8).

$$r_{real,premium} = \frac{1 + r_{nominal}}{(1 + i_s) * (1 + i_g)} - 1 = \frac{1 + 0.06}{(1 + 0.03) * (1 + 0.02)} - 1 = 0.0089 = 0.89\%, \tag{8}$$

The premium value ($C_{t,prémio}$) is obtained by equalling Equations (5) and (7). The value obtained is converted to Euros paid per household by multiplying the premium by the cladding's area and then dividing by the number of individual apartments. The final definition of the premium value should also consider the insurance company's profit margin. The different models proposed allow transforming the probability of the claddings reaching the end of their service life into profit margins.

4. Application of the Methodology to a Real Case Study

One of the most important aspects in the insurance premium calculation is price discrimination. A bonus-malus premium calculation principle must be used, ensuring that not all the individual policies in the portfolio present the same premium, because the level of risk is distinct [21,22]. This concept is widely adopted in car insurance, in which a series of variables affecting the risk level are analyzed to establish the premium, e.g., the age and the gender of the driver, his/her living address, time since the driving licence was issued, among others. A similar principle should apply to insurance policies for claddings, in which the characteristics of these elements should be contemplated when the premium is determined.

For that purpose, the proposed models are analyzed using a real building as a case study. The first step was the visual inspection of the cladding in situ, collecting the relevant information to apply the service life prediction models. This inspection was complemented by an interview with the building's architect. This case study is in Lisbon, Portugal, and has 15 apartments and 606 square meters of stone cladding. The cladding is 14 years old, presents medium-sized limestone plates (0.24 m^2), is located less than 5 km from the sea, and has a high exposure to damp.

With the building's characteristics, the application of the multiparameter models for the case study is feasible and the comparison between the resulting cladding age values and the ones obtained through the single-parameter models. That knowledge, paired with the costs previously presented, allows calculating the risk premium for each model.

Table 7 presents the case study's remaining years until the proposed maintenance actions, for each model, and considering four insurance risks for which different premiums must be designed.

Table 7. Case study's remaining time (in years) before action for the defined degradation levels, by model.

		Deterministic		Single-Parameter Stochastic				Multiparameter Stochastic			
		Single-Parameter	Multiparameter	Risk Margins							
				5%	10%	20%	50%	5%	10%	20%	50%
	10%	38	34	20	25	29	37	19	24	28	36
S_w	20%	54	54	39	43	47	54	50	52	55	59
	40%	74	73	60	66	71	80	70	72	76	81

The values were obtained by subtracting from the different estimated cladding age values the present age of the cladding (14 years). Table 6 typifies some aspects deemed important:

i.   The cladding's characteristics do not produce significant deviations between the remaining time before action obtained by the deterministic models. The comparison between the remaining time before action obtained by the single-parameter stochastic model and multiparameter stochastic model, for the threshold $S_w = 10\%$, for any risk margin is also inconclusive;

ii. In the multiparameter stochastic model, the risk's reduction is equivalent to an anticipation of the maintenance action's schedule. This effect is less noticeable in the second and third degradation levels;

iii. For the thresholds $S_w = 20\%$ and $S_w = 40\%$, in both stochastic models, when a lower level of risk is assumed, naturally a lower remaining time before action is obtained. This conclusion is more evident for the single-parameter stochastic model, for replacement actions ($S_w = 40\%$);

iv. For a fixed risk margin, in both the second and third degradation thresholds, the expected remaining time before action is higher in the multiparameter stochastic model than in the single-parameter one. The difference is substantial for lower risk margins (5%) and for more profound maintenance actions, such as the replacement.

Based on these values, the premium for each model is determined. The method is the same for every model, only changing the years in which the maintenance actions occur and the maximum duration of the policy until renewal. This duration corresponds to the period until the materialization of the last maintenance action associated with a $S_w = 40\%$. The procedure will be exemplified with the single-parameter deterministic model, as shown in Equation (9).

$$PV, premium = C_{t,premium}\left(\frac{1}{r_{real,premium}} - \frac{1}{r_{real,premium}\left(1 + r_{real,premium}\right)^{t_{40}}}\right), \tag{9}$$

Equation (10) presents the present value of the future payment.

$$PV, cost = \frac{33.05}{1.029^{38}} + \frac{48.90}{1.029^{54}} + \frac{59.83}{1.029^{74}} = 28.63 \ /m^2, \tag{10}$$

Finally, it is possible to obtain the annual premium in €/m², as shown in Equation (11).

$$28.63 = C_{t,premium}\left(\frac{1}{0.0089} - \frac{1}{0.0089(1 + 0.0089)^{74}}\right) \Leftrightarrow C_{t,premium} = 0.53/m^2, \tag{11}$$

This result is the risk premium rate for this case study, when the insurance is subscribed in 2017 based on this model. The risk premium by household is obtained through Equation (12).

$$\frac{C_{t,premium} * Area\ of\ stone\ cladding}{N.o\ of\ households} = \frac{0.53 * 606}{15} = 21.45 \tag{12}$$

The commercial premium is obtained by multiplying the result of Equation (10) by some coefficients that represent the margins for administrative expenses and expected profit. In this methodology, a fixed coefficient of 1.3 universal to all models is considered first, intending to cover the fixed costs present in any financial institution. A safety margin of 1.2 for the single-parameter and 1.1 for the multiparameter models is also applied, accounting for the estimation errors of the models. The commercial premium by household of the case study analyzed, for the single-parameter deterministic model, is 33.47€. Table 8 summarizes the same calculations for the remaining models.

Table 8. Summary table with estimated times before action, costs, risk premium rates, risk premiums, risk coefficients, and commercial premiums per household for the different models.

		Deterministic		Single-Parameter Stochastic				Multiparameter Stochastic			
		Single-parameter	Multiparameter	5%	10%	20%	50%	5%	10%	20%	50%
S_w	10%	38	34	20	25	29	37	19	24	28	36
	20%	54	54	39	43	47	54	50	52	55	59
	40%	74	73	60	66	71	80	70	72	76	81
$VP,cost$ (€/m^2)		28.63	30.19	45.26	39.35	34.85	27.82	38.81	35.15	31.63	26.59
$C_{t,premium}$ (€/m^2)		0.53	0.56	0.98	0.79	0.67	0.49	0.75	0.66	0.58	0.46
Risk premium by household (€)		21.45	22.84	39.55	32.02	26.90	19.75	30.26	26.86	23.26	18.72
Risk coefficients		1.3*1.2	1.3*1.1			1.3*1.2				1.3*1.1	
Commercial premium per household (€)		33.47	32.66	61.69	49.96	41.96	30.80	43.28	38.41	33.26	26.77
Final premium per household (€)		-	-	33.89	32.72	31.92	30.80	28.42	27.93	27.42	26.77

Table 8 shows the relation between the estimated service lives related with each maintenance action and the resulting insurance's premium. The deterministic model, although simpler, has the same probability of not estimating correctly the service life as the stochastic model for a risk margin of 50%. The simplification of using the average values leads to a slightly higher risk premium.

5. Results

In this study, the knowledge of the loss of performance of stone claddings, based on the evolution of degradation of these elements in real exposure conditions, is used to develop an insurance product. According to the results obtained, the following main conclusions can be drawn:

i.   An insurance product based on a deterministic model will not be able to compete against an insurance based on a stochastic model with a 50% risk margin, because it provides the same coverage and risk charging, with a higher premium;

ii. In the deterministic approach, the commercial premium's difference between single-parameter and multiparameter is below 2€, which reveals the lack of preponderance of the discrimination of the cladding's characteristics in these models;

iii. In the stochastic models, the lesser the assumed risk margin, the greater the premium. This aspect is more evident in the single-parameter model;

iv. For a fixed risk margin, the premium is lower in the multiparameter stochastic model than in the single-parameter one. The difference is more expressive for smaller risk margins (5%);

v.   These results reveal that the design of an insurance product knowing the characteristics of the insured object allows reducing the costs for the policyholder, which also increases the possibility of acquiring this insurance policy.

The risk premium is higher for an insurance product that predicts with greater prudence the expected intervention periods because the cost's increase is passed on to the client. During the subscription to the product "multiparameter stochastic 5%", the insurance company considers that the client belongs to the 5% whose cladding will need a cleaning at year 19, a major intervention at year 50 and a total replacement at year 70. In reality, 95% of the clients will have those maintenance needs at a later stage of the cladding's life, which results in lower costs for the insurer.

In the real world, it is well-known that the prices reflect elasticities of demand just as much as costs [23]. If the premium prices are not well-adjusted to the market and to the insured object, customers will naturally choose competing companies with a better, risk adjusted pricing system [24]. Therefore, in this study, in order to provide a greater equilibrium to the commercial premiums, increasing their competitiveness and motivating a risk reduction, the final premium is adjusted. The insurer obtains some advantages through the application of the service life prediction models, i.e., knows, for a given cladding, the risk of failure with a known margin of risk. This knowledge must beneficiate not only the insurer, but also the policyholder. For that, the premium calculations must be made with the stochastic model's results, and for models with risk margins lower than 50%, the insurance company can decrease up to 90% of the difference between the premium of the considered model and the premium of the 50% risk margin model, in order to make the product more competitive. Equation (13) presents an example of the determination of the final premium, adopting the multiparameter stochastic model with 5% risk margin.

$$Final\ premium\ by\ household = 43.28 - 0.9 * (43.28 - 26.77) = 28.42 \tag{13}$$

The commercial premium for the policyholders of the case study varies between 25€ and 35€ per year. Considering the average apartment's real estate value, the yearly premium corresponds to less than 0.1% of the cost of an apartment. The same cannot be said about vehicle insurance, the most popular nonlife insurance branch, where it is unexpected that the premium equals 0.1% the value of the vehicle. This difference shows the asymmetry in the way society values the building stock compared to other tangible assets.

6. Discussion

As mentioned by Kunreuther et al. [25], consumers and landlords usually misperceive the insurance products, mainly because of their feelings about losses they may or may not experience. The consumers usually want to pay low premiums, which do not cover all the possible losses, and feel disappointed when they suffer a loss; on the other hand, consumers see the insurance product as an unwise investment when they pay an insurance premium for several years and the loss does not occur. The authors [25] refer that when purchasing an insurance product, a costumer's mantra should be *the best return is no return at all*, i.e., the insurance product should protect the costumer from potential losses, at an acceptable premium, while it is expected that the loss or an extreme event would never occur.

The potential losses due to the future degradation of buildings' components are usually neglected by insurance companies, and most homeowners do not voluntarily adopt cost-effective measures to reduce the impact and costs of those losses [26]. These attitudes led to a built park with clear signs of degradation and a high economic burden associated with their maintenance and repair. In this sense, the role of the insurance companies is extremely important, since the policyholders can transfer the risk of degradation of the construction elements to insurance companies, which will defray the cost of maintenance and replacement of the buildings' components in case of failure [27].

Currently, insurers do not have enough information to propose realistic risk-based pricing, incorporating mitigation activities [28,29]. The insurance business is highly dynamic and competitive and the clients' majority is driven by basic insurance products with the lowest premium possible. However, the insurer must ensure the definition of well-designed insurance products, even if it costs more. In this sense, it is extremely important to have accurate knowledge regarding the object insured and the risk and cost of losses if a given event occurs.

Past studies [27,30] have discussed the benefits of using insurance as a mechanism for risk mitigation. In fact, several studies [31–34] discuss how insurance can be efficiently used to decrease litigation for defective construction. Moreover, some studies [35] also analyze the risks associated with the serviceability of the building and its components. However, the literature does not provide much evidence or insights on the particular design of such insurance policies for a specific maintenance/performance risk. This study has an innovative character, providing the methodological basis for the design of such insurance policies. It follows the standard approach for designing insurance (probability analysis, incident cost, and risk premium calculation), but specifically adapted to natural stone facade claddings.

More studies are needed in order to properly evaluate the aging and deterioration, and the consequent probability of failure of the facade elements. Some standard methods can be found in the literature, addressing the analysis of accelerated climate aging of building materials under different laboratory conditions [36,37]. These methods are extremely relevant for a deeper knowledge of the materials characteristics and their behaviour under different exposure conditions, allowing predicting an expected lifetime for a specific building material under a specific set of conditions. Accelerated artificial aging exposure experiments, such as the methods proposed by [36,37], save both time and cost; nevertheless, the extrapolation of these analyses to the real degradation processes can be a challenging task since the deterioration process in-service conditions is ruled by the simultaneous and synergic occurrence of several degradation factors. As mentioned by Silva et al. [15], each facade is a unique prototype subjected to unrepeatable conditions. In this sense, this study suggests the adoption of the analysis of real case studies, under natural exposure conditions, to evaluate the future trend of degradation of stone claddings.

Currently, there is an urgent demand for methodologies to evaluate the probability of failure of buildings and their components. Some probabilistic-based methodologies have been proposed for the evaluation of the degradation of building materials [38]. Structural approaches to evaluate the risk of failure can also be applied for modelling the deterioration of facade claddings. For example, Silva et al. [15] proposed the adoption of artificial neural networks, fuzzy systems, and Markov chains to model the deterioration and service life of external claddings. Markov chains are usually adopted for the stochastic analysis of the performance prediction, service life, and maintenance management of bridges. This model was successfully adopted for modelling the transition between degradation condition of facades' claddings, and as well, other similar structural approaches can be used to model the service life prediction of facades' claddings, thus providing extremely useful information related with the risk of failure of a given cladding, according to its age and specific conditions of exposure and use. In this study, a probabilistic approach is also proposed for the definition of the insurance policy, revealing that a better knowledge of the risk of failure of the cladding benefits not only the insurer, but also the policyholder.

The methodology for the premium's valuation presented in this study needs to be adjusted based on the insurer's portfolio and on the market conditions. In this study, a proposal that fits the observed context was presented, although it needs to be readjusted over time.

In future research, it would be relevant to evaluate similar insurance products covering different facade materials, such as ceramic cladding or renderings, and other coverings of building components, such as roofs and structure. Other aspects should also be analyzed in the future and adapted to each specific context, such as variation in the type, periodicity and effects of the maintenance actions, flexible scheduling of the maintenance actions in accordance with the owners' needs and expectations, and the measurement of the severity of degradation of the cladding before defining the insurance policy and estimating the premium cost. Finally, future research should also include an analysis of the transaction and administrative costs associated with this methodology, considering that they will be higher for insurers when compared with the "business as usual" approach.

7. Conclusions

In this study, an innovative approach for the design of an insurance policy model, applied to natural stone facade claddings, is proposed. In this model, the insurance rates are scientifically and effectively determined, based on technical knowledge related with the real loss of performance of the stone claddings over time and according to their characteristics (collected through an extensive fieldwork). The insurance premiums are thus designed based on the information obtained by deterministic and stochastic service life prediction models. The service life prediction models allow estimating, in a more precise and scientific manner, the expected claims and the related risk load.

The results obtained reveal that stochastic models produce more relevant information for the definition of the insurance policies. The knowledge of the probability of failure of the building's components over time and according to their characteristics is a very useful information to calculate the fair price of an insurance contract. Therefore, the stochastic models allow reducing the risk assumed by the insurance company by evaluating whether the planned maintenance activities must be anticipated or delayed, according to the corresponding degradation thresholds and the risk of the cladding reaching those thresholds.

The segmentation of the claddings' characteristics allows discriminating the price of the insurance policy for different policyholders. In this sense, multiparameter models are more precise and useful. For an equal risk margin, and for more intrusive maintenance actions (also more expensive), such as the cladding's replacement, a more accurate knowledge regarding the cladding's characteristics leads to a postponement of the claims (i.e., the execution of the maintenance actions). An increased knowledge of the insured object allows decreasing the insurance premiums paid by the policyholders, as well as increasing the insurer's safety margins.

Therefore, the use of multiparameter stochastic models in the design of the insurance premium is beneficial for both parties. The application of the insurance policy proposed in this study presents several advantages. For the policyholder, it allows altering the nature of the risk and its allocation, shifting increased risks to the insurance company. Moreover, this insurance product promotes the conservation of the built heritage, increasing the patrimonial value of the asset (which is beneficial for the image of the cities), reducing the risk of failure and the uncertainty of the cost of maintenance over time. For residential condominiums, this type of insurance can be very profitable, since the risks are shared by the households, resulting in a lower price for the insurance premium. For the insurance company, the adoption of accurate information regarding the probability of the claims and the related risks allows reducing the premium. On one hand, in the long-term, a lower insurance premium can affect the profits per insured. However, lower premium prices, once well-adjusted as described in this study, with safe and well-known risk margins, will expectedly lead to a greater number of clients in the portfolio, thus leading to a global reduction of the risks assumed by the insurance company.

Author Contributions: Data curation, A.S.; Formal analysis, C.O.C.; Investigation, M.M.; Methodology, J.d.B.; Writing—original draft, M.M.; Writing—review & editing, J.d.B., A.S. and C.O.C.

Funding: This research was funded by the FCT (Foundation for Science and Technology) through the projects SLPforBMS (PTDC/ECM-COM/5772/2014) and BestMaintenance-LowerRisks (PTDC/ECI-CON/29286/2017).

Acknowledgments: The authors gratefully acknowledge the support of the CERIS Research Institute, Instituto Superior Técnico (IST), University of Lisbon.

Conflicts of Interest: The authors declare no conflict of interest.

References

1. Dorfman, M.S. *Introduction to risk management and insurance*, 6th ed.; Prentice Hall International, Inc.: Upper Saddle River, NJ, USA, 1998.
2. Mossin, J. Aspects of rational insurance purchasing. *J. Political Economy* **1968**, *76*, 553–569. [CrossRef]
3. Bruno, M.G.; Grande, A.; Oliva, C.F. Anomalous demand and supply in cat risks insurance market. *Annali del dipartimento di metodi e modelli per l'economia il territorio e la finanza* **2015**, *1*, 51–62.
4. Blong, R. Residential building damage and natural perils: Australian examples and issues. *Build. Res. Inf.* **2004**, *32*, 379–390. [CrossRef]
5. Silva, A.; de Brito, J.; Gaspar, P. *Methodologies for service life prediction of buildings: With a focus on facade claddings*; Springer International Publishing: Zurich, Switzerland, 2016.
6. Khanduri, A.C.; Morrow, G.C. Vulnerability of buildings to windstorms and insurance loss estimation. *J. Wind Eng. Ind. Aerodyn.* **2003**, *91*, 455–467. [CrossRef]
7. American Insurance Association—Property-Casualty Insurance Basics: A Look inside the Fundamentals and Finance of Property & Casualty Insurance. Available online: http://www.aiadc.org/File%20Library/Resources/Industry%20Resources/Insurance%20101/Ins_101.pdf (accessed on 2 May 2017).
8. Ewald, F. Insurance and risk. In *The Foucault Effect: Studies in Governmentality*; Burchell, G., Gordon, C., Miller, P., Eds.; University of Chicago Press: Chicago, IL, USA, 1991; Chapter 10; pp. 197–210.
9. Dickson, D. *Insurance risk and ruin*; International Series on Actuarial Science, Cambridge University Press: Cambridge, UK, 2010.

10. Pollner, J.D. *Managing catastrophic Disaster Risks Using Alternative Risk Financing and Pooled Insurance Structures*; Volumes 23-495, World Bank Technical Paper no. 495; The International Bank for Reconstruction and Development, The World Bank: Washington, DC, USA, 2001.

11. Gaspar, P.; de Brito, J. Limit states and service life of cement renders on facades. *J. Mater. Civ. Eng.* **2011**, *23*, 1396–1404. [CrossRef]

12. Silva, A.; de Brito, J.; Gaspar, P. Service life prediction model applied to natural stone wall claddings (directly adhered to the substrate). *Constr. Build. Mater.* **2011**, *25*, 3674–3684. [CrossRef]

13. Mousavi, S.H.; Silva, A.; Brito, J.; Ekhlassi, A. Service life prediction of natural stone claddings with an indirect fastening system. *J. Perform. Constr. Facil* **2017**, *31*, 04017014. [CrossRef]

14. ISO 15686-1. *Building and constructed assets: Service life planning—Part 1: General principles*; International Standard Organization: Geneva, Switzerland, 2011.

15. Silva, A.; Gaspar, P.L.; de Brito, J. Comparative analysis of service life prediction methods applied to rendered facades. *Mater. Struct.* **2016**, *49*, 4893–4910. [CrossRef]

16. Macedo, M.; de Brito, J.; Cruz, C.; Silva, A. Methodological proposal for the development of insurance policies for building components. *J. Build. Eng.* **2018**. under review.

17. MacDonald, D.N.; Murdoch, J.C.; White, H.L. Uncertain hazards, insurance, and consumer choice: Evidence from housing markets. *Land Econ.* **1987**, *63*, 361–371. [CrossRef]

18. Madureira, S.; Flores-Colen, I.; de Brito, J.; Pereira, C. Maintenance planning of facades in current buildings. *Constr. Build. Mater.* **2017**, *147*, 790–802. [CrossRef]

19. CYPE®Price Generator. Available online: http://www.geradordeprecos.info/ (accessed on 9 November 2016).

20. Brealey, R.A.; Myers, S.C.; Allen, F. *Principles of corporate finances*; McGraw-Hill Publishers: New York, NY, USA, 2011; ISBN 9788580552386.

21. Rolski, T.; Schmidli, H.; Schmidt, V.; Teugels, J.L. Stochastic processes for insurance and finance. In *Wiley Series in Probability and Statistics*; John Wiley & Sons Ltd.: Chichester, UK, 1999.

22. Denuit, M.; Marechal, X.; Pitrebois, S.; Walhin, J.F. *Actuarial Modelling of Claim Counts: Risk Classification, Credibility and Bonus-Malus Systems*; John Wiley & Sons, Ltd.: Chichester, UK, 2007.

23. Ungern-Sternberg, T. The limits of competition: Housing insurance in Switzerland. *Eur. Econ. Rev.* **1996**, *40*, 1111–1121. [CrossRef]

24. Mahmoudvand, R.; Aziznasiri, S. Bonus-malus systems in open and closed portfolios. In *Modern Problems in Insurance Mathematics*; Silvestrov, D., Matin-Löf, Eds.; Springer: Cham, Switzerland, 2014; Chapter 16; pp. 261–271.

25. Kunreuther, H.C.; Pauly, M.V.; McMorrow, S. *Insurance & behavioral economics. Improving decisions in the most misunderstood industry*; Cambridge University Press: Cambridge, MA, USA, 2013.

26. Kleindorfer, P.R.; Kunreuther, H. The complementary roles of mitigation and insurance in managing catastrophic risks. *Risk Anal.* **1999**, *19*, 727–738. [CrossRef]

27. Odeyinka, H.A. An evaluation of the use of insurance in managing construction risks. *Constr. Manage. Econ.* **2000**, *18*, 519–524. [CrossRef]

28. Arnell, N.W.; Clark, M.J.; Gurnell, A.M. Flood insurance and extreme events: The role of crisis in prompting changes in British institutional response to flood hazard. *Appl. Geogr.* **1984**, *4*, 167–181. [CrossRef]

29. Lamond, J.E.; Proverbs, D.G.; Hammond, F.N. Accessibility of flood risk insurance in the UK: Confusion, competition and complacency. *J. Risk Res.* **2009**, *12*, 825–841. [CrossRef]

30. El-Adaway, I.H.; Kandil, A.A. Construction risks: Single versus portfolio insurance. *J. Manag. Eng.* **2009**, *26*, 2–8. [CrossRef]

31. Dekker, D.; Green, D.; Palley, S. The expansion of insurance coverage for defective construction. *Constr. Law* **2008**, *28*, 19.

32. National Association of Home Builders (NAHB). Insurance Coverage for Claims of Latent Defects: What Protection Is a Builder Buying? Available online: file:///C:/Users/Ana/Downloads/INSURANCE-COVERAGE-FOR-CLAIMS-OF-LATENT-DEFECTS-2015.pdf (accessed on 28 April 2019).

33. Ostrager, B.R.; Newman, T.R. *Handbook on insurance coverage disputes*, 19th ed.; Wolters Kluwer: New York, NY, USA, 2019.

34. Tronquet, M.C. There's no place like home—Until you discover defects: Do prelitigation statutes relating to construction defect cases really protect the needs of homeowners and developers. *Santa Clara Law Rev.* **2004**, *44*, 4.

35. Bunni, N.G. *Risk and insurance in construction*; Taylor & Francis Group: London, UK, 2003.
36. Jelle, B.P.; Nilsen, T.N.; Hovde, P.J.; Gustavsen, A. Accelerated climate aging of building materials and their characterization by Fourier transform infrared radiation analysis. *J. Build. Phys.* **2011**, *36*, 99–112. [CrossRef]
37. Nosrati, R.; Berardi, U. Long-term performance of aerogel-enhanced materials. *Energy Procedia* **2017**, *132*, 303–308. [CrossRef]
38. Gradeci, K.; Berardi, U.; Time, B.; Köhler, J. Evaluating highly insulated walls to withstand biodeterioration: A probabilistic-based methodology. *Energy Build.* **2018**, *177*, 112–124. [CrossRef]

Article

Pathology and Rehabilitation of Vinyl and Linoleum Floorings in Health Infrastructures: Statistical Survey

Cláudia Carvalho [1], Jorge de Brito [2,*], Inês Flores-Colen [2] and Clara Pereira [2]

[1]   Instituto Superior Técnico, Universidade de Lisboa, Av. Rovisco Pais, 1049-001 Lisbon, Portugal; claudia.m.s.carvalho@ist.utl.pt

[2]   CERIS, Instituto Superior Técnico, Universidade de Lisboa, Av. Rovisco Pais, 1049-001 Lisbon, Portugal; ines.flores.colen@tecnico.ulisboa.pt (I.F.-C.); clareira@sapo.pt (C.P.)

*   Correspondence: jb@civil.ist.utl.pt; Tel.: +351-218-497-650

Received: 15 April 2019; Accepted: 5 May 2019; Published: 7 May 2019

Abstract: A statistical survey on the pathology and rehabilitation of linoleum and vinyl floorings is presented. It is based on the visual inspection of 101 vinyl and linoleum floorings, in six health infrastructures in Lisbon, Portugal, which enabled the validation of the classification/nomenclature previously proposed, as well as the corresponding correlation matrices. It was also possible to identify the most common types of anomalies, their probable causes, the most adequate in situ diagnosis methods, and the most useful repair techniques. Anomalies, diagnosis methods, and repair techniques files were also validated. The obtained data enabled anomalies to be related to their causes, in situ diagnosis methods, and respective repair techniques e.g., a high number of scratches/wear were detected associated with dragging of equipment. The conclusions drawn intend to raise awareness among the industry actors and minimize the development of anomalies and their causes at the design and application stages. Furthermore, the main sensitive issues of the cladding system during its service life were revealed, highlighting the importance of a correct maintenance plan to minimize the surface's susceptibility to various degradation mechanisms.

Keywords: linoleum and vinyl floorings; inspection; pathology; statistical survey; healthcare infrastructures

---

1. Introduction

Vinyl and linoleum are generally used in floorings and wall claddings. Technological and scientific evolution has played an important role on this type of flooring, allowing the development and marketing of a wide and innovative range of vinyl and linoleum floorings (VLF). It led to an increase in use and a broader scope of application, as well as easier manufacturing and new application methods [1]. In order to obtain better results, the application of VLF must be founded on scientific, physical, and economic criteria. Hence, the choice of flooring should start with an adequate and demanding prescription, based on factors such as the materials' properties, location of the flooring, and the conditions it will be subjected to throughout its service life [2].

The design and execution stages are crucial for the quality of VLF, since, when these stages are not given the proper attention, there is an increase in pathological manifestations and the service life decreases [3]. There is a need to prevent the occurrence of anomalies, along with accurately assessing them, in order to make a correct diagnosis. It is also important to recommend the adequate repair techniques, considering the anomalies' causes. An expert inspection system is the adequate tool to perform these methodological changes in assessment approach and decision-making [4].

However, anomalies that arise in VLF often deny the expectations of a cladding with good performance and durability. In fact, pathological situations in VLF have been occurring more often in the first years in service, showing a direct relation with factors like maintenance, use, and substrate. This may be due to the lack of selection criteria for the type of flooring to apply [5], suited to the surrounding environment, allied with the lack of an economic comparative analysis of materials. Tight schedules may also play an important role at the design and execution stages, as well as the lack of training and awareness of non-specialized workmanship, disregard for technical issues, and exclusive consideration of aesthetics. Moreover, the frequent lack of coordination among different designers results in the incompatibility of elements of the prescribed solution (e.g., the adhesion procedure and the substrate) that are only detected at the execution stage.

Therefore, it is of high importance to study the durability of VLF in sets of buildings in various locations, subjected to different actions, in order to understand the pathology according to the characteristics of VLF. Inspections can be systematized and should be included in a maintenance plan, to prevent degradation of the flooring and its substrate.

The American Society for Testing Materials (ASTM) has several publications [6–13] covering distinct aspects, especially in the project and application stages of VLF. The standard tolerances and guides [14–18] from the American Concrete Institute (ACI), should also be taken into account, since the substrate plays such a big role, and it is usually made of concrete. However, apart from these publications, there is hardly any literature (e.g., manuals) regarding the inspection, diagnosis, and rehabilitation in this field. Hence, a system to support the inspection, diagnosis, and repair of VLF was developed by the authors [19].

Regarding the health sector, several studies emphasize the importance of adequate maintenance, in order to prevent hospital-acquired infections [20–30].

The purpose of this paper is thus two-fold, as it intends to show how the expert system of inspection was validated, as well as present and analyze the statistical data acquired during that process.

Finally, though the expert system of inspection was built based on the pathology detected in healthcare facilities, it can be extrapolated to any other facilities (e.g., schools, sports facilities, airports). Still, data on healthcare facilities pathology should not be applied directly for decision-making in other types of facilities. Systems created outside the scope of healthcare cannot be used in healthcare facilities, due to specific criteria that must be considered (e.g., infection control).

Research Significance

During the inspection campaign, in July 2017, 101 floorings were inspected in six healthcare facilities in the Lisbon area, Portugal; two of them private and four public. These healthcare facilities are representative of health infrastructures in Portugal. Healthcare facilities were chosen as the object of the inspection campaign due to the common use of VLF in this type of facility and to the maintenance specificity of healthcare facilities. Thirteen floorings were of linoleum and 88 of vinyl, influencing the analysis of results. Following the methodology implemented by the same research team in other cladding systems [31–42], the inspections' data enabled the comparison with the theoretical correlation matrices, validating them. The matrices that relate anomalies and causes, anomalies and diagnosis methods, anomalies and repair techniques, and anomalies with each other are presented by Carvalho et al. [19]. Detailed files with information on each type of anomaly, diagnosis method, and repair technique were also created and validated.

After validating the expert inspection system, the inspection data were statistically analyzed, and conclusions were drawn. To the best of the authors' knowledge, there is no other work in the literature with the same scope and objectives considering VLF.

2. Pathology and Classification System

The classification of anomalies proposed by Carvalho et al. [19] is divided in three groups, according to the location of the anomaly, and has a total of ten anomalies, listed in Table 1. The first group includes anomalies on the surface or in depth of the flooring (A-A), the second anomalies that occur on the substrate (A-B), and the third group anomalies on flooring welding joints (A-C). Anomalies are coded with an A for anomaly, a hyphen, and a sequential capital letter for the group reference (i.e., A for surface and depth, B for substrate, and C for joints), followed by sequential numbering. The probable causes of anomalies, diagnosis methods, and repair techniques are coded using similar labelling.

Table 1. Proposed classification of flooring anomalies, their probable causes, diagnosis, and rehabilitation techniques [19].

Code	Anomaly	Code	Anomaly
A-A	*on the surface or in depth*	A-A4	prickles/punctures
A-A1	scratches/wear	A-A5	cracking
A-A2	staining/dirt/color changes	A-A6	loss/fracture/rotting
A-A3	brightness changes	-	-
A-B	*on the substrate*	A-B2	peeling
A-B1	blistering	A-B3	depression/settlement
A-C	*on the joints*	A-C1	faulty welding joints

Code	Probable cause	Code	Probable cause
C-A	*design errors*	C-A3	deficient detail of singular areas
C-A1	choice of incompatible or unsuitable materials	C-A4	defective design of the coating system
C-A2	areas inaccessible for cleaning	-	-
C-B	*execution errors*	C-B4	inadequate substrate preparation
B-B1	use of materials not available and/or inadequate and/or incompatible with each other	C-B5	inadequate application of material
C-B2	application under extreme environmental conditions	C-B6	poor interpretation of the project
C-B3	non-compliance with the pauses the between the several phases of execution	C-B7	non-compliance with the setting time until the use of flooring
C-C	*external mechanical actions*	C-C3	substrate deformation
C-C1	fall of objects	C-C4	dragging of equipment
C-C2	concentration of stress in the substrate	C-C5	punching actions
C-D	*environmental actions*	C-D3	presence of moisture
C-D1	sun exposure	C-D4	natural ageing
C-D2	biological action/chemically aggressive agents	-	-
C-E	*maintenance errors*	C-E1	insufficient/incorrect cleaning of the flooring
C-F	*change of the original use conditions*	C-F2	changing the use of space
C-F1	excessive loads	-	-

Code	Diagnosis method	Code	Diagnosis method
M-A	*assisted visual analysis*	M-A2	crack meter (ND)
M-A1	visual inspection (ND)	M-A3	crack ruler (ND)
M-B	*mechanical*	M-B2	percussion hammer (ND)
M-B1	sphere impact test (SM)	M-B3	pull-off adhesion test (D)
M-C	*thermal*	M-C1	infrared thermography (ND)
M-D	*water content and temperature*	M-D3	hygrometer (ND)
M-D1	inner moisture and temperature measurement (ND)	M-D4	speedy moisture test (D)
M-D2	superficial moisture measurement (ND)	-	-

Code	Repair techniques	Code	Repair techniques
R-A	*superficial zones*	R-A3	stripping (*) (cr; pr; m)
R-A1	cleaning the cladding (cr; m)	R-A4	mixture of white glue and linoleum scraps (*) (cr)
R-A2	applying a surface protector (cr; pr; m)	R-A5	application of welding (cr)
R-B	*whole flooring*	R-B2	partial replacement of the flooring (cr; pr)
R-B1	total replacement of the flooring (cr; pr)	R-B3	glue injection (cr; pr)
R-C	*substrate*	R-C1	replacement of the levelling layer (cr; pr)
R-D	*flooring surrounding area*	R-D1	repair of anomalies in footers (cr; pr)

D—destructive; SM—semi-destructive; ND—non-destructive. (*) The R-A3 and R-A4 techniques only concern linoleum floorings. cr—curative repair technique; pr—preventive repair technique; m—maintenance work.

Carvalho et al. [19] also proposed a classification of probable causes of anomalies, to adequately decide how to control and repair detected anomalies (Table 1). It groups the causes of anomalies in VLF into six categories, organized chronologically: C-A design errors, C-B execution errors, C-C external mechanical actions, C-D environmental actions, C-E maintenance errors, C-F change of the original use conditions. As for the classification of diagnosis methods, it considers the technology used in the tests to group methods. The classification of repair techniques takes the location of the anomaly into account to group the techniques.

3. VLF Inspection Plan

Planning of VLF inspections is essential, as inspection and validation forms must be prepared, as well as specific equipment (rulers, measuring equipment, photographic equipment, and other accessories, if necessary). Inspection and validation forms were created based on previous research [39, 41,42]. The scope of the inspection program was to validate the system proposed by Carvalho et al. [19].

3.1. Inspection Forms

In order to make VLF inspections more effective, inspection forms were created to characterize the buildings and the materials and procedures related to the flooring's execution. These forms also contain important information on the building's surroundings. Validation forms include data on the detected anomalies and all related data, as well on the VLF areas that should be given special attention in the future.

In each inspection form, the following information should be filled out: In the heading, the number of the inspection form, date, surveyor's name and role, and the inspection's main purpose; for each building, the location, main use, year of construction, location of the floor under analysis, exposure to polluting agents, and solar radiation, and any contacts made; for each inspected flooring, linoleum or vinyl type, type of application (tile or roller), type of glue used, area (in m^2), type of substrate, existence of a vapor barrier, welding cord, type of finish/skirting (and height), and any singularities of the floor; maintenance operations recorded, namely, type and frequency of maintenance actions, and details about them (date, type of materials, and techniques used).

3.2. Validation Forms

For each VLF inspected, one validation form was filled out, in which the detected anomalies were identified according to the classification described in Section 2. It should be noted that the number of anomalies identified in each flooring was always equal to or greater than one, since only VLF with anomalies were considered. Each anomaly is characterized in terms of location within the VLF, affected area or length (nominal and relative), and the degradation of the laying material or the substrate itself. Other data of interest for the characterization of the repair urgency were also collected, such as the aesthetic value of the affected areas and the possible recurrent occurrence.

For each detected anomaly, probable causes (direct or indirect), adequate diagnosis methods, and repair techniques were registered on site, based on visual observation and according to the classification systems mentioned in Section 2.

4. Data analysis and Statistical Characterization

As shown in Figure 1, of the 101 VLF inspected, approximately one third was over 20 years, one third was in the remaining age gaps, and the other third did not have information on the construction year (NI). The oldest VLF inspected, of those with a known age, was 32 years old. That flooring was quite worn due to consecutive waxing and stripping. However, it did not show anomalies other than wear, scratching, and welding joints between tiles requiring replacement. According to Gorrée et al. [43], stripping and resealing a linoleum floor should be done six times in 20 years (approximately every three years) in public buildings, but the specificity of the inspected floors may have required a different periodicity of this activity.

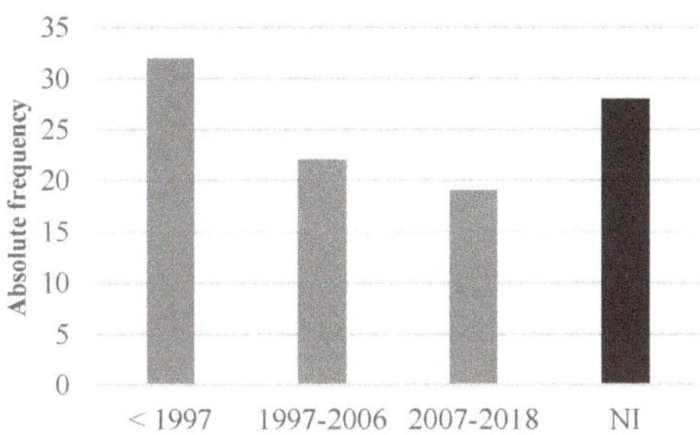

Figure 1. Absolute frequency of the inspected vinyl and linoleum floorings (VLF) according to the construction year (NI—not identified).

Even though the sample's floorings are only divided in vinyl and linoleum materials, as no laboratory tests were performed, nor sufficient information was available, the analysis of the age of the flooring system shows that more recent floorings are likely to have improved materials and application techniques that result from the constant innovation in the sector.

4.1. Anomalies Observed in the Sample

A total of 344 anomalies were detected in the inspected floorings, resulting in an average of 3.4 anomalies per flooring. As for the probable causes, 537 were recorded, resulting in an average of approximately 1.6 causes per anomaly. As to diagnosis methods and repair techniques, a total of 898 and 739, respectively, were recommended, which results in an approximate average of 2.6 diagnosis methods and 2.1 repair techniques per anomaly. The absolute and relative frequencies of detected anomalies were analyzed (Figure 2). Note that the relative frequency was calculated by dividing the absolute frequency by the number of inspected VLF.

Anomalies "A-A1 scratches/wear", "A-A2 staining/dirt/color changes", and "A-C1 faulty welding joints" occurred in more than half of the inspected floorings. The most frequent anomaly was A-A1, occurring in 76 floorings. Considering how easy it is to scratch a VLF [1], such frequency is not surprising, as careless dragging of equipment is enough to leave a scratch, and the wheels of most trolleys have inadequate or worn-out material. An even higher frequency could be expected, but it should be noted that recent floorings, with only months of age, were also considered in the inspections campaign.

Anomaly "A-B1 blistering" was the least frequent one (in approximately 10% of the sample). Nevertheless, it has considerably more severe consequences than others. The low frequency of A-B1 may be associated with its seriousness, since, when blistering occurs, it can condition the mobility of people and equipment, as well as affect the aesthetic value. If it is due to infiltration, the absence of repair actions may aggravate the situation. Thus, it is assumed that, when blistering occurs, it is usually repaired, resulting in a low frequency of the anomaly.

Anomaly A-C1 faulty welding joint is the second most frequent anomaly, occurring in 53% of floorings. This result may be associated with poor execution, leading to the accumulation of dirt, and fractures in smaller areas, when subjected to stress.

As for anomaly "A-A2 staining/dirt/color changes", 52 occurrences were detected, but a higher frequency was expected, according to the results of Personen-Leinonen et al. [44]. However, since the inspected infrastructures belong to the health sector, they involve very high cleaning requirements, justifying the lower than expected frequency. In general, dirt can be eliminated with cleaning. As for indelible stains and color changes in linoleum floorings, they can only be removed by stripping. In vinyl or recent linoleum floorings, they can be removed by polishing. In these cases, the VLF is only replaced when the affected area is of exceptional aesthetic value.

Comparing the incidence of anomalies in VLF with that of wood floorings, it is found that, in wood floorings, scratches, or wrinkles (above 80% of floorings) are also the most common detected anomaly [33]. In wood floorings, scratches are even more predominant, as the next frequent anomaly is only detected in about 25% of inspected floorings. These results raise the question whether scratching may be a generalized problem of floorings, possibly associated with the inadequate functional classification of space and rooms.

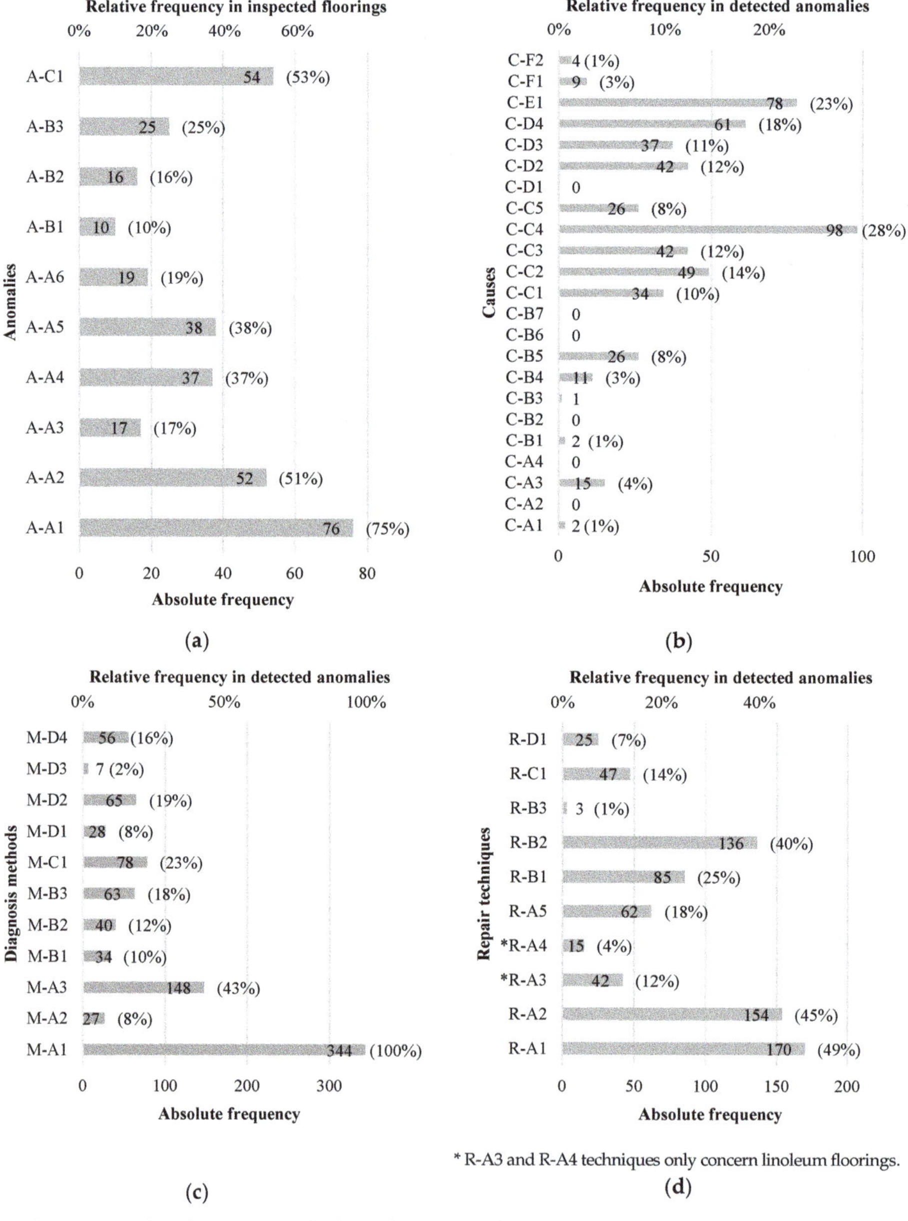

Figure 2. Absolute frequency (and relative frequency) of: (**a**) Detected anomalies; (**b**) probable causes; (**c**) adequate diagnosis methods; (**d**) recommended repair techniques.

In regard to urgency of repair, approximately 47% of anomalies belong to intervention level 1, 40% to level 0 (the most urgent), and 13% to level 2 (Figure 3). Anomalies on the substrate (A-B) usually require a higher repair urgency, as the affected area is prone to rapidly increase. It is also the case of anomalies "A-A5 cracking" and "A-C1 faulty welding joints", if there is a possibility of infiltration. In the health sector, aseptic conditions and functionality must be guaranteed. However, functionality becomes impossible when blistering and depressions occur, as they can un-calibrate instruments and hinder the stability of a patient being transported on a stretcher.

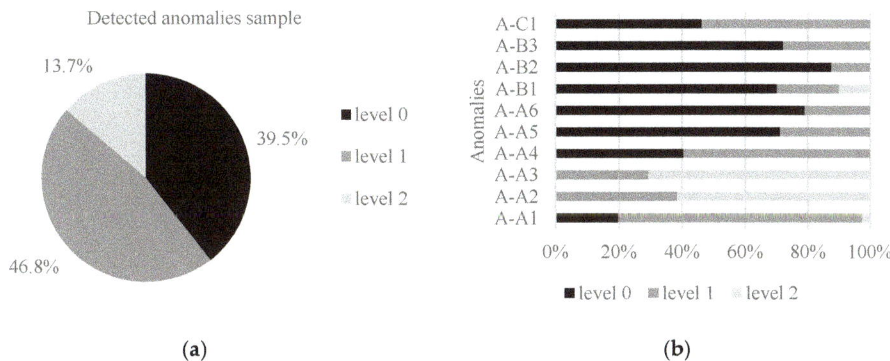

Figure 3. Urgency of repair: (**a**) Relative frequency of anomalies in each level; (**b**) relative frequency of each level for each type of detected anomalies.

Figure 3 shows that the anomalies that most require short-term intervention were, in descending order of priority, "A-B2 peeling" (88%), "A-A6 loss/fracture/rotting of material" (79%), "A-B3 depression/settlement" (72%), "A-A5 cracking" (71%), and "A-B1 blistering" (70%). In group A-A anomalies on the surface or in depth, anomalies "A-A4 prickles/punctures" (41%) and "A-A1 scratches/wear" (20%) could also be highlighted. As for anomaly "A-C1 faulty welding joints", approximately half of the sample was at level 0 and the other half at level 1. The absence or deficiency of the welding joint, allied with inadequate cleaning, creates a singularity where water may enter, triggering detachment of the flooring. This shows the importance of timely maintenance interventions at welding joints to guarantee the expected service life of the VLF. Regarding anomalies "A-A2 staining/dirt/color changes" and "A-A3 brightness changes", most belonged to level 2 (less urgent), with 62% and 71%, respectively. This is related to the mentioned high cleaning requirements in health infrastructures.

As for location of anomalies, in some situations, a specific type of anomaly is predominant. It is the case of anomaly "A-A1 scratches/wear", which is often seen in poorly executed transition areas or at the entrance of the rooms (Figure 4a,b). It is also the case of anomaly "A-A3 brightness changes", which often occurs immediately below the alcohol-based disinfectant dispensers (Figure 4c).

As mentioned, out of the six inspected health infrastructures, two were private and four were public, resulting in 24 and 77 analyzed floorings, respectively. Thus, a comparison between anomalies in private and public infrastructures was made. First of all, a significant age difference was noted, as all VLF inspected in Private-1 were more than 20 years old, while those in Private-2 were up to 10 years old. The frequency of anomalies in Private-1 (12 cases), Private-2 (12 cases), and the inspected Public sector facilities (77 cases) is shown in Figure 5.

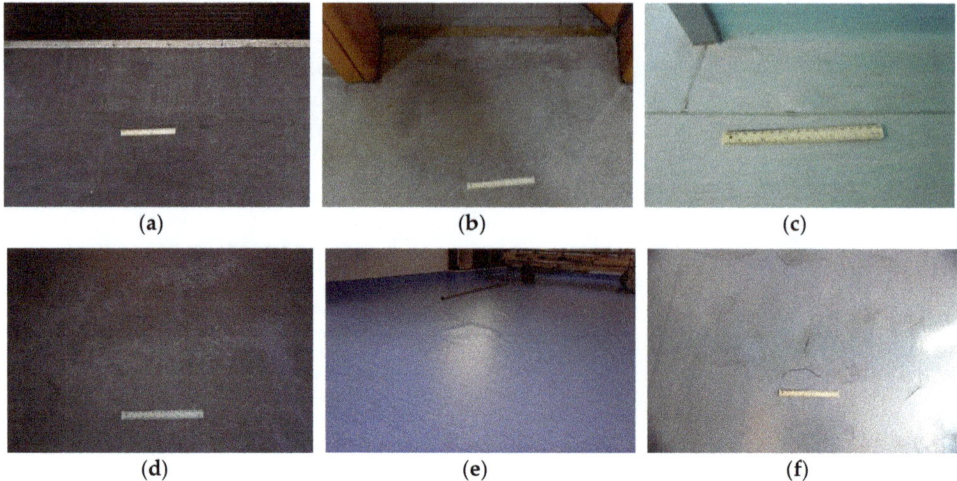

Figure 4. Anomalies: (**a**) A-A1 scratches; (**b**) A-A1 wear; (**c**) A-A3 brightness changes; (**d**) A-A3 brightness changes on a linoleum flooring due to an inadequate maintenance; (**e**) A-B2 peeling, requiring the use of repair technique R-B2 partial replacement; (**f**) A-B3 depression/settlement, requiring the use of repair technique R-B1 total replacement.

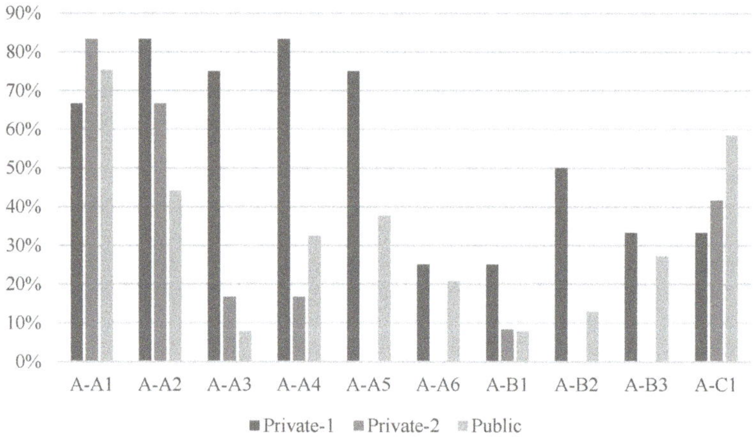

Figure 5. Relative contribution of each type of anomaly for the degradation of VLF in each type of inspected health facility.

Considering the age of Private-1 floorings, in most cases it had a greater incidence of anomalies. In fact, when visiting Private-1 facility, a need for major repair works was detected, in view of the high number of anomalies. Moreover, inspected Private-1 floorings were exclusively in linoleum. That explains the high relative frequency of anomaly "A-A4 prickles/punctures", associated with the greater residual indentation of the material and natural ageing. As for "A-A3 brightness changes", it was confirmed in situ that the high relative frequency is due to inadequate maintenance.

Compared with inspected Public infrastructures, Private-2 floorings had a higher incidence of "A-A1 scratches/wear", and the group of private infrastructures floorings had a higher incidence of

"A-A2 staining/dirt/color changes", "A-A3 brightness changes", "A-A4 prickles/punctures", "A-B1 blistering", and "A-B2 peeling".

4.2. Probable Causes of Anomalies

The frequency of probable causes is shown in Figure 2. During the inspection program, causes C-A1, C-A2, and C-A4 (design errors), most "C-B execution errors" causes (except for C-B4 and C-B5), and cause "C-D1 sun exposure" were never observed, or observed only once. This is probably due to the lack of information on the design and execution stages, specifically about application conditions and methodologies used. It results in the difficult establishment of relationships between design and execution causes and anomalies in VLF more than one year old. Cause "C-C4 dragging of equipment" was the most frequent, as was indicated in approximately 28% of anomalies, probably due to a close correlation with the occurrence of scratches (A-A1).

The relative frequency of each of group of causes in the sample (Figure 6) shows that "C-C external mechanical actions" (46%) and "C-D environmental actions" (26%) were the ones that contribute the most to the occurrence of anomalies. In fact, each anomaly was caused by 0.72 mechanical actions, on average.

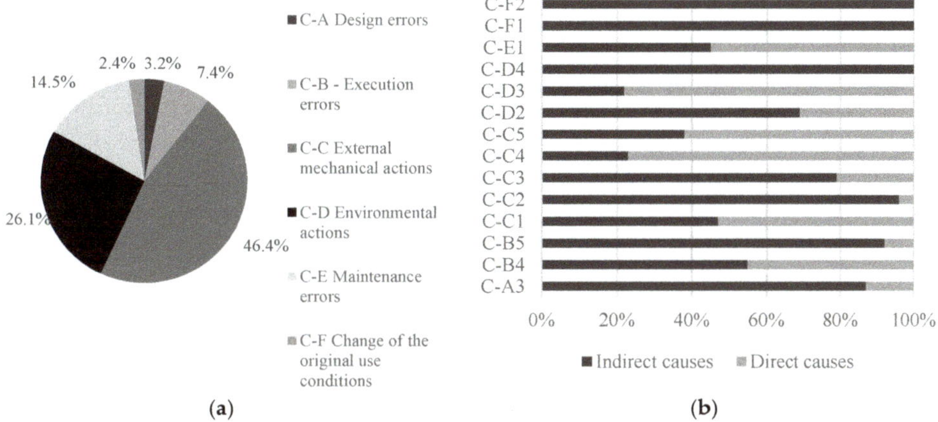

Figure 6. Probable causes of anomalies: (**a**) Relative frequency of each group of causes; and (**b**) direct vs. indirect causes.

In Figure 6, the distribution of causes in direct and indirect causes is shown, not considering causes with two occurrences or less. The causes most often identified as direct were "C-D3 presence of moisture" (78%) and "C-C4 dragging of equipment" (77%). On the other hand, the causes most frequently identified as indirect (in this case, with a frequency of 100%) were "C-D4 natural ageing", "C-F1 excessive loads", and "C-F2 changing the use of space".

To analyze the contribution of each group of causes to each anomaly, Figure 7 was drawn. As to anomaly "A-A4 prickles/punctures", it resulted almost entirely from causes in group "C-C external mechanical actions" (on average, 1.51 causes per anomaly), and, within the group, the most influential causes were "C-C1 fall of objects" (81%) and "C-C5 punching actions" (68%). The latter occurs because furniture, like chairs and tables, have worn or absent protections. Punching may also be related with the accidental fall of objects, such as oxygen bottles.

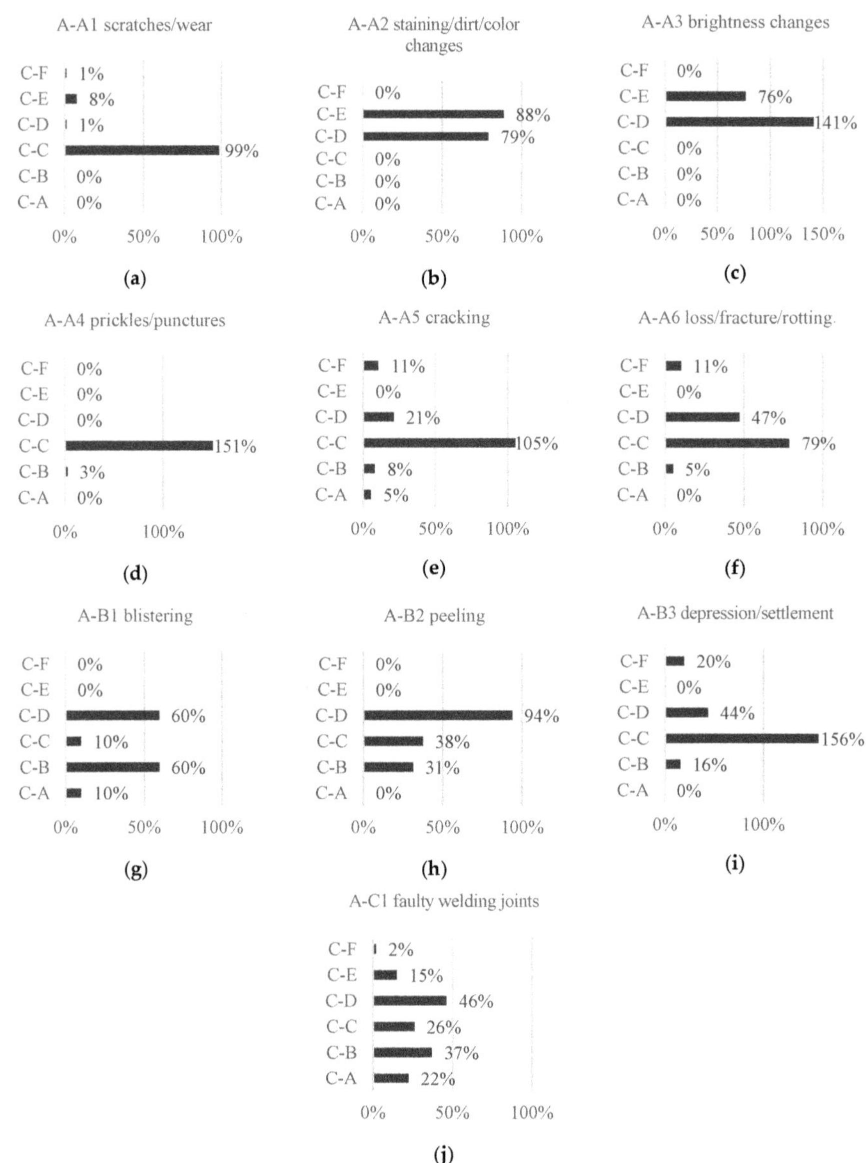

Figure 7. Contribution of each group of causes to each type of anomaly in the sample: (**a**) A-A1 scratches/wear; (**b**) A-A2 staining/dirt/color changes; (**c**) A-A3 brightness changes; (**d**) A-A4 prickles/ punctures; (**e**) A-A5 cracking; (**f**) A-A6 loss/fracture/rotting; (**g**) A-B1 blistering; (**h**) A-B2 peeling; (**i**) A-B3 depression/settlement; (**j**) A-C1 faulty welding joints.

As for anomalies "A-A2 staining/dirt/color changes" and "A-A3 brightness changes", they resulted exclusively from causes in groups "C-D environmental actions" and "C-E maintenance errors", the latter represented only by "C-E1 insufficient/incorrect cleaning of the flooring". In both A-A2 and A-A3, the predominant environmental causes were "C-D2 biological action/chemically aggressive agents" (50% and 94%, respectively) and "C-D4 natural ageing" (21% and 47%, respectively). Such frequencies were expected, as the use of inappropriate cleansers or spilling of aggressive substances, such as antiseptic drugs, causes blemishes, changes in brightness, or indelible stains. Comparing these relationships with those in wood floorings, staining and color changes were mainly related with an inadequate finishing, natural ageing, and poor maintenance works [33]. So, only natural ageing was identified in both types of flooring, as the expected wear signs in floorings can only normally be eliminated with large investment (major repair works or replacement).

Anomaly "A-A1 scratches/wear" was almost exclusively influenced by group "C-C external mechanical actions" (99%), due solely to "C-C4 dragging of equipment" (a direct cause). This cause represents the use of improper wheel material, which easily leaves a mark in VLF, or moving of furniture without adequate protections. In wood floorings [33], the circulation on floorings is also identified in all scratch's anomalies.

Anomaly "A-A5 cracking" was strongly influenced by the group of causes "C-C external mechanical actions", and, within this group, the most influential causes were "C-C3 substrate deformation" (50%), "C-C2 concentration of stress in the substrate" (34%), and "C-C4 dragging of equipment" (18%). When deformation occurs on the substrate, such as depression, the VLF follows the deformation shape. Over time, the VLF loses flexibility and ends up cracking in the uneven zone.

Concerning anomaly "A-A6 loss/fracture/rotting", it was influenced by "C-D environmental actions" (47%), specifically by causes "C-D3 presence of moisture" (37%) and "C-D4 natural ageing" (11%). But "C-C external mechanical actions" had an even stronger influence on anomaly A-A6 (79%).

As for anomalies on the substrate, "A-B1 blistering", "A-B2 peeling", and "A-B3 depression/settlement", they were all influenced by "C-D environmental actions" (60%, 94%, and 44%, respectively). Once more, the only influential causes of group C-D were "C-D3 presence of moisture" and "C-D4 natural ageing". Group "C-B execution errors" also had a strong influence on anomalies A-B1 (60%) and A-B2 (31%), whereas group "C-C external mechanical actions" had a strong influence on anomalies A-B2 (38%) and A-B3 (1.56 causes per anomaly, on average). Hence, the two main causes of substrate anomalies were moisture and natural ageing of the VLF allied with execution errors and mechanical actions, such as stress concentration in the substrate.

Finally, anomaly "A-C1 faulty welding joints" was influenced by all groups of causes, although group "C-F change of the original use conditions" was only identified once. It is an expected result, as joints can easily be degraded if incorrectly designed, executed, submitted to mechanical actions and aggressive agents, or poorly maintained. A similar variety of types of causes may be verified in wood floorings' [33] defects in joints (change of joint size).

4.3. Diagnosis Methods

The absolute and relative frequencies of the diagnosis methods identified in the sample are shown in Figure 2. It should be noted that, by default, all anomalies were associated with the diagnosis method "M-A1 visual inspection". The most frequent method, not considering M-A1, was "M-A3 crack ruler", adequate to 43% of the anomalies, probably due to its easy use in superficial anomalies and joints. The least chosen method was "M-D3 hygrometer" (seven cases), which may be related with the legal requirements, in terms of ambient moisture content, healthcare infrastructures are subjected to. Comparing with diagnosis methods that were found adequate for anomalies in wood floorings, moisture measurement was much more recommended (almost in 30% of anomalies), as vinyl floorings are not so sensitive to dampness and the wood floorings sample did not include healthcare infrastructures [33].

The frequency of the diagnosis methods recommended for each anomaly is shown in Figure 8. Anomaly "A-A1 scratches/wear" was only associated with methods "M-A1 visual inspection" (100%) and "M-A3 crack ruler" (97%), as expected. By measuring the thickness of the scratch, one can more easily diagnose the probable cause of the anomaly. While dragging of equipment generally causes a finer scratch, the movement of inadequate cart wheels leaves a thicker scratch. Anomalies "A-A2 staining/dirt/color changes", and "A-A3 brightness changes" were exclusively associated with method M-A1.

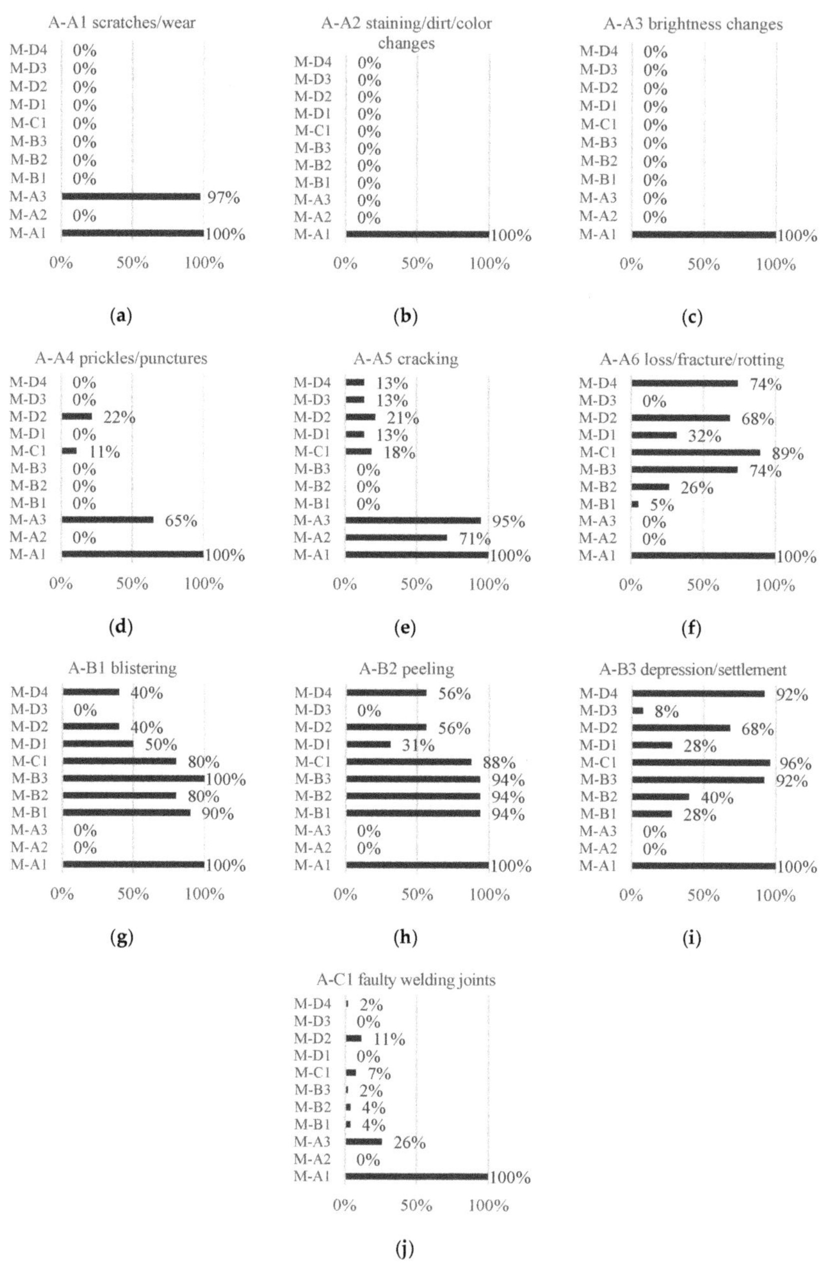

Figure 8. Contribution of each diagnosis method to each type of anomaly in the sample: (**a**) A-A1 scratches/wear; (**b**) A-A2 staining/dirt/color changes; (**c**) A-A3 brightness changes; (**d**) A-A4 prickles/ punctures; (**e**) A-A5 cracking; (**f**) A-A6 loss/fracture/rotting; (**g**) A-B1 blistering; (**h**) A-B2 peeling; (**i**) A-B3 depression/settlement; (**j**) A-C1 faulty welding joints.

As for anomaly "A-A4 prickles/punctures", aside from method M-A1, methods "M-A3 crack ruler" (65%), "M-D2 superficial moisture measurement" (22%), and "M-C1 infrared thermography"

(11%) were recommended. These results were expected, since the crack ruler may assist in measuring the anomaly, while measuring moisture may show changes in moisture content, whose causes and extent may be better determined with infrared thermography.

Anomaly "A-A5 cracking" was strongly associated with methods "M-A2 crack meter" (71%) and "M-A3 crack ruler" (95%), aside from M-A1. These tools, allied with the visual inspection, were the best at diagnosing this type of anomaly. A-A5 was also characterized by a weak relationship with all other methods, except for mechanical methods (M-B).

For anomaly "A-A6 loss/fracture/rotting", diagnosis methods "M-A1 visual inspection" (100%), "M-B3 pull-off adhesion test" (74%), "M-C1 infrared thermography (89%), "M-D2 superficial moisture measurement" (68%), and "M-D4 speedy moisture test" (16%) were strongly recommended to better assess its causes.

Regarding anomalies on the substrate, "A-B1 blistering", "A-B2 peeling", and "A-B3 depression/ settlement" were all associated with diagnosis methods "M-B1 sphere impact test", "M-B2 percussion hammer", "M-B3 pull-off adhesion test", "M-C1 infrared thermography", "M-D1 inner moisture and temperature measurement", "M-D2 superficial moisture measurement" and "M-D4 speedy moisture test", besides method "M-A1 visual inspection". For anomaly A-B1, methods M-B3 (100%) and M-B1 (90%) should be highlighted as they were advised in circumstances in which the adhesion conditions have changed. As for anomaly A-B2, diagnosis methods M-B1, M-B2, and M-B3 were equally recommended in 94% of detected cases. For anomaly A-B3, methods M-C1 (96%), M-B3 (92%), and M-D4 (92%) were the most recommended, given the probable association of A-B3 with dampness and the usefulness of determining adhesion conditions.

Finally, anomaly "A-C1 faulty welding joints" was mostly associated with method "M-A1 visual inspection", but "M-A3 crack ruler" (26%) may also play an important role in determining the size of the joint and of any detected cracks in its welding.

4.4. Repair Techniques

The absolute and relative frequencies of repair techniques were analyzed (Figure 2). On the one hand, the most advised technique was "R-A1 cleaning the cladding", recommended for approximately 49% of detected anomalies, which can be due to its immediate effect on the aesthetic value of the affected area. On the other hand, technique "R-B3 glue injection", was the least chosen (three cases), which is probably associated with being only advised in special circumstances of anomaly "A-B1 blistering". In wood floorings [33], the injection of voids with resin was also not recommended for many of the detected anomalies. Specific repair works do not tend to be used frequently, regardless of the flooring material.

Concerning the frequency of suggested repair techniques for each anomaly (Figure 9), "A-A5 cracking", "A-A6 loss/fracture/rotting", "A-B2 peeling", and "A-B3 depression/settlement" were highly associated with repair technique "R-B2 partial replacement of the flooring", as it was considered in more than 60% of anomalies of these types. Anomalies "A-A1 scratches/wear", "A-A2 staining/dirt/color changes", and "A-A3 brightness changes" were associated with repair technique "R-A2 applying a surface protector" in more than 75% of the cases.

Anomaly "A-A1 scratches/wear" should be repaired by "R-A1 cleaning the cladding" (96%) and "R-A2 applying a surface protector" (87%), to reduce its aesthetic impact by smoothing the surface. As stripping (R-A3) is exclusively used in linoleum floorings, it has a small incidence in the sample (12%), since, in the inspection campaign, only 13 of 101 analyzed floorings were linoleum. When A-A1 occurs, partial (R-B2) or total replacement (R-B1) only should take place when the area has a high aesthetic value, hence these techniques were only scarcely recommended in this context.

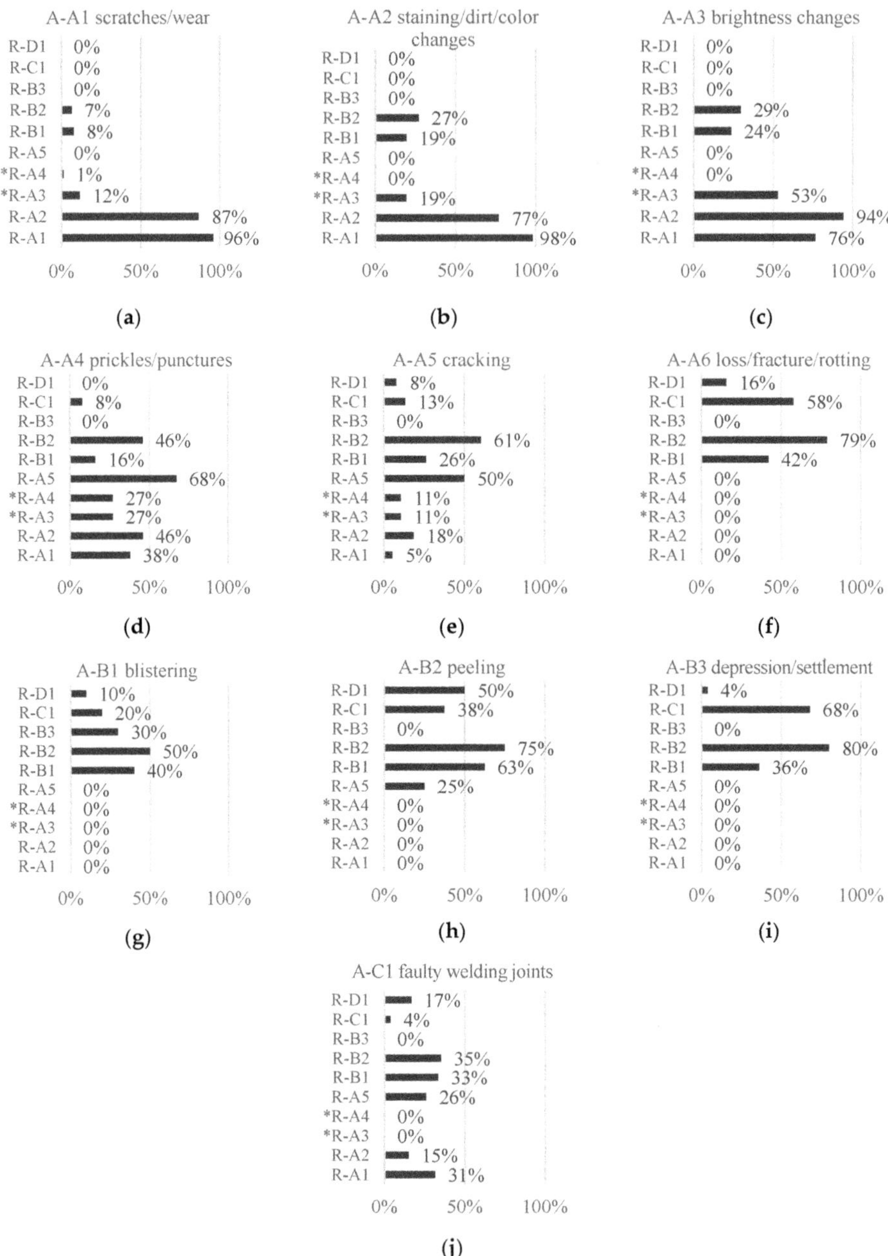

Figure 9. Contribution of each repair technique to each type of anomaly in the sample: (**a**) A-A1 scratches/wear; (**b**) A-A2 staining/dirt/color changes; (**c**) A-A3 brightness changes; (**d**) A-A4 prickles/ punctures; (**e**) A-A5 cracking; (**f**) A-A6 loss/fracture/rotting; (**g**) A-B1 blistering; (**h**) A-B2 peeling; (**i**) A-B3 depression/settlement; (**j**) A-C1 faulty welding joints.

Anomalies "A-A2 staining/dirt/color changes" and "A-A3 brightness changes" were both repaired by "R-A1 cleaning the cladding" (98% and 76%, respectively), "R-A2 applying a surface protector"

(77% and 94%), "R-A3 stripping" (19% and 53%), "R-B1 total replacement of the flooring" (19% and 24%), and "R-B2 partial replacement of the flooring" (27% and 29%, respectively). Although the inspected rooms were in health facilities, and the cleaning requirements were extremely high, the results on the replacement techniques were acceptable, since most A-A2 and A-A3 anomalies were stains, discoloration, and brightness changes with some aesthetic value, but not accumulated dirt. As for stripping (R-A3), in linoleum floorings, it was less associated with stains than with brightness changes. It was noticed that floorings were cleaned and waxed regularly, but stripping was only partial and not in the required timings, often resulting in alternated areas with different wax thicknesses, as seen in Figure 4d.

Anomaly "A-A4 prickles/punctures" was only not associated with repair techniques R-B3 and R-D1. The application of the remaining seven repair techniques depends on the area affected by the anomaly. For small areas, cleaning (R-A1) and applying a sealant (R-A2) or a weld bead was sufficient (R-A5). In the case of a linoleum flooring, it can be stripped (R-A3), or the punctures can be covered with a mixture of white glue and linoleum scraps (R-A4).

Anomaly "A-A5 cracking" was mostly associated with repair techniques "R-A5 application of welding" (50%), "R-B1 total replacement of the flooring" (26%), and "R-B2 partial replacement of the flooring" (61%). The use of these techniques depends on the anomaly's depth, width, and affected area.

As for anomaly "A-A6 loss/fracture/rotting", techniques "R-B1 total replacement of the flooring" (42%) and "R-B2 partial replacement of the flooring" (79%) were repeatedly recommended, as well as "R-C1 replacement of the levelling layer" (58%), the latter acting on the causes of the defect.

Regarding the anomalies on the substrate, "A-B1 blistering", "A-B2 peeling", and "A-B3 depression/ settlement" may all be repaired by techniques "R-B1 total replacement of the flooring", "R-B2 partial replacement of the flooring", "R-C1 replacement of the levelling layer", and "R-D1 repair of anomalies in footers", or skirting. Anomaly A-B1 may also be repaired with "R-B3 glue injection" (30%), which is used in specific cases of blistering. This happens especially when, in the months following the application, it is found that the blistered area has no glue and consists of a small extent, as seen in Figure 4e. As for anomaly A-B2, in 25% of the cases it could also be repaired by "R-A5 application of welding". In addition, A-B2 is the type of anomaly most associated with repair technique R-D1, in relative terms (50% of the cases). In order to be well laid, the VLF must be heated, to become flexible and follow the curve of the cove former, and then compressed, to ensure adhesion. This process may be difficult, as the surface is not smooth or horizontal. If these steps are not carefully followed, the detachment of the coving may occur.

Anomaly A-B3 may occur both in small and large areas, influencing the choice of repair techniques R-B1 (36%) or R-B2 (80%). Figure 4f illustrates a corridor, where the constant passage of trolleys and a possible poor levelling layer caused depressions, which later evolved to cracks.

Finally, anomaly "A-C1 faulty welding joints" may be repaired by all techniques except R-A3, R-A4, and R-B3. The most commonly recommended techniques were "R-A1 cleaning the cladding" (31%), "R-A5 application of welding" (26%), "R-B1 total replacement of the flooring" (33%), and "R-B2 partial replacement of the flooring" (35%).

5. Conclusions

The gathered data, based on a significant sample of 101 vinyl and linoleum floorings, enabled the validation of the classification system of anomalies, probable causes, in situ diagnosis methods, and repair techniques for vinyl and linoleum floorings, as well as each correlation matrix proposed by Carvalho et al. [19].

The main problems detected in VLF consisted of scratches, staining, color changes, some dirt, and anomalies on the joints, which were detected in more than a half of the inspected floorings. Accordingly, the dragging of equipment and the incorrect cleaning of the flooring were frequently identified as causes of anomalies in VLF. To assist the diagnosis of detected anomalies, the use of a crack ruler was found useful for its versatility, including the measurement of scratches. To correct the detected defects, cleaning, applying a surface protector, and even partially replacing the flooring were frequently recommended. It is essential to train cleaning companies, since inadequate maintenance directly contributes to accelerate the degradation of VLF. It is also necessary to ensure the quality control of the application of VLF through regular inspections and skilled manpower.

In the sample of inspected healthcare facilities, most detected anomalies have an average urgency of repair. However, a significant percentage of anomalies were still considered in need of urgent repair. In absolute terms, anomalies "A-A5 cracking" and "A-C1 faulty welding joints" present the highest number of cases in level 0 of urgency of repair, although, in relative terms, a higher percentage of cases of "A-B2 peeling" in level 0 was detected.

Considering the private and public healthcare facilities samples, in relative terms, all types of anomalies show a higher relative frequency in private facilities, except for "A-C1 faulty welding joints", detected in higher percentage in public healthcare facilities. In the inspected private facilities, "A-A1 scratches/wear", "A-A2 staining/dirt/color changes", and "A-A4 prickles/punctures" were the most frequently detected. In public facilities, anomaly A-A1 was also detected in the most relevant number of floorings.

This paper contributes to the dissemination of knowledge on the degradation of VLF, specifically in health infrastructures. With the presented data in mind, maintenance plans may be improved and more effective. Data may also be useful, for instance, at the design stage, as future defects, such as those caused by dragging, may be prevented if the type of vinyl or linoleum flooring is adequately chosen, restricting the functional requirements to more demanding ones. Additionally, at the design stage, a maintenance manual or a concise user's manual for the building may be developed and delivered to the client. That measure may potentially improve maintenance operations, such as cleaning. At the application stage, several measures may be implemented to avoid defects, always starting with the use of specialized labor and better communication between contractors. Specialized labor is more likely to apply floorings according to the best practices, hence avoiding defects caused by execution errors. As for communication between contractors, it may contribute, for instance, to an adequate substrate preparation, according to the requirements for VLF. It is also expected that the efficiency and effectiveness of maintenance in VLF increases if the inspection system proposed by Carvalho et al. [19] is used. In the future, to improve the system, a wider inspections campaign should be made.

Author Contributions: Conceptualization, J.d.B.; methodology, J.d.B. and I.F.-C.; formal analysis, C.C. and C.P.; investigation, C.C.; resources, C.C.; writing—original draft preparation, C.C.; writing—review and editing, C.P. and J.d.B.; visualization, C.C. and C.P.; supervision, J.B. and I.F.-C.

Funding: This research received no external funding.

Acknowledgments: The authors gratefully acknowledge the support of the CERIS Research Institute; IST, University of Lisbon; and FCT, Foundation for Science and Technology.

References

1. Douglas, J.; Noy, E.A. *Building Surveys and Reports*, 4th ed.; Wiley-Blackwell: Chichester, UK, 2011; pp. 203–204.
2. Dejaco, M.C.; Re Cecconi, F.; Maltese, S.; Spagnolo, S.L.; Kamara, J.M. Requirements compliance checking for existing buildings. In Proceedings of the CIB World Building Congress, Tampere, Finland, 30 May–3 June 2016; pp. 706–718.
3. Aktas, C.B.; Bilec, M.M. Service life prediction of residential interior finishes for life cycle assessment. *Int. J. Life Cycle Assess.* **2012**, *17*, 362–371. [CrossRef]
4. CIB W86. *Building Pathology: A State-of-the-Art Report*; CIB: Delft, The Netherlands, 1993.
5. Dixit, M.K.; Singh, S.; Lavy, S.; Yan, W.; Pariafsai, F.; Ostadalimakhmalbaf, M. Floor finish selection in the design of healthcare facilities: A survey of facility managers. *Facilities* **2019**. [CrossRef]

6.   ASTM E96/E96M-16. *Standard Test Methods for Water Vapor Transmission of Materials*; ASTM International: West Conshohocken, PA, USA, 2016.

7.   ASTM E1745-17. *Standard Specification for Plastic Water Vapor Retarders Used in Contact with Soil or Granular Fill under Concrete Slabs*; ASTM International: West Conshohocken, PA, USA, 2017.

8.   ASTM F710-17. *Standard Practice for Preparing Concrete Floors to Receive Resilient Flooring*; ASTM International: West Conshohocken, PA, USA, 2017.

9.   ASTM F1482-15. *Standard Practice for Installation and Preparation of Panel Type Underlayments to Receive Resilient Flooring*; ASTM International: West Conshohocken, PA, USA, 2015.

10.   ASTM F1516-13. *Standard Practice for Sealing Seams of Resilient Flooring Products by the Heat Weld Method (When Recommended)*; ASTM International: West Conshohocken, PA, USA, 2013.

11.   ASTM F1869-16a. *Standard Test Method for Measuring Moisture Vapor Emission Rate of Concrete Subfloor Using Anhydrous Calcium Chloride*; ASTM International: West Conshohocken, PA, USA, 2016.

12.   ASTM F1913-04. *Standard Specification for Vinyl Sheet Floor Covering without Backing*; ASTM International: West Conshohocken, PA USA, 2014.

13.   ASTM F2170-17. *Standard Test Method for Determining Relative Humidity in Concrete Floor Slabs Using In Situ Probes*, ASTM International: West Conshohocken, PA, USA, 2017.

14.   ACI 117-10. *Specification for Tolerances for Concrete Construction and Materials*; ACI Committee 117; American Concrete Institute: Farmington Hills, MI, USA, 1990.

15.   ACI 223R. *Standard Practice for the Use of Shrinkage-Compensating Concrete*; ACI Committee 223; American Concrete Institute: Farmington Hills, MI, USA, 2010.

16.   ACI 302.1R. *Guide for Concrete Floor and Slab Construction*; ACI Committee 302; American Concrete Institute: Farmington Hills, MI, USA, 2015.

17.   ACI 302.2R. *Guide for Concrete Slabs that Receive Moisture-Sensitive Flooring Materials*; ACI Committee 302; American Concrete Institute: Farmington Hills, MI, USA, 2006.

18.   ACI 360R. *Guide to Design of Slabs-on-Ground*; ACI Committee 360; American Concrete Institute: Farmington Hills, MI, USA, 2010.

19.   Carvalho, C.; de Brito, J.; Flores-Colen, I.; Pereira, C. Inspection, diagnosis and rehabilitation system for vinyl and linoleum floorings in health infrastructures. *J. Perform. Constr. Facil.* **2018**, *32*, 04018078. [CrossRef]

20.   Lent, T.; Silas, J.; Vallete, J. Chemical hazards analysis of resilient flooring for healthcare. *HERD-Health Environ. Res.* **2010**, *3*, 97–117. [CrossRef]

21.   Carling, P.C.; Bartley, J.M. Evaluating hygienic cleaning in health care settings: What you do not know can harm your patients. *Am. J. Infect. Control* **2010**, *38*, S41–S50. [CrossRef]

22.   Brito, F. Controlo da contaminação de superfícies no meio hospitalar (Contamination control of surfaces in the hospital environment). *TecnoHospital* **2017**, *79*, 10–11. (In Portuguese)

23.   Toreki, W.; Kanga, R.S. Regeneration of Antimicrobial Coatings Containing Metal Derivates upon Exposure to Aqueous Hydrogen Peroxide. U.S. Patent 9,549,547 B2, 24 January 2017.

24.   Harris, D.D.; Pacheco, A.; Lindner, A.S. Detecting potential pathogens on hospital surfaces: An assessment of carpet tile flooring in the hospital patient environment. *Indoor Built Environ.* **2010**, *19*, 239–249. [CrossRef]

25.   Harris, D.D.; Detke, L.A. The role of flooring as a design element affecting patient and healthcare worker safety. *HERD-Health Environ. Res.* **2013**, *6*, 95–119. [CrossRef]

26.   Harris, D.D.; Fitzgerald, L. A life-cycle cost analysis for flooring materials for healthcare facilities. *J. Hosp. Adm.* **2013**, *4*, 92–100. [CrossRef]

27.   Schwartz, P.H.; Wesselschmidt, R.L. Essential requirements for setting up a stem cell laboratory. In *Stem Cell Technologies in Neuroscience*; Srivastava, A., Snyder, E., Teng, Y., Eds.; Humana Press: New York, NY, USA, 2017; pp. 225–237.

28.   Ayçam, İ.; Yazici, A. Evaluation of operating room units within the context of green design criteria. *GU J. Sci.* **2017**, *30*, 1–15.

29.   Heisterberg-Moutsis, G.; Heinz, R.; Wolf, T.F.; Harper, D.J.; James, D.; Mazzur, R.P.; Kettler, V.; Soiné, H.; Peoples, R. Floor coverings. In *Ullmann's Encyclopaedia of Industrial Chemistry*; Wiley-VCH Verlag GmbH & Co. KGaA: Weinheim, Germany, 2017. [CrossRef]

30.   Harris, D.D. A material world: A comparative study of flooring material influence on patient safety, satisfaction, and quality of care. *J. Interior Des.* **2017**, *42*, 85–104. [CrossRef]

31. Amaro, B.; Saraiva, D.; de Brito, J.; Flores-Colen, I. Statistical survey of the pathology, diagnosis and rehabilitation of ETICS in walls. *J. Civ. Eng. Manag.* **2014**, *20*, 511–526. [CrossRef]
32. Conceição, J.; Poça, B.; de Brito, J.; Flores-Colen, I.; Castelo, A. Data analysis of inspection, diagnosis, and rehabilitation of flat roofs. *J. Perform. Constr. Facil.* **2019**, *33*, 04018100. [CrossRef]
33. Delgado, A.; Pereira, C.; de Brito, J.; Silvestre, J.D. Defect characterization, diagnosis and repair of wood flooring based on a field survey. *Mater. Constr.* **2018**, *68*, 1–13. [CrossRef]
34. Gaião, C.; de Brito, J.; Silvestre, J. Technical note: Gypsum plasterboard walls: Inspection, pathological characterization and statistical survey using an expert system. *Mater. Constr.* **2012**, *62*, 285–297. [CrossRef]
35. Garcez, N.; Lopes, N.; de Brito, J.; Silvestre, J. Pathology, diagnosis and repair of pitched roofs with ceramic tiles: Statistical characterisation and lessons learned from inspections. *Constr. Build. Mater.* **2012**, *36*, 807–819. [CrossRef]
36. Neto, N.; de Brito, J. Validation of an inspection and diagnosis system for anomalies in natural stone cladding (NSC). *Constr. Build. Mater.* **2012**, *30*, 224–236. [CrossRef]
37. Pereira, A.; Palha, F.; de Brito, J.; Silvestre, J.D. Diagnosis and repair of gypsum plaster coatings: Statistical characterization and lessons learned from a field survey. *J. Civ. Eng. Manag.* **2010**, *20*, 485–496. [CrossRef]
38. Pires, R.; de Brito, J.; Amaro, B. Statistical survey of the inspection, diagnosis and repair of painted rendered façades. *Struct. Infrastruct. E* **2015**, *11*, 605–618. [CrossRef]
39. Sá, G.; Sá, J.; de Brito, J.; Amaro, B. Statistical survey on inspection, diagnosis and repair of wall renderings. *J. Civ. Eng. Manag.* **2015**, *21*, 623–636. [CrossRef]
40. Santos, A.; Vicente, M.; de Brito, J.; Flores-Colen, I.; Castelo, A. Analysis of the inspection, diagnosis, and repair of external door and window frames. *J. Perform. Constr. Facil.* **2017**, *31*, 04017098. [CrossRef]
41. Da Silva, C.; Coelho, F.; de Brito, J.; Silvestre, J.; Pereira, C. Statistical Survey on Inspection, Diagnosis, and Repair of Architectural Concrete Surfaces. *J. Perform. Constr. Facil.* **2017**, *31*, 04017097. [CrossRef]
42. Silvestre, J.D.; de Brito, J. Inspection and repair of ceramic tiling within a building management system. *J. Mater. Civ. Eng.* **2010**, *22*, 39–48. [CrossRef]
43. Gorrée, M.; Guinée, J.B.; Huppes, G.; Oers, L.V. Environmental life cycle assessment of linoleum. *Int. J. Life Cycle Assess.* **2002**, *7*, 158–166. [CrossRef]
44. Personen-Leinonen, E.; Redsven, I.; Neuvonen, P.; Hurme, K.R.; Pääkkö, M.; Koponen, H.K.; Pakkanen, T.T.; Uusi Rauva, A.; Hautala, M.; Sjöberg, A.M. Determination of soil adhesion to plastic surfaces using a radioactive tracer. *Appl. Radiat. Isotopes* **2006**, *64*, 163–169. [CrossRef]

Article

Diagnoses in the Aging Process of Residential Buildings Constructed Using Traditional Technology

Beata Nowogońska

Faculty of Civil Engineering, Architecture and Environmental Engineering, University of Zielona Góra, Szafrana 1, 65-516 Zielona Góra, Poland; b.nowogonska@ib.uz.zgora.pl; Tel.: +48-68-3282290

Received: 18 April 2019; Accepted: 14 May 2019; Published: 20 May 2019

Abstract: The perspective of maintaining residential buildings in adequate technical condition is one of the most important problems over the course of their service life. The aim of the work is to present issues connected with the methods of predicting the process of changes in performance characteristics over the entire period that a building, constructed using traditional technology, is operational. Identification of the technical situation consists of a prognosis based on the analytical form of the distribution function and probability density of building usability. The technical condition of a building results from its past, while familiarity with the condition is necessary to determine how the building will behave in the future. The presented predictive diagnostics of the performance characteristics of an entire building and its elements is an original methodology of describing the lifespan of a building. In addition to identifying the technical condition, its aim is also to aid in making decisions regarding maintenance works. The developed model of predicting changes in the performance characteristics of buildings, the Prediction of Reliability according to Exponentials Distribution (PRED), is based on the principles applied for technical devices. The model is characterized by significant limitations in its application due to the negligible influence of wear processes. In connection with the above, the Prediction of Reliability according to Raleigh Distribution (PRDD) was developed, where the carried-out processes of changes in the performance characteristics are described using Rayleigh's distribution, and the building is a multi-element system. Model development would be incomplete without subjecting it to verification. Predicting the degree of the technical wear of load-bearing walls of a building is a form of checking the proposed PRED and PRRD models on the basis of data derived from periodical inspections of the research material. The developed model of the time distribution of the proper functioning of a building, presented as an image of the forecast of changes in the technical condition, can be applied to solving problems occurring in practice. The targeted approach to predicting the occurrence of damage will allow for optimal planning of maintenance works in buildings during their entire service life.

Keywords: technical condition; performance characteristics; degree of wear; service life; preventive maintenance

---

1. Introduction

In many fields, such as biology, technology, or management, there is a need for understanding the prediction of the aging process of an analyzed object. Prospective familiarity with the degradation process over the entire cycle of the existence of an object will enable the analysis of repair works. The problem of maintaining residential buildings in adequate technical condition also imposes providing for optimal planning of maintenance works, while the proper determination of the scope and program of refurbishment requires a diagnosis of the technical condition to be carried out. Accurate determination of the state of the building, the reasons behind damage, and above all, the prediction of unfavorable changes make it possible specify refurbishment needs.

Diagnosis comprises the basis for properly carried out refurbishment activity in every technical object. It pertains to both issues of assessing the technical state and its causes, as well as the prediction of the degradation process. The diagnosis of the technical condition of buildings can be carried out using two methods. The one most frequently applied is the assessment of unfavorable changes in the objects on the basis of site inspections, nondestructive testing, measurements, and calculations. Another course of action that can be taken is the predictive method, relying on predicting the degradation of a building.

Predicting the aging process of residential buildings carried out using traditional technologies is necessary when planning maintenance works in these buildings [1–7]. A better understanding of the service life of buildings results in more efficient building maintenance and reduced environmental costs [8,9]. Service life analysis aims at establishing and explaining the performance-over-time functions [10–13]. Reference [14] developed three design proposals that target different service lives (30 years, 50 years, 100 years) based on the building's expected life and uses BIM technology to simulate the life cycle cost and design performance as based on the renovation scenario analysis of the building's life cycle.

In accordance with the recommendations of ISO 7162:1992 standards [15], an assessment of the performance characteristics of a building should be carried out. The changes in these characteristics should be predicted over time using a method simulating the predicted degradation of a good over time known as Predicted Service Life Distribution of the Component (PSLDC). The ISO standards "Planning the Service Life of a Building" [16,17] provide general standards regarding the issues of predicting the service life of a building. These guidelines contain an introduction to predicting performance characteristics, but lack details regarding forecasting. These standards highlight difficulties in indicating degradation, even in the case of similar buildings, as there are many variables influencing the service lives in practice. The variety of buildings, environments, quality of construction works, and future conditions of maintenance lead to uncertainty in forecasting service life. In one of the ISO standards "Planning the Service Life of a Building" [17], there is an entry that the course of the service life runs in accordance with the Weibull distribution and its possible modifications.

References [8,18,19] were given the classification of methods for predicting the service life of buildings (PSLDC). The classification is comprised of deterministic, probabilistic, and simulation methods. Deterministic methods based on the ISO standard provide merely approximated results. However, probabilistic methods based on indicating variables of the service model as random values with a known probability distribution are labor-intensive. The other group of probabilistic methods for describing the course of the destruction process are methods containing Markov chains. Markov chains are often applied for describing the destruction process of bridges and technical infrastructure. The example of applying discrete Markov [18] chains is the description of the destruction process of a residential building described in four states.

The simulation methods [8,18] are comprised of methods for determining the service life of a building, somewhere between the not-very-accurate deterministic methods and probabilistic methods, requiring a high quantity of data. The simulation methods are based on developing a mathematical model using probability distributions for determining the individual variables of the model. Reference [20] presented problems related to exploitation reliability, which was defined as the probability of exploiting a building without any interference (failure) in a given period of time. The author evaluated the construction technologies of residential buildings in terms of their reliability, and buildings erected in traditional technologies occurred to be the best.

Service Life Prediction is necessary to evaluate Life Cycle Assessment for a sustainable construction process [21]. In Reference [22], the time behavior of building components is simulated using time dependent environmental models thanks to the correlation between performance characteristics' decay (measured by lab programs) and the conditions established by users' requirements. The analysis in Reference [23] shows that the environmental performance of buildings is affected by the service life of a building and the replacement intervals of building components.

The study on service life prediction of building components is presented in References [24–27]. Artificial Neural Network, with its back-propagation learning algorithm, was used in developing model of service life prediction. The simulation results using trained neural network model indicated that implementation of the neural network model is important for future application of service life prediction of building components.

Building management [28] is a particular economic activity comprising a set of property maintenance, operation, and repair. This is a technical set of operations required for building maintenance and preservation of usable condition, as well as functionally required for the maintenance of the land to ensure that property is used in accordance with the purpose.

Reference [29] proposes a new stochastic dynamic programming model where optimality conditions are derived through the Hamilton-Jacobi-Bellman equations. The model defined the joint production and repair, which are major maintenance switching strategies that minimize the total cost over an infinite planning horizon. Two probabilistic approaches are described in Reference [30], one approach using a mathematical function (Weibull) to describe the performance of a component over time and one approach using discrete Markov chains.

The purpose of the research reported in Reference [31] was to develop a model that allows for the identification of the owner's needs in all phases of the building life cycle. A six-level classification system for the information required in the project and a two-dimensional model that maps the life cycle were corroborated and improved by applying the Delphi technique to a panel of 10 experts in two rounds. Reference [32] concluded that the planning and the application of the condition-based maintenance strategy its significant characteristics and make reference to the resulting prediction model.

Assessments of materials degradation require that methods be available to aid prediction of service life. Researchers in advanced aerospace, nuclear, electronics, and medicine have been more successful than researchers in building and construction technology in responding to the need for reliable predictions of service life [33].

In Reference [34], a proposed method combines the use of failure mode and effect analysis to permit identifying likely failure modes from which maintenance actions could be planned and the limit states method to assess the durability of the given retrofit action. In Reference [35], the model of the lifespan curve for a residential building was presented, where the Pareto principle was applied as the strategy for undertaking maintenance works.

Obsolescence presents a serious threat to built property [36], as it rarely accounts for the immobile, long-lasting, and (financial and natural resource) capital-intensive characteristics of property, nor for its societal and cultural significance. Minimizing obsolescence and extending longevity are therefore indispensable for maintaining the physical, economic, and societal investments.

The research aim is to present the method of diagnosing the aging process of a building and its components. Ensuring an adequate technical condition of residential buildings is one of the most important problems over the course of their service life. Prognostic familiarity of the aging process of buildings over the entire period of their existence is helpful when analyzing repair works. The proper development of the range and program of maintenance works requires the diagnosis of the technical condition to be carried out. Proper determination of the condition of the building, the reasons behind damage, and above all, the forecast of unfavorable changes will make it possible to determine repair needs.

2. Methods of Functions Describing the Aging Process of a Building

The problem of ensuring an adequate the technical condition of a building occurs over the entire period it is in service. The technical condition of a building changes as a result of the aging process. In solving problems connected with developing a prediction of changes in performance characteristics of a residential building, using algorithms to determine changes in the reliability of technical devices is proposed. The prognosis of unfavorable processes will make it possible to determine the time frame in which the technical condition of a building will be unsatisfactory in the future and will necessitate repair works.

The problem of the wear of objects is an increase in damage and partial defects [37]. A model presented in Reference [38] allows for different kinds of obsolescence to be characterized and distinguished. To improve the representation of service life in life cycle assessments and the evaluation of environmental impacts, building service life prediction modelling was examined.

The object's reliability is defined as the ability to fulfil the task resulting from the purpose it was intended [39]. The object is demanded to fulfill a determined function in determined time t in determined conditions of operation. The measure of the reliability of an object is the probability of the task completing. Such defined reliability measure is a function of time of the building's reliable performance and is called reliability function [40].

For modeling situations in survival analysis, when the probability of failure changes over time, the Weibull distribution is most often applied as the distribution of the random variable of the time a building is fit for service. This distribution has been applied for many years, as a strength distribution as well as a distribution of the time of the proper operation and durability of analyzed goods [41–47].

The probability density function for the Weilbull distribution is determined with relation:

$$f(t) = \alpha \, \beta^{\alpha} \, t^{\alpha-1} \, \exp\left[-(\beta t)^{\alpha}\right] \qquad \text{for } t \geq 0, \tag{1}$$

where:

t   the exploitation period,

α scale parameter (a real number) $\alpha > 0$,

β the shape parameter (real number), $\beta > 0$.

Parameter α of the distribution determines the probability of a breakdown in time:

- for $\alpha < 1$, the probability of breakdown decreases in time, which indicates the fact, when the object breakdown is modeled, that some specimens may have production defects and slowly fall out of the population,
- for $\alpha = 1$ (exponential distribution), the probability is constant, which suggests that that breakdowns are caused by external random events,
- for $\alpha > 1$, the probability grows in time, it indicates the fact time-related technical wear of elements is the main cause of breakdowns,
- for $\alpha = 2$ (the Rayleigh distribution), the probability grows in time.

The distributions function for the Weilbull distribution:

$$F(t) = 1 - \exp\left[-(\beta t)^{\alpha}\right], \tag{2}$$

A measure of the reliability of technical devices is the R(t) function, also referred to as the survival function. The distribution function is called the probability of damage, a destruction function, breakdown, or a failure function, and is determined with the relation:

$$F(t) = P(t < T_R) = 1 - R(t), \tag{3}$$

T_R period of object durability,

R(t)   reliability function, also called the probability of proper operation, or durability function.

The specific example of the Weibull distribution, when the shape parameter is $\alpha = 1$, is the exponential distribution. The characteristic for the exponential distribution is a constant intensity of damage throughout the whole period of the object exploitation [42,45,47]. Exponential distribution is frequently used in the examination of a proper performance time of mechanical and electronic devices.

3. Assumptions Made Regarding the PRRD Method

In business planning of repairing the buildings, one should determine the scope of works. There are no reliable mathematical models that allow the estimation of the operational reliability of a building, often referred to as changes in the building performance. In the case of technical appliances (mechanical and electronic), attempts are undertaken to determine the prediction of their operational reliability. However, for buildings, only indicative graphs of changes in performance are presented.

Literature [41–47] gives the general form of Weibull as a model for describing the prognoses of the aging. In the case of mechanical and electronic devices, a particular case of Weibull distribution—exponential distribution—is used. The distribution function of exponential distribution is treated as a function of the damage of a device. The damage function, also referred to as unreliability, is a complement of reliability, which is the R(t) function.

The developed model of predicting changes in the performance characteristics of buildings—Prediction of Reliability according to Exponentials Distribution (PRED) [46,48]—is based on the principles applied for technical devices. However, large approximations are accepted in exponential distribution. The negligible influence of wear processes is assumed. The exponential model of degradation curves of objects is based on a constant intensity of damage, which, according to the author, is not precise in the case of building structures.

Another specific case of Weibull distribution is Rayleigh distribution. In this distribution, the wear of the object along with the passing of time is considered to be a significant cause of failure. The author proposes using this distribution for the mathematical description of the aging process of a residential building. In the aging process of a building, its wear increases continuously, which is why the author believes that, in assessing the changes in the performance characteristics of residential buildings, formulas based on Rayleigh distribution are more accurate relationships.

The general form of the function describing the course of changes in the performance characteristics over time t based on the Weibull distribution is expressed by the relationship (1). In Rayleigh distribution, parameters α and β are equal to: $\alpha = 2$, $\beta = 1/T$.

The proposed [48–50] model PRRD (Prediction of Reliability according to Raleigh Distribution) of changes in the performance characteristics $R_i(t)$ of the i-th building component in time t, based on the Raleigh distribution, and using durability periods of component T_{Ri} from data found in literature [51–54], is described by the relationship [45,46]:

$$R_i(t) = \exp\left[-(t/T_{Ri})^2\right], \tag{4}$$

The disabilities of building components are given in literature [51–54] as ranges of periods which result from the different conditions concerning location, maintenance, variability in properties of materials, the level of workmanship, and the level of design.

The selected results of calculations are presented in Figure 1.

Buildings are made up from many elements with a difficult to define and determine homogeneity of materials. They have a complex structure and are subject to wear as a result of many processes. In addition to this, building structures are exposed to outside factors, often with a very diverse intensity. Indicating changes in the technical state over the entire course of the use of a building is a process requiring many generalizations to be made.

Each building is made up of many components. These elements serve various functions and are created from dissimilar construction materials, with each being characterized by different properties and different service lives. The initial stage of the diagnostic analysis of a building is its decomposition.

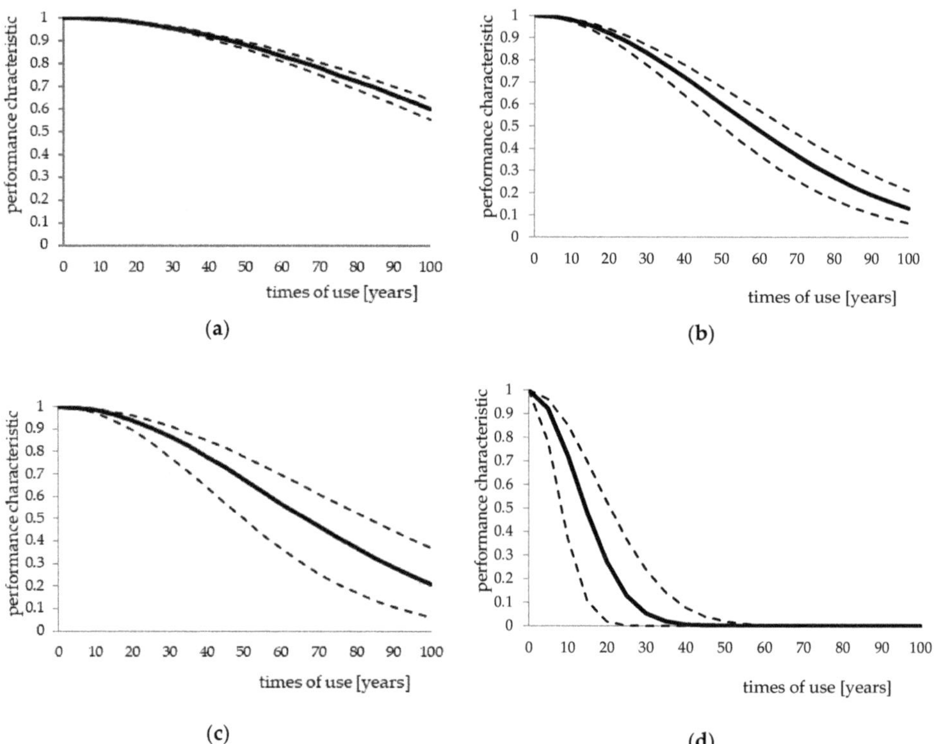

Figure 1. Changes in the performance values of building components: (**a**) Masonry walls, (**b**) wooden floors (**c**) wooden rafter, (**d**) gutters, and drain pipes.

A building constructed using traditional technology with exemplary material and construction solutions was subjected to decomposition into diagnostic levels. The depth of diagnosis, as well as level p of probing into the structure of the building, were assumed. The components of level p were listed in Table 1. Each component is characterized by its own service life. Studies pertaining to the service lives of building components were carried out at research centers in Poland. As a result of the analysis of literature data pertaining to the service lives minimum T_{Rmin}, maximum T_{Rmax}, and average T_{Rsr} values were established. These values are compiled in Table 1. Under comprehensive analysis of changes in the service properties of a building, the problem ought to be looked into accounting for the phenomena accompanying the long-term use of the object. Among them are: Wear, frequency of carrying out maintenance works, manner of use, the type of use, the influence of outside factors, the correctness of the design process, the quality of the applied materials, and the precision with which it had been made. In the case of buildings which have been used for a few dozen years, there is no possibility of checking the conditions derived from the design and construction period of the building. The long period of use to date points to the fact that design and construction processes were correct and high-quality building materials were used. The assumption was also made that the type of use of the objects (all are residential buildings), as well as external factors (similar atmospheric conditions due to the same climate, close geographical location), for all objects are at an equal level and do not influence the diversification of their performance characteristics.

Each building component serves a specific function. Structural elements have the most significant influence on the service life of a building. Other elements influence the performance characteristics of a building to a lesser degree, with their influence resulting from the fact that damaged auxiliary elements can lead to changes in the parameters of basic elements. The intensities of the influence of performance characteristics of the i-th components (Figure 1) in the form of a scale of weights of elements A_i (importance weight) based on scale serving to assess the quality of a building were accounted for when determining the performance characteristics of the entire building, which is a collection of n components. Changes in the performance characteristics of building $R_A(t)$ in time t are determined by the relationship [48]:

$$R_A(t) = \sum A_i R_i(t) \tag{5}$$

Table 1. Periods of durability of individual building components from selected building materials [51–54].

Component	Time Periods for Establishing the Dates of Maintenance Works	
	min [years]	max [years]
Brick foundations	63	110
Masonry brick walls	80	140
Masonry partition walls	57	100
Wooden beam ceilings	40	70
Wooden stairs	20	35
Roof rafter	45	80
Tail caver	40	70
Gutters and drain pipes	10	18
Internal plasters	32	55
External plasters	26	45
Windows	29	50
Doors	51	90
Glazing	22	40
Wooden floor	28	50
Wall coatings	3	4
Woodwork oil coatings	3	5
Cores of ceramic cookers	20	35
Tiled stove	26	45
Central heating pipes	20	35
Boilers and heaters for c.h.	28	50
Water supply and sewage pipes	21	38
Water supply and sanitation fittings	17	30
Gas pipes	21	38
Electrical installations	34	60

4. Verification of the Mathematical Model

The examined material comprises 592 residential buildings performed in the traditional technology, situated within the area of the town of Zielona Gora [55]. The buildings were built between 1915—2015. The buildings were divided into subgroups research buildings 5, 10, 15, ..., 100 years.

The buildings covered by the analysis are characterized by similar solutions in terms of their construction and materials used. All of them are two stories high with full basements. The walls of the analyzed buildings were made of full brick and the floors over the basements are brick infill floors, with the remaining floors constructed on wooden beams construction. The stairs and roof trusses are wooden with a collar beam roof structure, and roof covering is comprised of plain type pantile or roofing felt.

The technical states of all the buildings were periodically (every five years) inspected by experts. The periodic monitoring, consisting in the examination of technical wear, resulted in the reports containing the information on the percentage wear of 25 components of the 592 buildings.

For each building element, it is possible to determine the prediction of the technical wear in any arbitrary exploitation period. The durability periods of building elements of determined material-structure solutions are given in [49–52], and after substituting them into Formula (4), the prediction of the degree of technical wear may be obtained according to the Rayleigh distribution.

The bibliography on reliability of electronic a device attributes the intensity of failure to technical wear as described in:

$$S_Z(t) = \int_0^t \lambda(t)dt \tag{6}$$

where $S_z(t)$—the rate of the product wear.

Often [41], a different definition of the intensity of damage $\lambda(t)$ is given, described as the speed at which unreliability $F(t)$ increases in relation to $R(t)$:

$$\lambda(t) = \frac{dF(t)}{dt}\frac{1}{R(t)} \tag{7}$$

Based on the above dependencies, the degree of consumption in the PRRD model is a function dependent on time:

$$S_Z(t) = \frac{t^2}{T^2} \tag{8}$$

Out of the results of periodic inspections (every five years) of building elements, the results for walls were selected. For each subgroup of buildings (5, 10, 15, ..., 100 years old), average values of wall wear were calculated. The results are presented in Table 2 and Figure 2.

According to the relationship (8), changes in the consumption of masonry brick walls in the full period of their use have been determined. For brick masonry walls, the durability period is determined within the limits 130–150 years. The degrees of technical wear were determined for the minimum (130) and the maximum (150) values, with the use of the according to Rayleigh distribution (Formula (8)). The obtained results are presented in Table 2 and Figure 2. The analysis adopts the same class of use for all buildings.

Table 2. Predictions of the degree of wear of masonry walls. Prediction of Reliability according to Raleigh Distribution (PRRD), as well as values of the degree of wall wear resulting from the periodic inspection of buildings located in Zielona Góra.

Times of Use [Years]	Degree of Walls Wear (Average Value)	Predictions of the Degree of Wear of Masonry Walls in Model PRRD (Formula (8))	
		Min	max
5	0.00	0.00	0.00
10	0.00	0.00	0.00
15	0.00	0.01	0.01
20	0.02	0.02	0.02
25	0.04	0.03	0.04
30	0.05	0.04	0.05
35	0.06	0.05	0.07
40	0.08	0.07	0.09
45	0.09	0.09	0.12
50	0.14	0.11	0.15
55	0.18	0.13	0.18
60	0.22	0.16	0.21
65	no data	0.19	0.25
70	no data	0.22	0.29
75	0.32	0.25	0.33
80	0.35	0.28	0.38
85	0.42	0.32	0.43
90	0.43	0.36	0.48
95	0.50	0.40	0.53
100	0.58	0.44	0.59

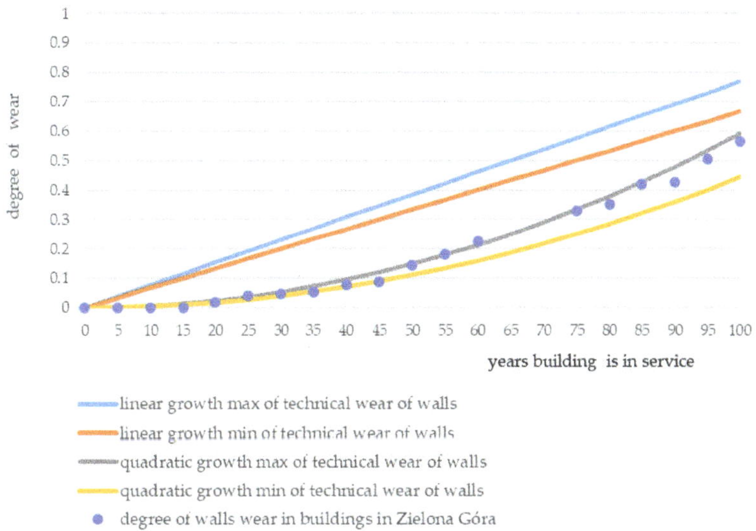

Figure 2. Predictions of the degree of wear of masonry walls as well as values of the degree of wall wear resulting from the periodic inspection of buildings located in Zielona Góra.

The values of the degree of wear of walls by the Rayleigh distribution during a 100-year period of building exploitation, as well as the average values of the degree of wear, obtained in periodic inspections of buildings five-year, 10, 15, ..., 100-year, were verified with the use of Student t-test for independent samples. The maximum probability of an allowable error, which could be made while drawing conclusions, was assumed to be equal $\alpha = 0.05$. For this level of significance, a critical area was determined. For buildings included in the analysis, the number of degrees of freedom equal to n = 19, the critical value of the test is $p = 2.0930$. In examining buildings with the t-test, the test result was 2.2957. This value is greater than $p = 2.0930$, which means that the results by the Rayleigh distribution are statistically significant. It can be assumed, therefore, that the reliability of predictions of buildings determined by the Rayleigh distribution is close to reality.

The prediction of the degree of wear of the walls is an example of the methodology of prediction of technical condition of building elements. In an analogous way, it is possible to elaborate predictive changes in the degree of wear of all building components.

5. Applying the PRRD Model in Planning Maintenance Works

Ensuring the proper technical condition of a building in over the course of its use can only be done by properly carried out refurbishment activities. The proposed PRRD model can be applied to modeling situations in the analysis of changes in the performance characteristics of a building which had not undergone refurbishment. The PRRD method of predicting changes in the performance characteristics can support activities aimed at avoiding an inadequate technical condition of a building. The accurate prediction of unfavorable changes and preventive repairs and maintenance works will make it possible to ensure the proper technical condition of the object.

The effective use of a building ought to be based on maintaining an adequate level of performance characteristics of the building, with refurbishment processes serving to fulfill this task. All types of refurbishment activities have a significant influence on the technical condition of a building over the course of its continued use. The full characterization of a refurbished object must account for the initial state and the scope of maintenance works. On this basis, the course of changes in the performance characteristics over time, prior to and after the refurbishment, can be determined.

The prediction of changes in the performance characteristics $R_M(t)$ of a building which had undergone refurbishment, where tp is the date of the refurbishment, is expressed by the Formula (9):

$$R_M(t) = \begin{cases} \sum_{i=1}^{m-r} A_i \exp\left(-\left(\frac{t}{T_{Ri}}\right)^2\right) & \text{if } t \in (0, t_{pi}) \\ \sum_{i=1}^{r} A_i \exp\left(-\left(\frac{t-t_{pi}}{T_{Ri}}\right)^2\right) & \text{if } t \in (t_{pi}, T), \end{cases} \tag{9}$$

where:

r—the number of refurbished components in a given inter-repair cycle,
m—number of all components,
A_i—weight of i-th component,
t_{pi}—time of carrying out refurbishment on component i,
T_{Ri}—life cycle of i-th component.

The following assumptions were made:

(1) The building is constructed using traditional technology, and the material and building solutions taken from Table 1;
(2) The frequencies of maintenance works for building components are:

    $c1 = 5$ five-year cycle—including wall and ceiling paint coats, oil paint coats on windows, and doors,
    $c2 = 15$ gutters and downpipes, electrical installations,
    $c3 = 30$ wooden stairs, plumbing pipes and pipe fittings, gas lines,
    $c4 = 60$ wooden ceilings, roof trusses, ceramic tile roof cover, outside plaster, window and door frames, glazing, floors and floor covers, furnaces and radiators, electrical installation cables.

(3) The times of carrying out maintenance works on building components stem from the life cycles of these elements nearing their end.

Proposals of times of carrying out planned-preventive maintenance works were initially indicated based on experiments and the life cycles of building components nearing their end. The proposed maintenance works aim to maintain the building in good, satisfactory, or average technical condition, and rely on prophylactic action and preventing the premature emergence of unfavorable changes.

The obtained results of predictions of changes in the performance characteristics of a renovated building according to Rule (9) for the earlier accepted assumptions are presented in Figure 3. For purposes of comparison, a dashed line is used to also present changes in the performance characteristics over the course of use of a building which had not undergone maintenance works.

Modeling various possible usage scenarios of a building according to the proposed method will allow for the optimal planning of building maintenance works undertakings. The changes in the performance characteristics of the building shown in Figure 3 illustrate the application of a sample strategy of maintenance works of selected building components at the assumed times.

The characteristics of various strategies of choosing refurbishment activities have an influence on the shape of the life curve of a building. The obtained data allow for the comparative analysis of the reliability effectiveness of a building and choice of the most adequate material and structural solutions.

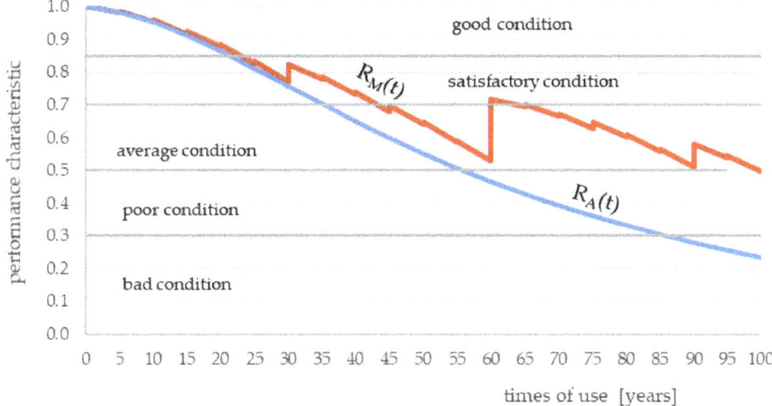

Figure 3. Prediction of changes in the technical condition of a renovated building $R_M(t)$ and unrenovated building $R_A(t)$.

6. Discussion of Proposal to Assess the Date of the Complete Maintenance Works

Changes in the performance characteristics of a building R(t), changes in the failure of a building (F(t), and its level of wear $S_Z(t)$ during subsequent years of its use are a picture of the changes in the technical condition of a building over the course of subsequent years of service.

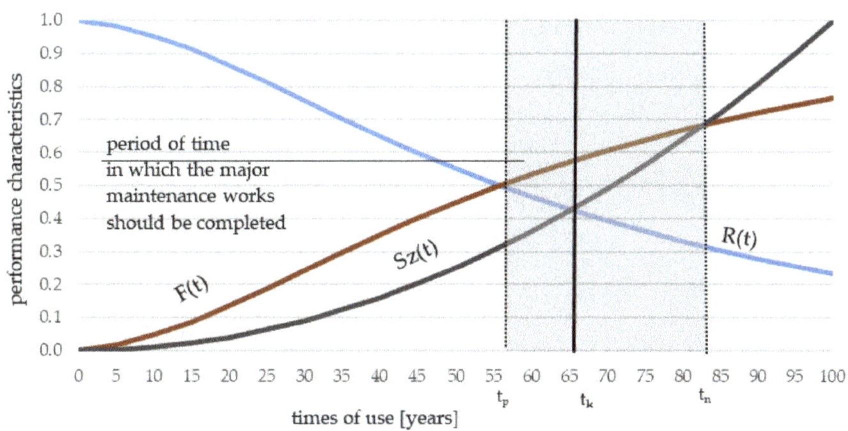

Figure 4. Functions describing the aging process of a building.

The performance characteristics of a building along with its failure over the course of its use can assume values from 0 to 1. The level of wear of a building, given as a percentage, increases over time, from 0 to the maximum value.

The performance characteristics of a building in which maintenance works had not been carried out decrease in subsequent years of use, whereas the failure of a building continually increases. It is assumed that an important moment is term t_p, when the performance characteristics are so poor, that they are equal to the failure of the building. Starting from term t_p, complete maintenance works of the building ought to be carried out due to its bad technical condition (Figure 4).

A more critical term is the moment t_k. The function of changes in the performance characteristics lessens and reaches lower values than the constantly increasing failure function. Term t_k is assumed as a critical term when the performance characteristics function is equal to the level of wear function. It is assumed that term t_k is a term requiring the unconditional carrying out of major maintenance works of the building.

Term t_n, on the other hand, indicates the limit of the unprofitability of carrying out any maintenance works. It is a date when the level of wear begins to exceed the failure of the building.

The proposed model of changes in the operational properties of a building and its components over time is a method of predicting unfavorable changes to the technical condition of a building. The aging process of residential buildings is an inevitable problem. Familiarity with the course of the damage process for the entire period a building is in use is helpful when planning maintenance works.

In this method, the value of the level of wear equal to 70% is given as the limit for carrying out maintenance works; due to the high costs of repairs, provided the demolition of the building is recommended if the level of wear is higher than 70%. The carried-out analysis does not account for the costs of remodeling and regards only the technical state of the building. Thanks to the indicated changes of the aging function over time, it can be stated that residential buildings whose level of wear is higher than 70% are to be demolished due to their bad technical state.

The exponential distribution is applied often to assess the distribution of the time of proper operation, however the exponential model of the reliability distribution does not exist in reality. In the exponential distribution, significant approximations are accepted, assuming the negligible influence of wear processes. Another specific example of the Weibull distribution is the Rayleigh distribution. This distribution occurs when the wear of the object with the passing of time is the main reason behind failure. It is the author's opinion that the choice of the Rayleigh distribution for modelling the aging process in buildings seems to be the most accurate. All buildings and their components undergo wear over the course of their use, with the Rayleigh distribution applied in cases when the wear of the building increases along with the passing of time.

The model of the distribution of the time during which a building operates properly presented as the prognosis of changes in the technical condition can be applied to solve problems occurring in practice. The approach directed at predicting the occurrence of damage will make it possible to plan maintenance works on buildings in an optimal manner over the entire course of their service lives.

The proposed method of determining the terms of complete maintenance works should be treated as a warning. The terms of maintenance work according to this method stem from the bad technical condition of the building. The additional operating costs accounted for result from the lack of maintenance works, and will surely lead the terms being moved to earlier.

The proposed prognostic models of the aging process of residential buildings can be effectively used by real estate managers for planning the maintenance of buildings.

7. Conclusions

Over the entire course of use, the performance characteristics of a building are increasingly lower, decreasing with usage and deteriorating as a result of technical wear. The increase in the performance characteristics can take place as a result of conservation, preventive measures, modernization, the replacement of components, or overhaul or rebuilding. In all periods of the life cycle of a technical object, the needs for predicting damage occur. Determining the changes in the performance characteristics of a building, in accordance with the recommendations set out by standards, requires the application of Predicted Service Life of a Component (PSLDC) danger curves as support tools in the planning the times of maintenance works. The proposed model of changes in the performance characteristics of a residential building in the Prediction of Reliability according to Rayleigh Distribution (PRRD) function of time facilitates the prediction of changes in the technical state of a building.

The next stages of research will concern the analysis of conservation strategies to control the degradation of the building. The decision-making process will include models of deterioration of conditions in the building and problems related to maintenance works costs.

The prediction of the degradation of a building should prove useful in processes of reacting to aging-related damage of buildings, while applying danger curves by administrators can be valuable as a means of supporting the planning of maintenance works.

Conflicts of Interest: The author declares no conflict of interest.

References

1. Masters, L.W. *Problems in Service Life Prediction of Building and Construction Materials*; Springer: Amsterdam, The Netherlands, 1985.
2. Crowder, J.R. Problems in service life prediction of building and construction materials. *Int. J. Fatigue* **1987**, *9*, 126–127. [CrossRef]
3. Lacasse, M.A. Advances in methods for service life prediction of building materials and components-final report-activities of the CIB W80. In Proceedings of the 10th International Conference on Durability of Building Materials and Components (DBMC), Lyon, France, 7–20 April 2005.
4. Lacasse, M.A. Advances in service life prediction-An overview of durability and methods of service life prediction for non-structural building components. In Proceedings of the annual Australasian Corrosion Association conference, Wellington Convention Centre, Wellington, New Zealand, 16 November 2008.
5. Sjöström, C. Durability of Building Materials & Components 7. In Proceedings of the Seventh Conference Durability of Building Materials and Components, Stockholm, Sweden, 4 July 1996.
6. Daniotti, B.; Spagnolo, S.L.; Paolini, R. Climatic Data Analysis to Define Accelerated Ageing for Reference Service Life Evaluation. In Proceedings of the 11DBMC International Conference on Durability of Building Materials and Components, Istanbul, Turkey, 11–14 May 2008. T32. [CrossRef]
7. Daniotti, B.; Spagnolo, S.L. Service Life Prediction Tools for Buildings' Design and Management. In Proceedings of the 11DBMC International Conference on Durability of Building Materials and Components, Istanbul, Turkey, 11–14 May 2008. T72. [CrossRef]
8. Silva, A.; de Brito, J.; Gaspar, P.L. *Methodologies for Service Life Prediction of Buildings*; Springer International Publishing: Cham, Switzerland, 2016; Volume VII, p. 432. [CrossRef]
9. Firląg, S.; Piasecki, M. NZEB Renovation Definition in a Heating Dominated Climate: Case Study of Poland. *Appl. Sci.* **2018**, *8*, 1605. [CrossRef]
10. Jernberg, P.; Sjöström, C.; Lacasse, M.A.; Brandt, E.; Siemes, T. Guide and Bibliography to Service Life and Durability Research for Buildings and Components. In *Joint CIB W080/RILEM TC 140—Prediction of Service Life of Building Materials and Components*; CIB W080/RILEM 175-SLM; International Council for Building Research, Studies and Documentation: Rotterdam, The Netherlands, 2004.
11. Masters, L.W. Prediction of service life of building materials and components. *Mater. Struct.* **1986**, *19*, 417–422. [CrossRef]
12. Varnier, D.J.; Lacasse, M.A.; Brown, W.C.; Chown, G.A.; Kyle, B.R. Applying service life and asset management techniques to roofing systems. In Proceedings of the Sustainable Low-Slope Roofing, Oak Ridge, TN, USA, 9 October 1996.
13. Zurbrügg, P. *Simulation des Phénomènes de Dégradation D'éléments de Construction*; Ecole Polytechnique Fédérale De Lausanne: Lausanne, Switzerland, 2010.
14. Juan, Y.K.; Hsing, N.P. BIM-Based Approach to Simulate Building Adaptive Performance and Life Cycle Costs for an Open Building Design. *Appl. Sci.* **2017**, *7*, 837. [CrossRef]
15. *ISO 7162:1992—Performance Standards in Building—Contents and Format of Standards for Evaluation of Performance*; ISO: Geneva, Switzerland, 1992.
16. *ISO 19208:2016—Framework for Specifying Performance in Buildings*; ISO: Geneva, Switzerland, 2016.
17. *ISO 15686-2:2012—Buildings and Constructed Assets—Service Life Planning—Part 2: Service Life Prediction Procedures*; ISO: Geneva, Switzerland, 2012.
18. Ortega, L.; Serrano, B.; Fran, J.M. Proposed method of estimating the service life of building envelopes. *Rev. Constr.* **2015**, *14*, 60–68.
19. Bucoń, R.; Sobotka, A. Decision-making model for choosing residential building repair variants. *J. Civ. Eng. Manag.* **2015**, *21*, 893–901. [CrossRef]

20. Sugier, J.; Anders, G.J. Modelling and evaluation of deterioration process with maintenance activities. *Maint. Reliab.* **2013**, *15*, 305–311.
21. Korentz, J.; Nowogońska, B. Assessment of the life cycle of masonry walls in residential buildings. In Proceedings of the MATEC Web of Conferences; 3rd Scientific Conference Environmental Challenges in Civil Engineering (ECCE 2018), Opole, Poland, 23–25 April 2018; 2018; Volume 174, p. 01025.
22. Lacasse, M.A.; Vanier, D.J. Durability of Building Materials and Components 8. Service life and Asset Management. In Proceedings of the Eighth Conference Durability of Building Materials and Components, Vancouver, BC, Canada, 30 May–3 June 1999.
23. Janjua, S.Y.; Sarker, P.K.; Biswas, W.K. Impact of Service Life on the Environmental Performance of Buildings. *Buildings* **2019**, *9*, 9. [CrossRef]
24. Usman, F. Service Life Prediction of Building Components. In Proceedings of the ICCBT—International Conference on Construction and Building Technology, Kuala Lumpur, Malaysia, 16–20 June 2008; pp. 287–296.
25. Knyziak, P. Estimating the Technical Deterioration of Large-Panel Residential Buildings Using Artificial Neural Networks. *Procedia Eng.* **2014**, *91*, 394–399. [CrossRef]
26. Caniato, M.; Bettarello, F.; Schmid, C.; Fausti, P. The use of numerical models on service equipment noise prediction in heavyweight and lightweight timber buildings. *Build. Acoust.* **2019**, *26*, 35–55. [CrossRef]
27. Radziszewska-Zielina, E.; Śladowski, G.; Sibielak, M. Planning the reconstruction of a historic building by using a fuzzy stochastic network. *Autom. Constr.* **2017**, *84*, 242–257. [CrossRef]
28. Pukite, I.; Geipele, I. Different Approaches to Building Management and Maintenance Meaning Explanation. *Procedia Eng.* **2017**, *172*, 905–912. [CrossRef]
29. Rivera-Gómez, H.; Oscar Montaño-Arango, O.; Corona-Armenta, J.R.; Garnica-González, J.; Hernández-Gress, E.S.; Barragán-Vite, I. Production and Maintenance Planning for a Deteriorating System with Operation-Dependent Defectives. *Appl. Sci.* **2018**, *8*, 165.
30. Rudbeck, K. Methods for Designing Building Envelope Components Prepared for Repair and Maintenance. Ph.D. Thesis, Department of Buildings and Energy, Technical University of Denmark, Lyngby, Denmark, 2000.
31. Alshubbak, A.; Pellicer, E.; Catala, J.; Teixeira, J. A Model for identifying owner's needs in the building life cycle. *J. Civ. Eng. Manag.* **2015**, *21*, 1046–1060. [CrossRef]
32. Chen, C.; Juan, Y.; Hsu, Y. Developing a systematic approach to evaluate and predict building service life. *J. Civ. Eng. Manag.* **2017**, *23*, 890–901. [CrossRef]
33. Masters, L.W.; Brandt, E. Systematic methodology for service life prediction of building materials and components. *Mater. Struct.* **1989**, *22*, 385–392. [CrossRef]
34. Morelli, M.; Lacasse, M.A. A systematic methodology for design of retrofit actions with longevity. *J. Build. Phys.* **2019**, *42*, 485–604. [CrossRef]
35. Nowogońska, B. Preventive Services of Residential Buildings According to the Pareto Principle. *IOP Conf. Ser.* **2019**, *471*, 112034. [CrossRef]
36. Cecconi, F.R.; Moretti, N.; Maltese, S.; Dejaco, M.C.; Kamarab, J.M.; Heidrich, O. A rating system for building resilience. *TECHNE J. Technol. Archit. Environ.* **2018**, *15*, 358–365.
37. Cempel, C.; Natke, H.G. Damage Evolution and Diagnosis in Operating Systems. In *Safety Evaluation Based on Identification Approaches Related to Time-Variant and Nonlinear Structures*; Springer: Cham, Switzerland, 1993; pp. 44–61.
38. Thomsen, A.; Flier, K. Understanding obsolescence: A conceptual model for buildings. *Build. Res. Inf.* **2011**, *39*, 352–362. [CrossRef]
39. Moubray, J. *RCM II-Reliability Centered Maintenance*; Industrial Press, Inc.: Oxford, UK, 2007.
40. Andrews, J.D.; Moss, T.R. *Reliability and Risk Assessment, Longman Scientific & Technical*; John Wiley: Hoboken, NJ, USA, 1993.
41. Cordeiro, G.; Ortega, M.; Lemonte, A. The Exponential-Weibull Lifetime Distribution. *J. Stat. Comput. Simul.* **2014**, *84*, 2592–2606. [CrossRef]
42. Walpole, R.E.; Myers, R.H. *Probability and Statistics for Engineers and Scientists*; Macmillan Publishing Company: London, UK, 1985.
43. Grant, A.; Ries, R. Impact of building service life models on life cycle assessment. *Build. Res. Inf.* **2013**, *41*, 168–186. [CrossRef]
44. Nowogońska, B. Proposal for determing the scale of renovation needs of residential buildings. *Civ. Environ. Eng. Rep.* **2016**, *22*, 137–144. [CrossRef]

45. Khelassi, A.; Theilliol, D.; Weber, P. Reconfigurability Analysis for Reliable Fault-Tolerant Control Design. *Int. J. Appl. Math. Comput. Sci.* **2011**, *21*, 431–439. [CrossRef]

46. Nowogońska, B. Model of the reliability prediction of masonry walls. In *Engineering Mechanics 2014-20th International Conference Svratka*; Brno University of Technology: Brno, Czechy, 2014; pp. 456–459.

47. Zaidi, A.; Bouamama, B.; Tagina, M. Bayesian reliability models of Weibull systems: State of the art. *Int. J. Appl. Math. Comput. Sci.* **2012**, *22*, 585–600. [CrossRef]

48. Nowogońska, B. The Life Cycle of a Building as a Technical Object. *Periodica Polytech. Civ. Eng.* **2016**, *60*, 331–336. [CrossRef]

49. Nowogońska, B.; Cibis, J. Technical problems of residential construction. *IOP Conf. Ser.* **2017**, *245*, 052042. [CrossRef]

50. Nowogońska, B. Performance characteristics of buildings in the assessment of revitalization needs. *Civ. Environ. Eng. Rep.* **2019**, *29*, 119–127. [CrossRef]

51. Winniczek, W. *Valuation of Buildings and Structures by Reconstruction*; CUTOB PZITB: Wrocław, Poland, 1993. (In Polish)

52. Lenkiewicz, W. *Repairs and Modernization of Building Objects*; Warsaw University of Technology: Warsaw, Poland, 1998. (In Polish)

53. Zaleski, S. *Renovation of Residential Buildings-a Guide*; Arkady: Warsaw, Poland, 1997. (In Polish)

54. Achterberg, G.; Hampe, K.H. Baustoffe und Bauunternehmungskosten—Wirtschaftlich günstige Relationen von Herstellung und Unterhaltungskosten der Gebäude. In *Schriftenreihe Bau und Wohnforschung des Bundesminister für Raumordnung, Bauwesen und Städtebau*; Heft 04.051, Nds MBL, 43; Amazon: Bonn, Germany, 1976.

55. *Periodic Inspection Protocol 2015*; Department of Municipal and Housing Administration in Zielona Góra: Zielona Góra, Poland, 2015. (In Polish)

 *buildings*

Article

A Cross-Domain Decision Support System to Optimize Building Maintenance

Nicola Moretti * and Fulvio Re Cecconi

Politecnico di Milano, Department of Architecture, Built Environment and Construction Engineering, 20133 Milano, Italy
* Correspondence: nicola.moretti@polimi.it; Tel.: +39-02-23995174

Received: 3 June 2019; Accepted: 2 July 2019; Published: 4 July 2019

Abstract: Operations and maintenance optimization are primary issues in Facility Management (FM). Moreover, the increased complexity of the digitized assets leads Facility Managers to the adoption of interdisciplinary metrics that are able to measure the peculiar dynamics of the asset-service system. The aim of this research concerns the development of a cross-domain Decision Support System (DSS) for maintenance optimization. The algorithm underpinning the DSS enables the maintenance optimization through a wiser allocation of economic resources. Therefore, the primary metric encompassed in the DSS is a revised version of the Facility Condition Index (FCI). This metric is combined with an index measuring the service life of the assets, one measuring the preference of the owner and another measuring the criticality of each component in the asset. The four indexes are combined to obtain a Maintenance Priority Index (MPI) that can be employed for maintenance budget allocation. The robustness of the DSS has been tested on an office building in Italy and provided good results. Despite the proposed algorithm could be included in a wider Asset Management system employing other metrics (e.g., financial), a good reliability in the measurement of cross-domain performance of buildings has been observed.

Keywords: Digitization; Key Performance Indicators; KPIs; Asset Management; Facility Management; Operations Maintenance & Repairs; Decision Support System; Facility Condition Index

---

1. Introduction

In Facility Management (FM), performance measurement is one of the primary issues to both ensure a high-quality built environment for owners and users, and to enhance competitiveness of the organization providing FM services. The performance measurement assumes a strategic role for the assessment and monitoring of critical factors within an organization [1], and therefore needs a particular attention in the context of improved decision-making.

Currently the FM context is changing. The physical asset should no longer be characterized only according to its presence and existence in the real estate; it should be conceived as strictly related to data and information which are used for its operation and that are generated through its use [2]. This trend can be named the digitization of the built environment. The digitization, together with the sustainability, is one of the driving forces in FM [3] and can be intended as those dynamics that are currently leading the research in FM and the development of new and more effective business processes in the industry. The digitization is shaping a different approach, which should be considered when dealing with physical assets [4]. The digitized asset is composed of the physical asset integrated with data related to the different phases of its life cycle. This strict integration is due to the contemporary need for collaboration and the interdisciplinary exchange of competences which requires new approaches for working on the digital built environment. Moreover, FM has always been related to the information management tools and took advantage of this strict interrelation from the first Computer Aided Facility Management (CAFM), followed by CAD tools and Integrated Workplace Management Systems (IWMS), up to Computerized Maintenance Management Systems (CMMS) [5].

More recently FM has been more and more strongly integrated within the Building Information Modeling (BIM), due to the wide adoption of this approach in design and construction phases [6]. A primary challenge in this field concerns how to maintain and effectively employ information coming from the previous phases in the use phase of assets, without a loss of reliability, in order to exploit the collaboration and integration capabilities offered by the BIM [7] approach.

Thanks to this dynamic, in FM new services can be delivered through the use of the digitized assets. Typically, FM has been conceived as a discipline composed of a set of theories and practices carried out for the reducing costs and inefficiencies of the non-core business, in order to increase the productivity of the core business of an organization. Today this trend is changing, and the development of FM services is more closely targeted towards the optimization of the core business of the organization [8].

The digitization brings new dynamics, enabled by the availability and accessibility of data, in all the life cycle phases of the assets. One of them concerns the servitization of the built environment. In manufacturing, the servitization is intended as a strategy for the improvement of the competitiveness of organizations [9] through the inclusion of a set of services, within the selling of a product. These services can be activated even after the transaction [10]. In management of the digital built environment, the demand for efficient and integrated spaces is a key enabler for success and competitiveness of the core business. Accordingly, the building is conceived not only as a physical asset but also as an integrated system composed of the physical asset and the services that can be provided to the users through its use [8].

The Aim of the Research

Traditional performance measurement categories (e.g., financial, functional, physical, user satisfaction [11]) are no longer suitable metrics for measurement of cross-domain performances offered by the digitized and servitized assets.

Accordingly, the research question concerns how effectively measure cross-domain performances of digitized assets. Moreover, once these performances have been measured, how is it possible to optimize them, in order to achieve an improved Facilities Management of digitized assets? Therefore, this research aims at proposing a cross-domain Decision Support System (DSS) for the allocation of maintenance budgets. The maintenance cost performance of the asset has been computed through the Facility Condition Index (FCI), one of the most acknowledged indicators measuring the technical performance of the building and its components through an economic metric. This indicator has been combined with a set of further parameters: the D index, measuring the service life of the component; the C index measuring the criticality as the consequence of the failure of the component for the effective execution of the core business of the organization occupying the asset; the P index measuring the preference of the asset owner for the execution of the core business. These indicators are combined together for each building component (asset) and are ranked on a scale from 0 to 1, through the calculation of the Maintenance Priority Index (MPI): the simple average of the four metrics.

Assuming a reduced budget (not enough to cover all maintenance costs), an optimization algorithm has been developed, thus the economic resources can be allocated to those components presenting the highest MPI value.

In order to test the robustness of the proposed methodology, the DSS has been applied to a case study involving an office building in Erba, Italy. A simulation with reduced maintenance budget has been carried out, demonstrating the possibility of effectively driving different maintenance policies, according to a variation of the C and P indexes.

2. State of the Art

In Asset and Facilities Management, performance measurement can be considered as a powerful mean for decision-making [12]. This activity can be seen both as a factor for enhancing the competitiveness of an organization and as a learning process within the organization [13], which allows the continuous improvement in the decision-making process, for the effective delivery of better services [14]. Performance measurement is always tied to the use of specific Key Performance Indicators (KPIs), which allow us to synthetize phenomena, without losing the systemic value of the information [15]. KPIs are obtained through the refinement of raw data, therefore they are employed at different decision-making levels, depending on the management needs. Typically, three decision-making levels can be recognized when dealing with performance measurement: strategic, tactical and operational [16,17].

Performance metrics are often aggregated to support a decision-making process and this aggregation can vary according to the disciplinary field, the type of organization and the decision-making level [18]. One of the most acknowledged methods for assessment of performances within an organization is the Balanced Score Card (BSC) framework: A model for measurement and management of strategic performances within an organization [1]. The balanced score card gave rise to the definition of specialized KPIs for FM [3].

In management of the built environment, metrics are usually domain-specific and related to technical, functional, economic/financial performances [19]. For a definition of effective DSS, metrics are seldom selected among the same domain, to avoid the excessive specialization of the performance assessment [14,20]. However, one of the most acknowledged indicators for measurement of technical performances through an economic dimension is the Facility Condition Index (FCI). This indicator was introduced for the first time by the National Association of College and University Business Officers (NACUBO) and Rush (1991) [21], and it is widely used for making informed decisions in AM and FM [11,22,23].

The Facility Condition Index

The FCI allows us to quantify on a scale from 0 to 100 the condition of the assets. The metric is a ratio between the Deferred Maintenance (DM) and the Current Replacement Value (CRV) of each component of an asset and can be calculated at different scales: from the single component in a building, until the whole building in a portfolio [24]. The basic equation used for the calculation of the metric is (1):

$$FCI = \frac{DM}{CRV} \, [\%] \tag{1}$$

where:

- DM is the cost of the deferred maintenance, defined as the costs of the maintenance operations that should have been done but it had been deferred,
- CRV is the current replacement value of the component or the same asset [25].

The result of the calculation is evaluated on three or four levels, on a scale from 0 to 100%, where 0 is the best value:

- good: 0%–5%,
- fair: 5%–10%,
- poor: 10%–30%,
- critical: 30%–100%.

The FCI, since its first publication, has been adopted for leading maintenance policies of assets portfolios of many departmental organizations in the United States of America (USA) [26]. Among them, the Department of the Interior (DoI) has published a set of documents and guidelines which contributed in the standardization of procedures and workflows related to the calculation of the metric [26–28]. However, a standardized calculation methodology has not been defined yet.

In literature, other definitions of the indicator can be found. These versions can be considered as improved variations of the original formula. Despite being a powerful mean for the calculation of the technical performances of the assets, this metric needs to be carefully calculated and often when combined with further indicators, is able to describe other performances not directly related to the technical and economic performance of the assets. More in general, recent versions of the indicator show differences in terms calculation of the DM and CRV. Some versions, for instance, consider the effect of obsolescence due to lack of compliance with codes in the calculation [29]. In some other cases, the DM is computed considering even the renewals and improvements [24]. Other approaches for the calculation of the DM are specific for the evaluation of historical assets [25]. Therefore, the USA governmental guidelines [25,30] suggest employing a combined version of the FCI for being more effective in the computation of the overall maintenance performance. This enhanced version encompasses in the calculation also major rehabilitation, replacement and maintenance recommendations registered during the periodic Condition Assessment (CA). These factors are used for the calculation of a composed index: the FCI_{comp} (2):

$$FCI_{comp} = \frac{DM + DM_{mr} + DM_{im}}{CRV} \, [\%] \tag{2}$$

where:

- DM is the deferred maintenance as defined for Equation (1),
- DM_{mr} is the cost of major rehabilitations and replacements,
- DM_{im} is the cost of the maintenance and repair recommendations highlighted during the periodic CA.

Finally, the FCI_{comp} is used for the calculation of a Condition Index (CI_{comp}) (3):

$$CI_{comp} = (1 - FCI_{comp}) \, [\%] \tag{3}$$

The CI_{comp} can be considered as one of the examples of how the FCI is modified in order to describe a wider array of phenomena. Other application of the FCI can be found in literature, the most of them stemming from the need for a more comprehensive performance measurement tools that are able to catch the complexity of the physical asset, while measuring performances belonging to different disciplinary fields [23,31]. A good example of how the FCI could be associated with other metrics involves coupling with the Asset Priority Index (API) [32]: a metric which measures the relevance of the asset according to the objective that the owning organization wants to achieve through its use. The API have been employed by many private and public organizations [33,34], being a powerful qualitative indicator that is able to correct the FCI according to the strategic importance of an asset, provided by the managing or owning organization.

In summary, the FCI can be considered as a powerful aggregated metric for the measurement of the maintenance performance of the assets, through an economic dimension. Nevertheless, some drawbacks can be highlighted. The most evident concerns its feature of being driven by the dimension of the denominator: Two components presenting the same DM but different CRV are ranked differently, since the ratio would penalize the component with the highest CRV. This issue does not allow us to use the FCI itself for prioritization of maintenance interventions, since the metric will always tend to penalize components with small CRV. A second drawback involves the evident technical nature of the indicator, not considering other phenomena but only the maintenance performance. Therefore, the FCI provides valuable information of the technical performance, despite needing to be associated with other metrics to be considered an effective tool for decision-making. Through this research these main two issues are addressed and a flexible and cross-domain DSS is proposed.

3. Materials and Methods

In this article a DSS for the optimization of maintenance budget is presented. The maintenance budget is allocated thanks to the inclusion of the FCI in the DSS to obtain the Maintenance Priority Index (MPI), thanks to the combination with 3 KPIs measuring the age of the assets (D), their criticality

(C) and the preference of the owner (P). The three metrics are described in detail in the following paragraphs. Figure 1 represents the process developed for the calculation of the algorithm.

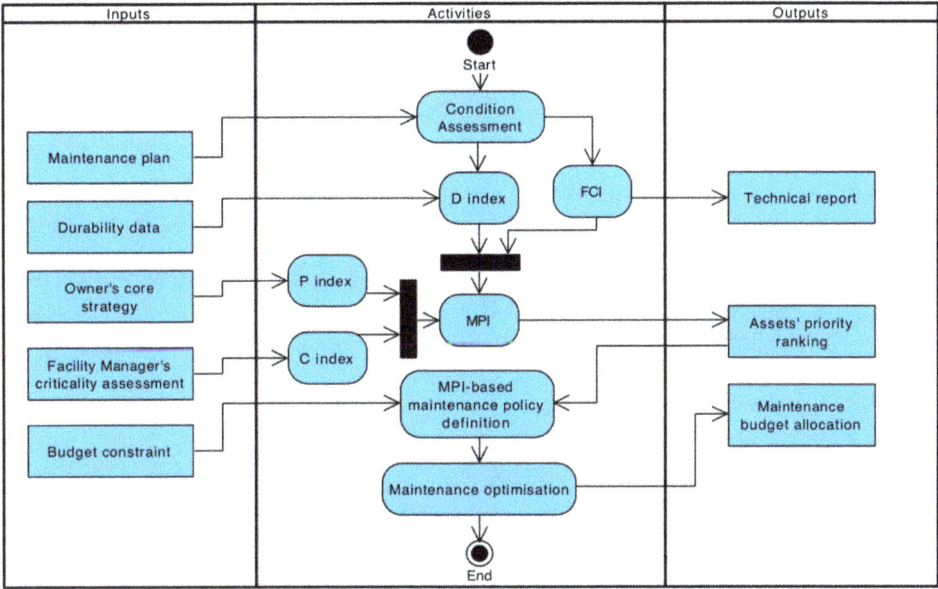

Figure 1. Process adopted for the development of the algorithm underpinning the Decision Support System (DSS).

The four KPIs were aggregated in a final equation which keeps the phenomena measured together through these different metrics and addresses cross-domain issues in supporting decisions on assets to be maintained. The power of this algorithm concerns the inclusion of metrics describing different phenomena in a synthetic MPI computed through Equation (4).

$$MPI_j = \frac{FCI_j^* + D_j + C_j + P_j}{4} \tag{4}$$

The metrics (FCI*, D, C and P indexes) are described through their equations and function in the next paragraphs.

For the calculation of these indicators, the first step concerns the definition of the entities that must be considered when the maintenance interventions are implemented. For this purpose, a peculiar breakdown criterion has been adopted: the elements or group of them on which to implement the maintenance interventions have been identified through the definition provided by buildingSMART international for IfcAsset. In the standard ISO 16739-1 [35], the ifcAsset is defined as the:

"level of granularity at which maintenance operations are undertaken. An asset is a group that can contain one or more elements" [35].

Thanks to this approach is possible to break down the building according to the maintenance management needs, adopting a standardized approach which allows an easier integration within the Asset Information Modeling (AIM) as defined in the ISO 19650-2 [36].

After a building condition assessment, the FCI (Equation (1)) of each asset (intended as ifcAsset) can be calculated. Noteworthy, the FCI is defined as a KPI ranging from 0 to 1 but it is very unlikely to find an asset with an FCI higher than 30% in a well-managed building. Even in an un-managed but lived-in building, a value higher than 50% is rare and represents an asset in a very critical condition. Thus, if directly used in Equation (8) to compute the MPI, the weight of FCI would be much smaller than the other 3 KPIs, with all of them actually ranging from 0 to 1. In order to compute MPI using 4 KPIs with the same weights, the FCI has been scaled through Equation (5).

$$FCI_j^* = \frac{FCI_j - FCI_{min}}{FCI_{max} - FCI_{min}} \tag{5}$$

where:

- FCI_j is the value of the metric calculated for the asset j;
- FCI_{min} is the minimum value of the FCI among the assets in the building;
- FCI_{max} is the maximum value of the FCI among the assets in the building.

The FCI_j^* loses its characterization as introduced through Equation (1), despite being more effective and suitable for the DSS and providing a contribution of 25% in the computation of the algorithm for the maintenance prioritization.

The building condition assessment allows us to also compute the D index: a metric relating to the service life of each asset. The D index is derived by the D^+ and D^- indexes developed by Dejaco et al. [16]. The D index, in this research, is computed through Equation (6)

$$D_j = \begin{cases} \frac{ASL_j}{RSL_j} \ if \ ASL_j \leq RSL_j \\ 1 \ if \ ASL_j > RSL_j \end{cases} \tag{6}$$

where:

- ASL_j is the Actual Service life of the asset, i.e., the period of time since the asset has been installed or built;
- RSL_j is the Reference Service Life of the same asset [37].

The D index weights for an additional 25% in the final calculation of the algorithm. The FCI and the D index are the two technical indexes. These metrics are combined with two more KPIs describing the criticality of the asset for the overall functioning of the building and the preference assigned by the owner of facility manager to the specific assets within the building.

The C index measures the criticality of the specific asset in term of the consequences of the damage on the overall functioning of the building and must be assigned by the Owner or the Facility Manager. This approach is similar to the one carried out for the definition of the Asset Priority Index (API): a metric mainly used by the US governmental organizations for the definition of the most critical assets within a mission [32]. In this case, the mission can be considered to be the correct and safe operation of the building. The C index is a qualitative metric, ranging from 0 to 1, computed standardizing a critical index C* that was defined according to 10 levels (Table 1).

Table 1. Severity classes adopted for the C* index. The classes are taken from the MIL-STD 1629(A) [38].

Severity	Rank	Criteria
No effect	1	No effect on the product or subsequent processes
Very slight effect	2	Very slight effect on the product performance
Slight effect	3	Slight effect on the product performance
Minor effect	4	Fault does not require repair
Moderate effect	5	Fault on non-vital part requires repair
Significant effect	6	Product performance was degraded, but operable and safe. Non-vital part inoperable
Major effect	7	Major effect on process, repair on parts necessary. Subsystem inoperable
Extreme effect	8	Extreme effect on process, equipment damaged. Product inoperable but safe
Serious effect	9	Potential hazardous effect.
Hazardous effect	10	Hazardous effect

The C index of an asset is calculated through the following Equation (7)

$$C_j = \frac{C_j^* - 1}{9} \tag{7}$$

where C* is the criticality assigned to each single asset by the facility manager.

The owner of the building or its facility manager is also asked to give his preference for which asset should be maintained first, since it could be crucial for business related issues. This preference is measured through a priority index P* on a 1-10 scale (Table 2). The P, then is normalized following the same approach applied to the C index.

Table 2. Preference scale values.

Preference	Rank
No impact on the business	1
Very slight impact	2
Slight impact	3
Minor impact on the business	4
Moderate impact	5
Significant impact	6
Major impact	7
High impact	8
Very high impact	9
Extremely high impact	10

The P index is eventually calculated with the following Equation (8):

$$P_j = \frac{P_j^* - 1}{9} \tag{8}$$

where the P* is the preference assigned to each asset by the owner of the building.

3.1. Sensitivity Analysis

A further investigation has been accomplished in order to understand better how the MPI is changing according to the variation of its four parameters. Therefore, 5 Montecarlo simulations have been accomplished (Table 3). The following are the parameters considered:

- Number of iterations: 25,000;
- FCI is random real number between 0 and 1 with a flat probability distribution;
- D is a random real number between 0 and 1 with a flat probability distribution;
- P is a random integer number between 1 and 10 with a flat probability distribution;
- C is a random integer number between 1 and 10 with a flat probability distribution.

Table 3. Results of the Montecarlo simulations.

MPI	Sim 1	Sim 2	Sim 3	Sim 4	Sim 5
Mean	0.499117	0.50098	0.499901	0.498664	0.500728
Std. Dev.	0.152626	0.152535	0.151193	0.152301	0.152611

In Table 3, the results of the Montecarlo simulations are presented.

Figure 2 and Table 4 present the results of one of the simulations that were carried out.

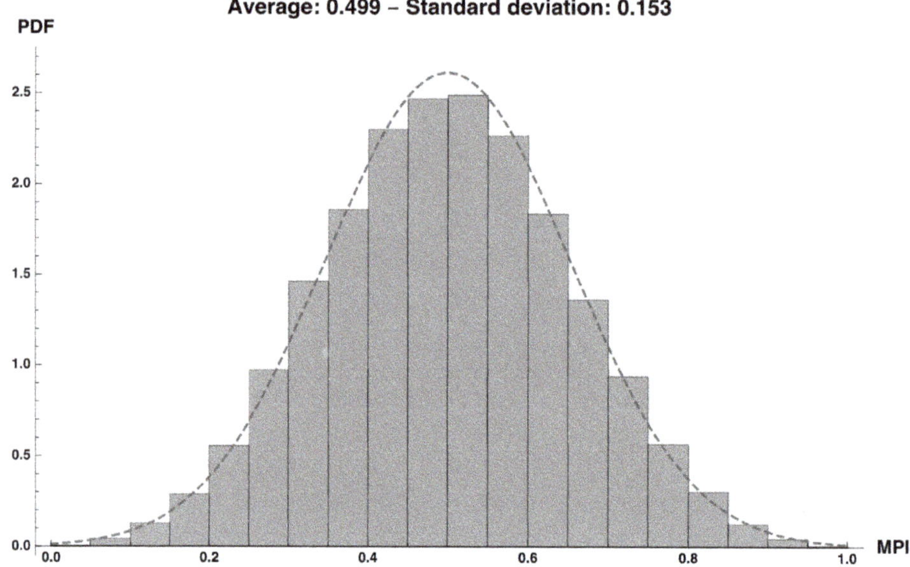

Figure 2. Results of the first of the five Montecarlo simulations compared to a Normal Distribution with the same average and standard deviation (Dashed line).

Table 4. Percentiles and Maintenance Priority Index (MPI) values.

%tile	5%	10%	15%	20%	25%	30%	35%	40%	45%	50%
MPI	0.2464	0.3011	0.3387	0.3681	0.3950	0.4197	0.4412	0.4608	0.4810	0.5007
%tile	55%	60%	65%	70%	75%	80%	85%	90%	95%	
MPI	0.5206	0.5420	0.5621	0.5836	0.6065	0.6321	0.6620	0.6989	0.7508	

The Montecarlo simulation has been carried out in order to be able to define better the thresholds in Table 5. The Probability Density Function (PDF) described in Figure 2 allows us to identify the most likely values according to the values assumed by the four coefficients, therefore it allows us to identify most suitable thresholds for the assessment of the assets.

Table 5. Thresholds defined for the four metrics of the Decision Support System (DSS).

Metric	Good	Fair	Poor
FCI	FCI $\leq$ 5%	5 < FCI < 10%	FCI > 10%
D	D $\leq$ 0.5	0.5 < D < 1	D = 1
P	P $\leq$ 3	3 < P $\leq$ 6	P > 6
C	C $\leq$ 3	3 < C $\leq$ 6	C > 6

3.2. The Facility Condition Index in the Decision Support System

The following step concerns the investigation of how the FCI behave in the DSS, since the FCI thresholds and equation have been revised in order to be more suitable for the calculation of the algorithm. Therefore, an evaluation of the variation of the MPI according to the variation of the values of the four parameters has been accomplished. In order to do this, some thresholds have been defined not only for the FCI but also for the other indicator. The thresholds have been defined thanks to the Montecarlo simulation (Figure 2 and Table 4).

For the FCI three levels have been defined: "good", "fair" and "poor" as suggested in many literature references [16]. For the D index, if the component presents an ASL below the half of the RSL, thus can be classified as in good conditions. If the ASL is between the half of the RSL and the end of the RSL, then the component is in a fair condition, otherwise it can be considered to be in a poor condition, since it means that it is operating with an ASL that is greater than expected. Therefore, it has been operating for a longer period than what is expected in standard conditions. The thresholds defined for the P and the C values are the same. Even in this case, the two parameters are following a similar logic in the evaluation of the priority and criticality assigned to the assets.

Once the thresholds have been identified, some reference values have been defined for the four metrics (Table 6).

Table 6. Reference values for the four metrics od the Decision Support System DSS.

Metric	Good	Fair	Poor
FCI	2.5%	7.5%	15%
D	0.25	0.75	1
P	2	5	7
C	2	5	7

The following step concerns the definition of the thresholds to be associated to the MPI. The index is composed of four metrics. The values of each metric can be organized in three classes: good, fair, poor (Table 6). For determination of the thresholds a minimum and maximum value of the FCI must be defined. These parameters have been set as described in Table 7.

Table 7. Facility Condition Index (FCI) and scaled Facility Condition Index (FCI*) values adopted for the definition of the Maintenance Priority Index (MPI) thresholds.

Parameters	Values
FCI_{min}	0
FCI^*_{min}	0
FCI_{max}	0.2
FCI^*_{max}	0.75

These values of the FCI have been combined with those related to the D, C and P indexes giving as results a combination of $3^4 = 81$ cases. These results have been organized in a table and the thresholds of the MPI have been agreed as shown in Table 8:

Table 8. Maintenance Priority Index (MPI) thresholds.

MPI Thresholds	Priority	Criteria
≥ 0.6	High	Maintain the asset as soon as possible
$0.3 > MPI > 0.6$	Medium	Maintain the asset if possible
$MPI \leq 0.3$	Low	No need for maintenance

Figure 3 represents an example of different combination of the four metrics computed for the calculation of the MPI. Additionally, the central red circle represents the MPI value obtained from the calculation of the four metrics. From top left and clockwise, the circles represent FCI, D, C and P.

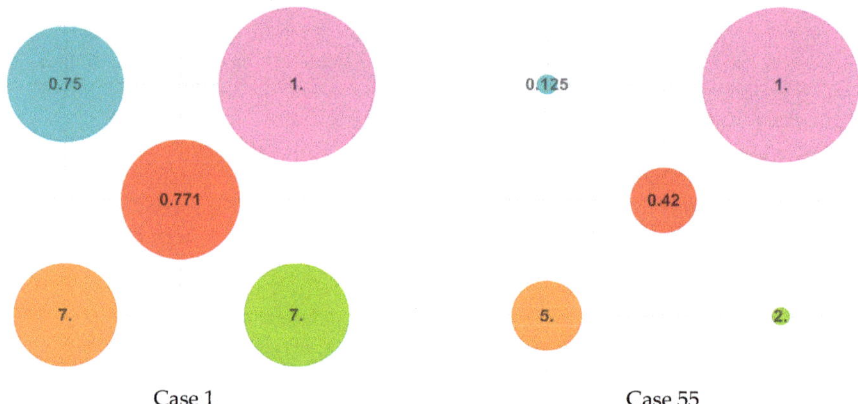

Figure 3. visualization of the combination of the four metrics composing the Maintenance Priority Index (MPI). The central circle represents the value of the Maintenance Priority Index (MPI).

Once the MPI has been computed for all the assets within the building, the budget allocation can take place. Therefore, assets are ranked in ascending order according to their MPI value. Economic resources are allocated first to those assets with a higher MPI. Maintenance interventions are assigned to assets as long as there is enough budget to implement them. The maintenance interventions and related costs are defined thanks to the calculation of the deferred maintenance (Equation (1)). The proposed algorithm sits in the core of the DSS for maintenance budget allocation.

4. Results

The proposed DSS has been tested and applied in a case study, in order to demonstrate the behavior of the metrics composing the algorithm and for testify its robustness. The following paragraphs describe the outcomes of these analyses and testing procedures.

4.1. The Influence of Criticality (C) and Preference (P) Indicators

The MPI allows us to combine the four indicators and to combine the evaluation of different phenomena in a single performance metric. This aggregation is used to overcome one of the main problems related to FCI and its use in maintenance prioritization, concerning the need for the combination with other metrics. If the priority of maintenance operations is defined only according to the FCI (i.e., an asset with higher FCI will be maintained sooner than the ones with al lower FCI), it will be very probable that assets with low CRV will be maintained first and those with very high CRV will not be considered properly. In fact, FCI is known to be a metric led by the CRV [39].

This gives rise to a question: are P and C indicators enough to change the priority of maintenance given by the only FCI? In order to answer this question a further investigation has been accomplished.

At first the ratio between to MPI of two different assets has been calculated through Equation (9), as function of the CRV ratio and the P ratio:

$$\frac{MPI_1}{MPI_2} = \frac{(9DM_1 + CRV_1(-2 + C_1 + 9D_1 + P_1))y}{9DM_2xy + CRV_1(P_1 + (-2 + C_2 + 9D_2)y)} \tag{9}$$

where:

- $x = \frac{CRV_1}{CRV_2}$ is the ratio between the current replacement values of two components;
- $y = \frac{P_1}{P_2}$ is the ratio between the priority indexes of the same two components.

It must be considered that the P and the C indexes impact on the function in the same way. Therefore, the *y* variable can be defined both considering the ratio between the P or the C index. In this case the ratio between the P indexes of two assets has been considered. The trend of the MPI ratio according to the variation of the ratios of CRV and P ratios is represented in Figure 4. The MPI ratio presents a negative variation when the CRV ratio increases, confirming that the FCI is a metric led by the denominator, which increases when the P ratio increases. This trend confirms that the P and C indexes are effectively impacting on the final outcome of the calculation of the algorithm.

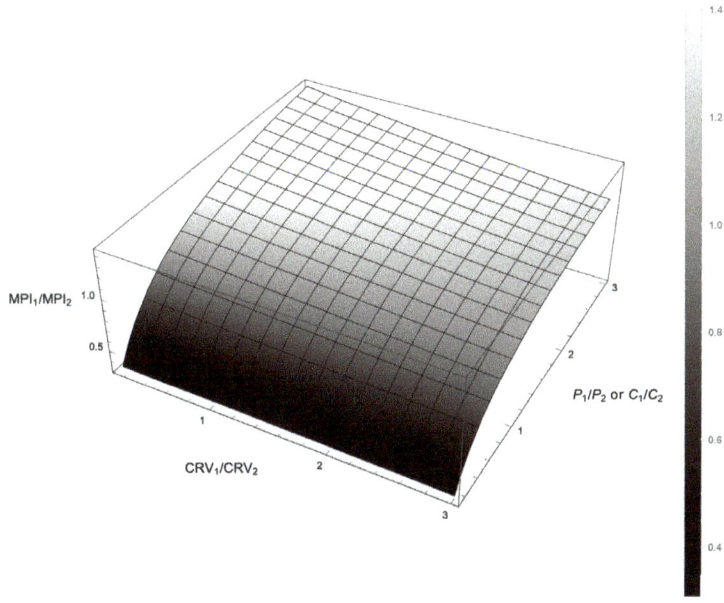

Figure 4. Ratio between the Maintenance Priority indexes (MPI) of two components according to the ratio of Current Replacement Values (CRV) and Priority index (or Criticality Index).

Table 9 presents the results of an example carried out on two assets. Even in the cases when the FCI would strongly orient towards the implementation of the maintenance interventions on the specific component (CRV1/CRV2 = 0.4 or 1.75), the Priority index (or Criticality index) can lead to completely different situations, compared to the results of the simple FCI. For instance, if two assets present $DM_1 = DM_2$ but a $CRV_1/CRV_2 = 0.5$, the prioritization of the maintenance interventions based on the FCI would lead to a resource allocation for maintaining the component 1 ($FCI_1 = 2\ FCI_2$), while the MPI ratio drive the allocation of the same resources on component 1 if $P_1/P_2 >= 1$, otherwise on the component 2 (see Table 9). The same considerations can be made using the ratio between the criticality of the two components instead of the one between the preference.

Table 9. Variation of the Facility Condition Index (FCI) ratio and Maintenance Priority Index (MPI) ratio according to Current Replacement Value (CRV) ratio and Priority index (P) ratio.

Replacement Cost Ratio [CRV$_1$/CRV$_2$]	Priority Ratio [P$_1$/P$_2$]							
	0.4	0.6	0.8	1	1.2	1.4	1.6	
0.5	2.00	2.00	2.00	2.00	2.00	2.00	2.00	FCI1/FCI2
	0.66	0.82	0.94	1.03	1.11	1.16	1.21	MPI1/MPI2
0.75	1.33	1.33	1.33	1.33	1.33	1.33	1.33	FCI1/FCI2
	0.65	0.81	0.93	1.02	1.09	1.14	1.19	MPI1/MPI2
1	1.00	1.00	1.00	1.00	1.00	1.00	1.00	FCI1/FCI2
	0.64	0.80	0.91	1.00	1.07	1.12	1.16	MPI1/MPI2
1.25	0.80	0.80	0.80	0.80	0.80	0.80	0.80	FCI1/FCI2
	0.63	0.79	0.90	0.98	1.05	1.10	1.14	MPI1/MPI2
1.5	0.67	0.67	0.67	0.67	0.67	0.67	0.67	FCI1/FCI2
	0.63	0.78	0.89	0.97	1.03	1.08	1.12	MPI1/MPI2
1.75	0.57	0.57	0.57	0.57	0.57	0.57	0.57	FCI1/FCI2
	0.62	0.77	0.87	0.95	1.01	1.06	1.10	MPI$_1$/MPI$_2$

4.2. Case Study Results

The proposed DSS has been applied to a case study to test its robustness and the possibility of effectively drive the budget allocation for maintenance interventions. The DSS has been applied to an office building in Erba, in the north of Italy. The building is occupied by a construction company and by a notary firm for a total surface of approx. 3000 sqm. The building was accomplished in 2007 and sits in a piedmont climatic condition (Figure 5). The building has been modeled through Revit with a low Level of Detail (LoD) and a high Level of Information (LoI). However, the DSS has not been completely integrated in the BIM model.

Figure 5. location (left) and Building Information Model (BIM) model of the office building on which the Decision Support System (DSS) has been implemented.

The first step for the calculation of the algorithm concerns the definition of list of assets to be maintained. This operation is crucial for effective information management framework since it defined the granularity at which maintenance operations take place. The assets have been defined considering the definition provided by the standard ISO 16739-1 [35]. Therefore, elements have been aggregated not only following the location and typological criteria but also following a functional criterion, grouping those elements on which maintenance operations can be carried out uniformly.

These assets have been employed for the development of a standard list of maintenance interventions, collected and used for the definition of their maintenance profiling. Accordingly, the CRV for each asset has been calculated thanks to the use of CAD drawings, technical specifications and 3D models available at the moment of the development of the case study. The CRV costs have been defined according to the Pricelist 2018 published by the Comune di Milano [40] and Camerca di Commercio Milano Monza Brianza Lodi [41]. Once the CRV values have been defined, a condition assessment has been carried out and the deferred maintenance interventions for each asset have been identified. This operation gave as an outcome a list of 246 components with the DM > 0, most of them with a small CRV (Figure 5).

172 assets out of the 246 present a DM smaller than 5000 € for a total amount of 253.908 €. The value of the DM for each of the 246 assets is represented in Figure 6. 69.9% of the 172 DM cost 7.15% of the total cost of DM, amounting to €351.632.

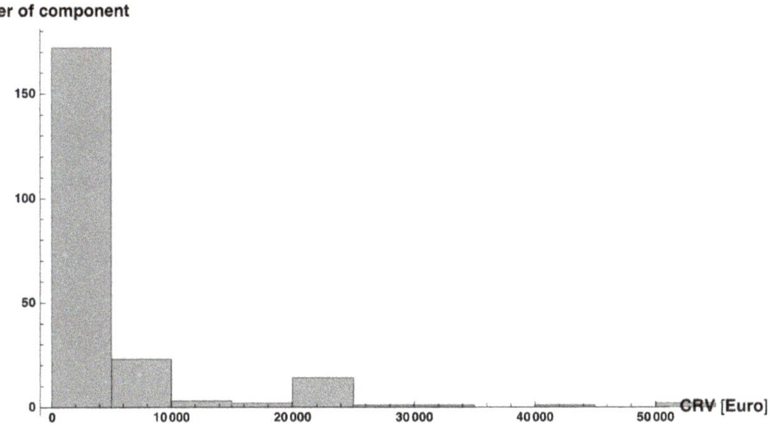

Figure 6. Histogram of Current Replacement Value (CRV) for the components with Deferred Maintenance (DM) > 0. Most of the component have small CRV.

After the condition assessment, carried out in this case checking the maintenance interventions performed against those defined in the maintenance plan, the MPI can be computed. Figure 7 shows the initial situation, after the condition assessment. For each asset, the colors display how much each of the four metrics (FCI*, D, P, C) contribute to the MPI.

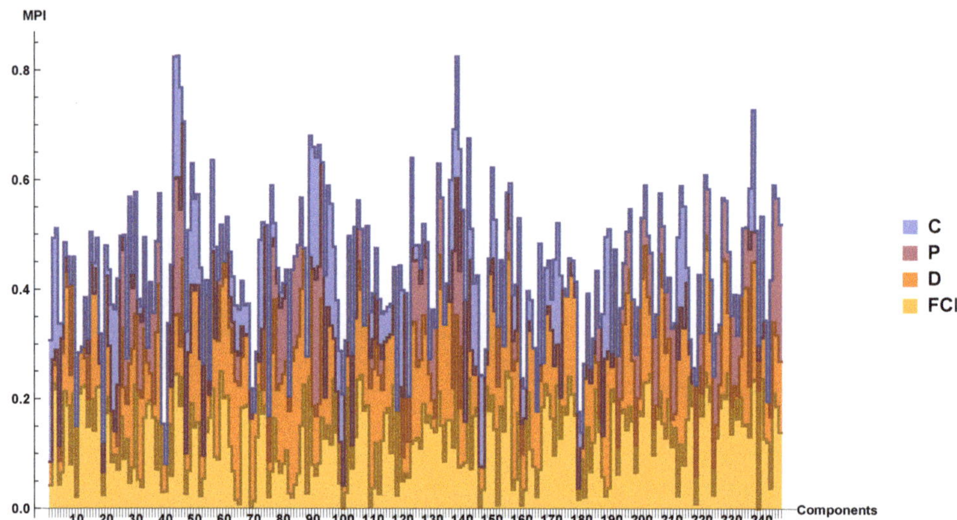

Figure 7. Maintenance Priority Index (MPI) of the components with Deferred Maintenance (DM) > 0. The colors show how the Facility Condition Index (FCI), D index (measuring the age of the assets), P index (measuring the preference of the owner) and C index (measuring the criticality of each asset) contribute to the MPI of each component.

The mean initial FCI of all the assets is $FCI_{init} = 10.8\%$ (poor) and the corresponding MPI_{init} is approximately 43.0% (medium). Figure 8 allows us to compare FCI_{init} and MPI_{init}. The vertical dashed lines represent the standard limits for poor and fair conditions used in FCI while the horizontal dashed lines are the limits for low and medium priority defined for MPI. It can be seen that some components with a very good FCI have a high MPI due to D, P or C.

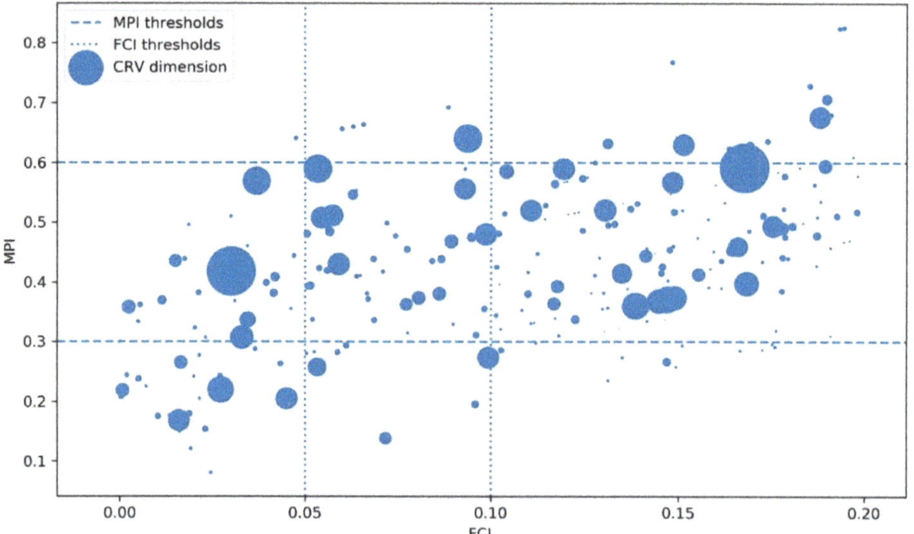

Figure 8. Comparison between Facility Condition Index (FCI) and Maintenance Priority index (MPI) of the components and dimension of the Current Replacement Value (CRV).

The following stage concerns the budget allocation of reduced resources for the optimization of maintenance operations. For this simulation, it is assumed a budget constrain (B) equal to 65% of the cost of all the DM detected during the condition assessment (Equation (10)).

$$B = \left(\sum_{i=1}^{n} C_{DM,i} \right) \times 65\% \tag{10}$$

Limited by this constrain, it is possible to perform maintenance activities only on 130 components chosen according to the MPI from the highest to the lowest possible, according to the budget constrain. The assets have been chosen confronting the cumulative sum of the deferred maintenance with the budget, until the budget is totally spent.

The budget set to 65% of the total DM cost, allows us to solve the deferred maintenance of approximately 130 assets (98% of the total DM of the 130 assets) identified through the MPI ranking. Maintenance interventions carried out, allowed to reduce the average MPI to 34.5%, namely 19.7% less than the initial MPI (Figure 9); whilst the average FCI assume the value of 3.9%, corresponding to a decrease of the FCI$_{init}$ of 63.7% (Figure 10). The breakdown of values assumed by each metric composing the MPI is shown in Figure 11.

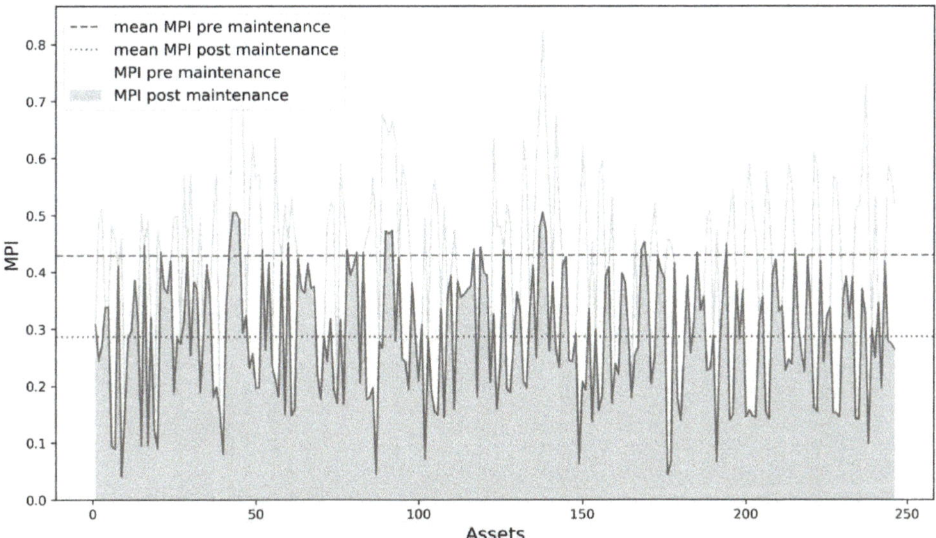

Figure 9. Maintenance Priority Index (MPI) before (ivory) and after (light grey) maintenance. The dashed line is the average MPI after maintenance, the dotted one is the average MPI before maintenance.

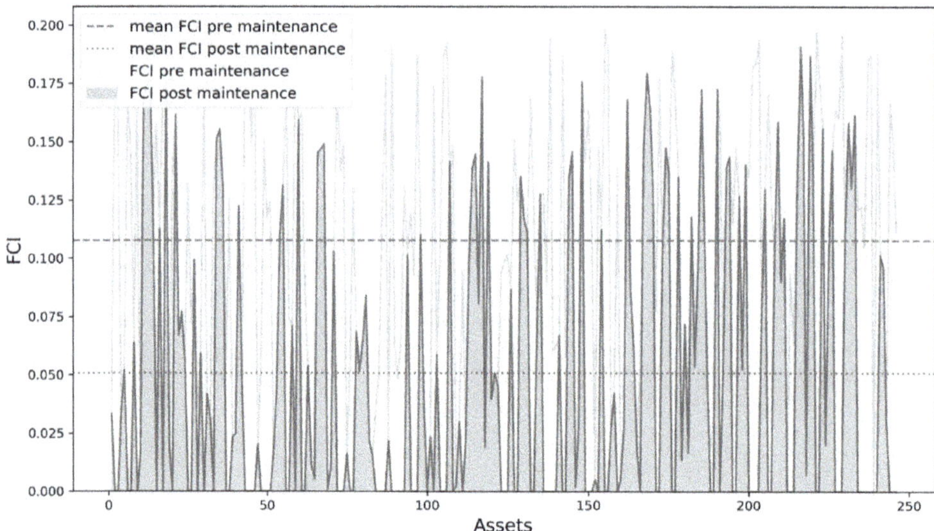

Figure 10. Facility Condition Index (FCI) before (ivory) and after (light grey) maintenance. The dashed line is the average FCI after maintenance, the dotted one is the average FCI before maintenance.

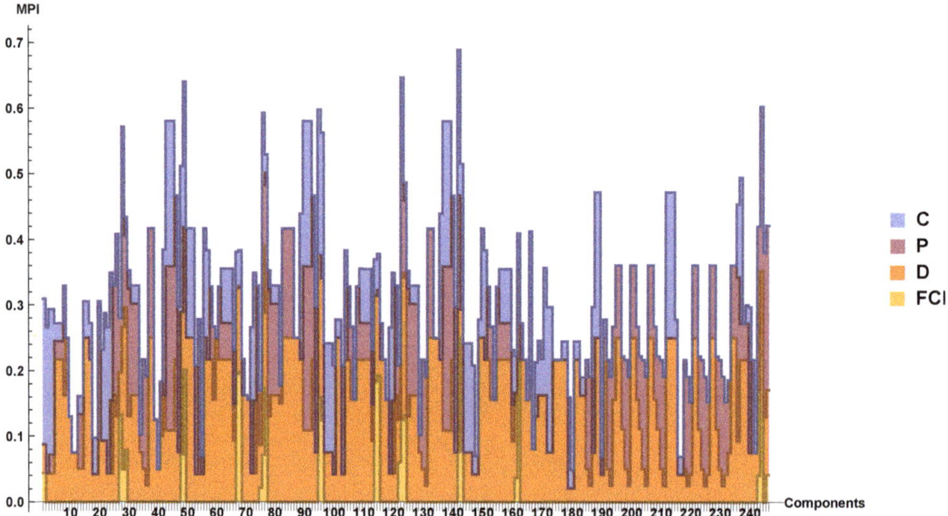

Figure 11. Maintenance Priority Index (MPI) of the components after the maintenance. The colors show how the Facility Condition Index (FCI), D index (measuring the age of the assets), P index (measuring the preference of the owner) and C index (measuring the criticality of each asset) contribute to the MPI of each component.

Comparing Figures 7 and 11 it is possible to visualize how the maintenance optimization affects $\frac{1}{4}$ of the MPI, sine it impacts on the reduction of the FCI. This is the only metric of the DSS that is drastically reduced after the prioritization of the maintenance interventions. Figures 11 and 12 demonstrate that some assets present, after the implementation of the maintenance interventions, an FCI = 0, despite the values of the remaining three metrics remain the same. This may cause a situation, as described in Figure 12, where some priority assets still maintain a high value of the MPI, despite the FCI is equal to 0.

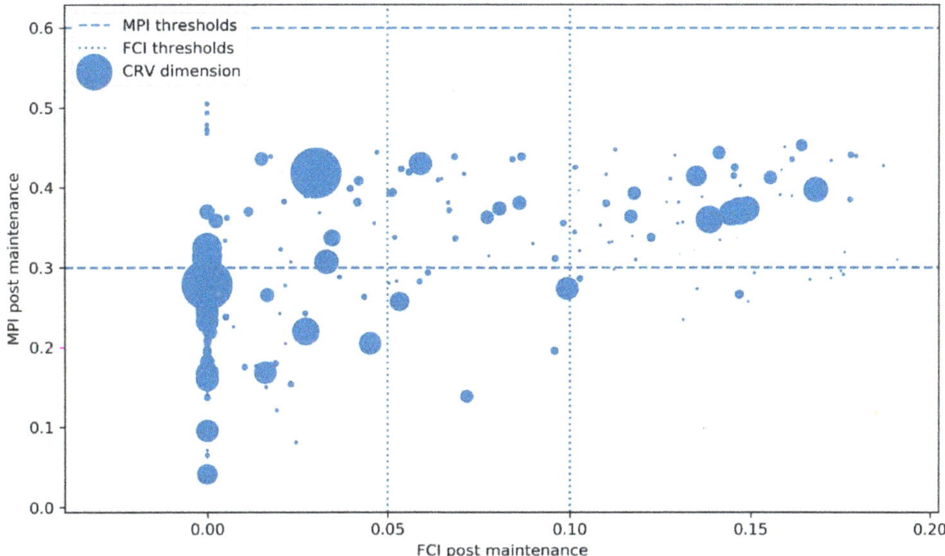

Figure 12. Comparison between Facility Condition Index (FCI) and Maintenance Priority Index (MPI) of the components after maintenance.

Using MPI or FCI to prioritize maintenance leads to completely different results. This because MPI considers also the age of components, their criticality and the priority of the owner / manager of the asset. Figure 13 compares two different maintenance prioritization strategies which can be applied to the same case study with the same budget constraint. One maintenance prioritization strategy is based on MPI, the other is based on the use of FCI. As it can be seen, there is a group of components, the green dots in Figure 13, characterized by high MPI and elevated FCI values that are going to be maintained regardless of the KPI used to set priority. There is another group of assets, identified by the red dots, that using either FCI or MPI, are not going to be maintained. The blue dots represent assets that are going to be maintained only if the priority is set using the FCI. Eventually, the yellow dots are components that need to be maintained if the priority is defined using MPI. In the case study presented in this article, with a maintenance budget set to 65% of the total maintenance needs, if the priority is set using MPI, 130 components out of 246 will be maintained. Otherwise, if FCI is used, 103 components are going to be maintained.

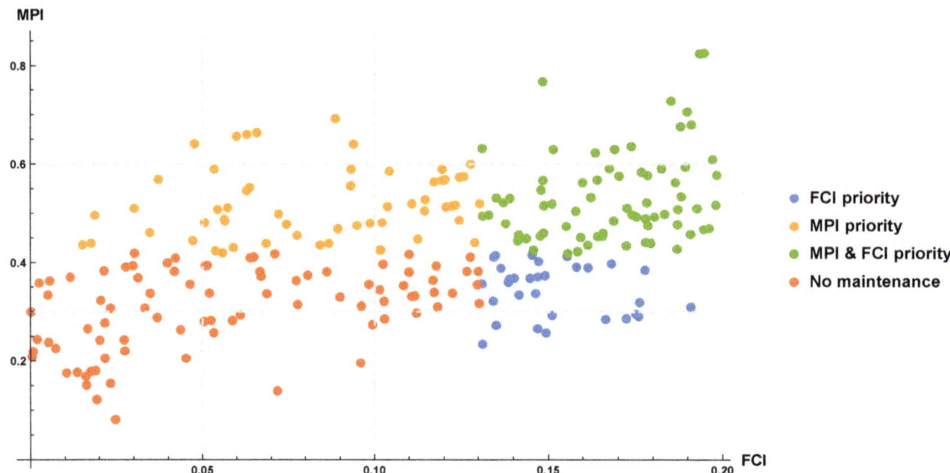

Figure 13. Assets to be maintained (or not) according to different maintenance prioritization strategies (Facility Condition Index and Maintenance Priority Index computed after the Condition Assessment are shown).

5. Discussion

Based on the results obtained, the DSS has been stressed considering different budget thresholds. A scenario analysis has been carried out, in order to investigate the behavior of the MPI when the budget constraint is changed. If the maintenance budget goes from 55% to 65% of the total cost of the DM the FCI decreases of 36% and the MPI reduction is 7%, while if the budget goes from 55% to 75% the FCI present a 49% reduction and the MPI is reduced by 10% (Figure 14).

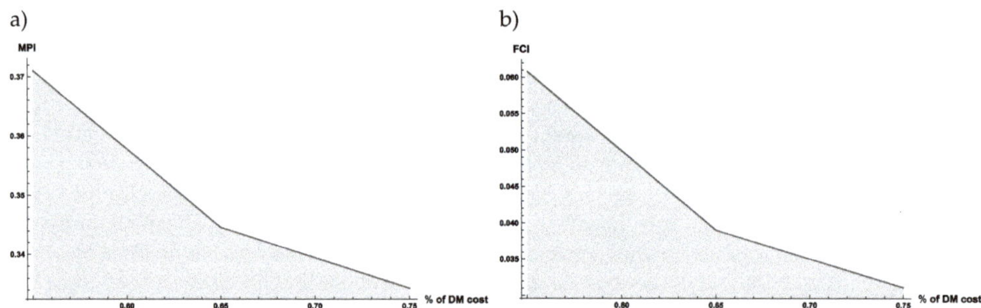

Figure 14. Scenario analysis. How Maintenance Priority Index (MPI) (**a**) and Facility Condition Index (FCI) (**b**) change according to the budget constrain expressed in percentage of the Deferred Maintenance (DM) costs.

The scenario analysis demonstrated that increasing the maintenance budget, the MPI decreases slower, since most critical assets are repaired first, thanks to the logic adopted in the development of the maintenance prioritization algorithm.

Influence of Uncertainty of Input Data on MPI

Computing replacement and maintenance costs is fraught with potential errors due to lack of knowledge of the actual condition of the asset to be maintained and of its surroundings. Noteworthy, most of the times these costs are not forecasted by the same person who made the survey for the condition assessment, this leads to estimates characterized by uncertainty. Many authors when forecasting maintenance and life cycle costs suggest dealing with uncertainty using Montecarlo (MC) simulation method: Tae-Hui, K. et al. [42] uses MC simulations to estimate maintenance and repair costs of hospital facilities, Rahman, S. et al. [43] for LCC of municipal infrastructure and Park, M. et al. [44] to analyze aged multi-family housing maintenance costs.

The correct choice of the stochastic inputs for the MC simulation is of paramount importance. In an analysis of building components life cycle costs, Di Giuseppe et al. [45] found that uncertainty in the estimation of the costs of the components (CRV of the assets in this case study), is well modeled using the uniform distribution, Equation (11), with the two limits a and b computed, respectively, as $-/+10\%$ of the deterministic costs.

$$u(x|a,b) = \begin{cases} \frac{1}{b-a} \; for \; a \leq x \leq b \\ 0 \; for \; x < a \; or \; x > b \end{cases} \tag{11}$$

Likewise, Di Giuseppe et al. [45] suggest using a Normal Distribution, Equation (12), for the maintenance costs (in this article the DM costs). The deterministic value of DM has to be used as the mean of the Normal distribution and 3% of the same cost as the standard deviation.

$$f(x|\mu,\sigma^2) = \frac{1}{\sqrt{2\pi\sigma^2}}e^{-\frac{(x-\mu)^2}{2\sigma^2}} \tag{12}$$

Eventually, the same source [45], suggests using a Uniform distribution with $-/+20\%$ of the deterministic ESL as a and b limit to model service life uncertainty while computing D, one of the four KPIs used in the proposed MPI. D, as shown in Equation (5), is the ratio of the actual service life of the component and the estimated one. While most of the times there is no uncertainty in ASL, ESL may be subject to a great variability that can be effectively modeled using the uniform distribution.

Main results of a 10,000 sample MC simulations are presented in Table 10, where mean and standard deviation of both FCI and MPI of the whole building (the mean of the MPI and FCI of all the components) are showed. These values have been compared to the deterministic values, respectively MPI = 0.3446 and FCI = 0.0390. The percentage difference between the mean of MC simulation and the deterministic values is 1.16% for MPI and 9.74% for FCI, demonstrating that the uncertainty on costs and service lives affects mostly the FCI forecast. This is not an unexpected result since costs only affect 25% of MPIs while FCI is totally dependent on them.

Table 10. Results of the Montecarlo simulation. Mean and standard deviation of Maintenance Priority Index (MPI) and Facility Condition Index (FCI) deterministic values obtained in the case study.

	Mean	Standard Deviation	Deterministic Values	% Difference Mean vs. Deterministic
MPI	0.3486	0.0047	0.3446	1.16%
FCI	0.0428	0.0034	0.0390	9.74%

Moreover, Figure 15 represents the probability density curve and cumulative one of the MPI, while Figure 16 represents the same outcome for the FCI. These images allow us to better compare deterministic values with the outcome of the MC simulations. As it can be seen also in Figure 16, the deterministic values computed for MPI and FCI are slightly optimistic, i.e., the uncertainty in costs and service lives will lead to higher values of both the KPIs (Figure 17). Noteworthy, even if the assumed uncertainty is quite high, $+/- 20\%$ for the ESL and $+/- 10\%$ for the CRV, the final result is slightly affected by it.

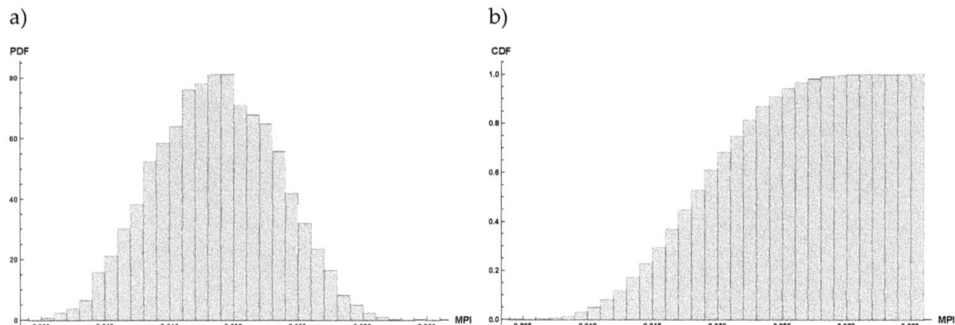

Figure 15. Montecarlo simulation results. Probability (**a**) and Cumulative (**b**) density function of Maintenance Priority Index (MPI).

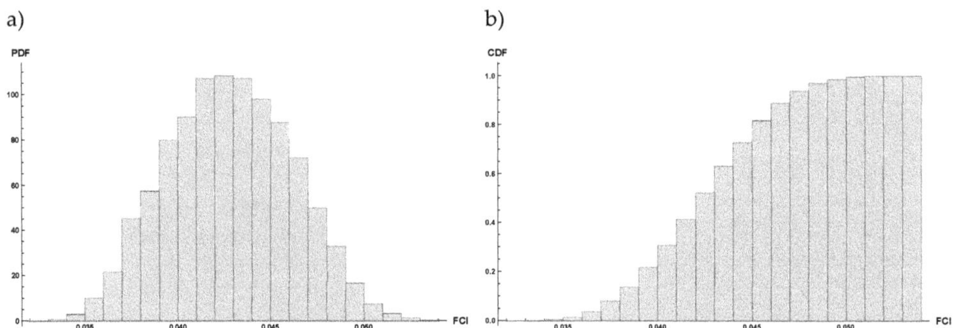

Figure 16. Montecarlo simulation results. Probability (**a**) and Cumulative (**b**) density function of Facility Condition Index (FCI).

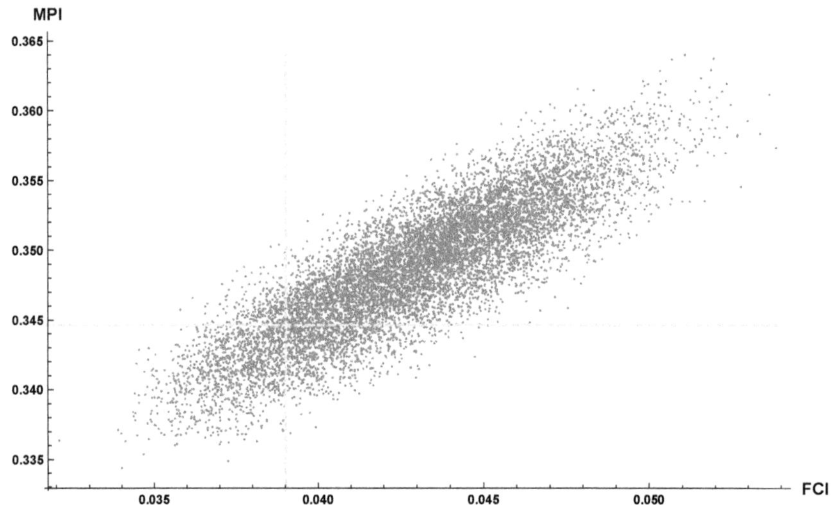

Figure 17. Combination of Maintenance Priority Index (MPI) and Facility Condition Index (FCI) for each one of the 10,000 simulations compared to the deterministic Maintenance Priority Index (MPI) and Facility Condition Index (FCI) (dashed lines).

The deterministic values of MPI and FCI are close to the average of MC simulations. This means that, with the uncertainty on CRV, DM and RSL assumed to be as described above, there is a 50% probability that the real value of MPI and FCI will be higher than expected. This is a very high probability of error. MPI and FCI values with a lower probability of error, can be taken from Table 11 which shows the quantile of the two KPIs. For example, MPI = 0.3519 and FCI = 0.0452 have a probability of been overcome, i.e., of error, of only 25%, meaning in 75% of the cases, the actual values of the two KPIs will be lower than these.

Table 11. Results of the Montecarlo simulation. Maintenance Priority Index (MPI) and Facility Condition Index (FCI).

p	5%	15%	25%	35%	45%	55%	65%	75%	85%	95%
MPI	0.3408	0.3435	0.3452	0.3467	0.3480	0.3493	0.3506	0.3519	0.3536	0.3564
FCI	0.0373	0.0391	0.0403	0. 0414	0.0424	0.0433	0.0442	0.0452	0.0464	0.0483

6. Conclusions

Through this research a DSS for maintenance budget allocation on physical assets has been developed and its robustness and applicability has been demonstrated through a case study. The technical assessment of the building has been carried out thanks to the FCI and the D indexes. The former describes, through an economic measure, the maintenance performance of the assets. The latter compares, in a scale 0-1, the ASL of each component to its ESL. Moreover, two non-technical performances related to the building are encompassed in the DSS through two indicators: the criticality (C) of the failure of each asset on the overall building and the maintenance preference (P) of the building owner for each single asset. This allows us to gather phenomena belonging to different domains within the Digital Asset Management context. This cross-domain characterization of the DSS allows us to address the contemporary trends in management of the built environment. The built environment should no longer be defined only through its physical properties, but also through the services that can be provided through its use. Therefore, the DSS is composed of metrics belonging to different domains of the AM discipline.

The definition of the level of aggregation of the components, composing the ifcAssets, on which maintenance interventions are carried out, is a critical issue to be considered in the development of the DSS. This phase is crucial, since it defines the level of aggregation of the information concerning the operation of a physical asset. This generalization of the information is a key factor for management of the asset during its life cycle, since it determines the level at which the information requirements for AM should be provided.

Moreover, some further development of the research can be highlighted. The final calculation of the MPI at the level of the single asset is obtained through the simple average among the four metrics of the MPI (FCI, D, C, P). This result could be different if a weighted average would be carried out. Nevertheless, the definition of the weighting system could lead to completely different results compared to those obtained through the current version of the proposed DSS, adding uncertainty to the DSS computation. The DSS, in fact, already comprehends some preference parameters (D and P), and therefore is partially driven by the choices made by specific subjects working for the correct operation of the building.

In this article the calculation of the MPI at the building level is presented. Once the MPI for each asset within the building has been computed, an average MPI for the whole building can be calculated. However, this value cannot be used for carrying out comparisons among buildings within a portfolio, since the FCI normalization do not allow us to compare same elements in different buildings. Therefore, further research should be accomplished to fill the scalability gap of the DSS.

Within the context of the digitization of the built environment, the DSS could be further integrated into the BIM approach, allowing a stronger integration for data collection and storage in the digital model. Therefore, further research will be done in the field of information requirements for using AM and FM to exploit the digital tools for information management. This could provide a further proof of the effectiveness in the implementation of BIM methodologies in AM and FM, towards the definition of a Digital Twin of the built environment. Moreover, the aggregation of the building components through the use of the ifcAsset definition, facilitate the integration of the DSS in a BIM environment [46,47]. Thanks to the flexibility in the definition of the level of aggregation at which maintenance interventions are applied, the DSS allows us to address the issue of portfolio management and infrastructural asset management. However, when the DSS is applied to a different scale, the inclusion of the location in the algorithm becomes a crucial issue. Therefore, further parameters could be included in the DSS, allowing the prioritization of the maintenance interventions, according to location-based metrics. The employment of the DSS at the portfolio or infrastructural level opens new research questions concerning how to integrate GIS and BIM for effective asset and facility management. Therefore, the scope of the research should be expanded to the GeoBIM Asset and Facility Management [48]. Further metrics could be included in the algorithm, as it is able to catch cross-domain issues, despite being based on only few KPIs.

Author Contributions: F.R.C.: Conceptualization, Methodology, Supervision, Visualization, Writing—original draft, Writing—review & editing. N.M.: Conceptualization, Methodology, Visualization, Writing—original draft, Writing—review & editing.

Funding: This research receives no external funding.

Acknowledgments: Authors would like to express their gratitude to Rigamonti Francesco S.p.A. for funding part of the Nicola Moretti's scholarship.

Conflicts of Interest: The authors declare no conflict of interest.

Abbreviations

AIM	Asset Information Model
AM	Asset Management
API	Asset Priority index
ASL	Actual Service Life
BIM	Building Information Modeling
C index	Criticality index is a KPI measuring the criticality as the consequence of the failure of the component, for the effective execution of the core business of the organization occupying the asset
CA	Condition Assessment
CAFM	Computer Aided Facility Management
CAD	Computer Aided Design
CI_{comp}	Composed Condition Index
CMMS	Computerized Maintenance Management Systems
CRV	Current Replacement Value
D index	KPI measuring the service life of the asset (building component)
DM	Deferred Maintenance
DoI	Department of Interior
DSS	Decision Support system
FM	Facility Management
FCI	Facility Condition Index
FCI_{comp}	Composed Facility Condition Index
IWMS	Integrated Workplace Management Systems
KPI	Key Performance Indicator
LoD	Level of Detail
LoI	Level of Information
MC	Montecarlo method

MPI	Maintenance Priority index
NACUBO	National Association of College and University Business Officers
P index	Priority index is a KPI measuring the preference of the asset owner, in the perspective of the execution of the core business
PDF	Probability Density Function
RSL	Reference Service Life
USA	United States of America

References

1. Amaratunga, D.; Haigh, R.; Sarshar, M.; Baldry, D. Application of the Balanced Score-card Concept to Develop a Conceptual Framework to Measure Facilities Management Performance within NHS Facilities. *Int. J. Health Care Qual. Assur.* **2002**, *15*, 141–151. [CrossRef]
2. Nical, A.K.; Wodynski, W. Enhancing Facility Management Through BIM 6D. *Procedia Eng.* **2016**, *164*, 299–306. [CrossRef]
3. Brochner, J.; Haugen, T.; Lindkvist, C. Shaping Tomorrow's Facilities Management. *Facilities* **2019**, *37*, 366–380. [CrossRef]
4. Saxon, R.; Robinson, K.; Winfield, M. *Going Digital. A Guide for Construction Clients, Building Owners and Their Advisers*; UK BIM Alliance: London, UK, 2018.
5. Volk, R.; Stengel, J.; Schultmann, F. Building Information Modeling (BIM) for Existing Buildings-literature Review and Future Needs. *Autom. Constr.* **2014**, *38*, 109–127. [CrossRef]
6. Chen, W.; Chen, K.; Cheng, J.C.P.; Wang, Q.; Gan, V.J.L. BIM-based Framework for Automatic Scheduling of Facility Maintenance Work Orders. *Autom. Constr.* **2018**, *91*, 15–30. [CrossRef]
7. Mcarthur, J.J. A Building Information Management (BIM) Framework and Supporting Case Study for Existing Building Operations, Maintenance and Sustainability. In Proceedings of the International Conference on Sustainable Design, Engineering and Construction, Chicago, IL, USA, 10–13 May 2015; Volume 118, pp. 1104–1111.
8. Moretti, N.; Dejaco, M.C.; Maltese, S.; Re Cecconi, F. The Maintenance Paradox. *ISTEA 2017-Reshaping Constr. Ind.* **2017**, 234–242.
9. Luoto, S.; Brax, S.A.; Kohtamaki, M. Critical Meta-analysis of Servitization Research: Constructing a Model-narrative to Reveal Paradigmatic Assumptions. *Ind. Mark. Manag.* **2017**, *60*, 89–100. [CrossRef]
10. Vandermerwe, S.; Rada, J. Servitization of Business: Adding Value by Adding Services. *Eur. Manag. J.* **1988**, *6*, 314–324. [CrossRef]
11. Lavy, S.; Garcia, J.A.; Dixit, M.K. KPIs for Facility's Performance Assessment, Part I: Identification and Categorization of Core Indicators. *Facilities* **2014**, *32*, 256–274. [CrossRef]
12. Alexander, K. Facilities Management Practice. *Facilities* **1992**, *10*, 11–18. [CrossRef]
13. Amaratunga, D.; Baldry, D. Assessment of Facilities Management Performance in Higher Education Properties. *Facilities* **2000**, *18*, 293–301. [CrossRef]
14. Yang, H.; Yeung, J.F.Y.; Chan, A.P.C.; Chiang, Y.H.; Chan, D.W.M. A Critical Review of Performance Measurement in Construction. *J. Facil. Manag.* **2010**, *8*, 269–284. [CrossRef]
15. Ladiana, D. *Manutenzione e Gestione Sostenibile Dell'ambiente Urbano*; Alinea Editrice: Firenze, Italy, 2007.
16. Dejaco, M.C.; Re Cecconi, F.; Maltese, S. Key Performance Indicators for Building Condition Assessment. *J. Build. Eng.* **2017**, *9*, 17–28. [CrossRef]
17. ISO. *ISO 41011:2017. Facility Management-Vocabulary*; ISO: Geneva, Switzerland, 2017.
18. Amaratunga, D.; Baldry, D.; Sarshar, M. Assessment of Facilities Management Performance. *Facilities* **2000**, *18*, 258–266. [CrossRef]
19. Amaratunga, D.; Baldry, D. A Conceptual Framework to Measure Facilities Management Performance. *Prop. Manag.* **2003**, *21*, 171–189. [CrossRef]
20. Parn, E.A.; Edwards, D.J.; Sing, M.C.P. The Building Information Modeling Trajectory in Facilities Management: A Review. *Autom. Constr.* **2017**, *75*, 45–55. [CrossRef]
21. Rush, S.C. *Managing the Facilities Portfolio: A Practical Approach to Institutional Facility Renewal and Deferred Maintenance*; National Association of College and University Business Officers: Washington, DC, USA, 1991.

22. Lavy, S.; Garcia, J.A.; Dixit, M.K. Establishment of KPIs for Facility Performance Measurement: Review of Literature. *Facilities* **2010**, *28*, 440–464. [CrossRef]

23. Lavy, S.; Garcia, J.A.; Dixit, M.K. KPIs for Facility's Performance Assessment, Part II: Identification of Variables and Deriving Expressions for Core Indicators. *Facilities* **2014**, *32*, 275–294. [CrossRef]

24. Maltese, S.; Dejaco, M.C.; Re Cecconi, F. Dynamic Facility Condition Index Calculation for Asset Management. In Proceedings of the 14th International Conference on Durability of Building Materials and Components, Ghent, Belgium, 29–31 May 2017.

25. U.S. Department of Interior. *U.S. Department of Interior Policy on Deferred Maintenance, Current Replacement Value and Facility Condition Index in Life-Cycle Cost Management*; U.S. Department of Interior: Washington, DC, USA, 2008.

26. U.S. Department of Interior. *U.S. Department of Interior Management Plan. Version 3.0*; U.S. Department of Interior: Washington, DC, USA, 2008.

27. U.S. Department of Interior. *U.S. Department of Interior National Park Service Park Facility Management Division Park Facility Management Division*; U.S. Department of Interior: Washington, DC, USA, 2012.

28. U.S. Department of Interior. *U.S. Department of Interior Deferred Maintenance and Capital Improvement Planning Guidelines*; U.S. Department of Interior: Washington, DC, USA, 2016.

29. IFMA. Asset Lifecycle Model for Total Cost of Ownership Management. Framework, Glossary Definitions. In *A Framework for Facilities Lifecycle Cost Management*; IFMA: Houston, TX, USA, 2008.

30. U.S. Department of Interior. *U.S. Department of Interior Site-Specific Asset Business Plan (ABP) Model Format Guidance*; U.S. Department of Interior: Washington, DC, USA, 2005.

31. Lavy, S.; Garcia, J.A.; Scinto, P.; Dixit, M.K. Key Performance Indicators for Facility Performance Assessment: Simulation of Core Indicators. *Constr. Manag. Econ.* **2014**, *32*, 1183–1204. [CrossRef]

32. U.S. Department of Interior. *U.S. Department of Interior Asset Priority Index Guidance*; U.S. Department of Interior: Washington, DC, USA, 2005.

33. Mills, C.D. *An Evaluation of U.S. Coast Guard Shore Facility Readiness Measures*; U.S. Army War College, Strategy Research Project: Carlisle, PA, USA, 2001.

34. NASA. *The NASA Deferred Maintenance Parametric Estimating Guide*; National Aeronautics and Space Administration: Washington, DC, USA, 2003.

35. ISO. *ISO 16739-1:2018-Industry Foundation Classes (IFC) for Data Sharing in the Construction and Facility Management Industries-Part 1: Data Schema*; ISO: Geneva, Switzerland, 2018.

36. ISO BS EN. *ISO 19650-2:2018. Organization and Digitization of Information about Buildings and Civil Engineering Works, Including Building Information Modeling (BIM)-Information Management using Building Information Modeling. Part 2: Delivery Phase of the Assets*; ISO: Geneva, Switzerland, 2018.

37. ISO. *ISO 15686-1:2011 Buildings and Constructed Assets-Service Life Planning-Part 1: General Principles and Framework*; ISO: Geneva, Switzerland, 2011.

38. US Department of Defense MIL-STD-1629A. *Military Standard. Procedures for Performing a Failure Mode, Effects and Criticality Analysis*; Department of Defense: Washington, DC, USA, 1980.

39. Re Cecconi, F.; Moretti, N.; Dejaco, M.C. Measuring the Performance of Assets: A Review of the Facility Condition Index. *Int. J. Strateg. Prop. Manag.* **2019**, *23*, 187–196. [CrossRef]

40. Comune di Milano. Listino Prezzi Comune di Milano. Edizione 2018. Available online: http://www.comune.milano.it/wps/portal/ist/it/amministrazione/trasparente/OperePubbliche/listino_Prezzi/Edizione%2B2018 (accessed on 17 March 2019).

41. Camera di Commercio Milano Monza Brianza Lodi Home-PiùPrezzi, Il portale dei prezzi della Camera di Commercio di Milano. Available online: http://www.piuprezzi.it/ (accessed on 13 May 2019).

42. Tae-Hui, K.; Jong-Soo, C.; Young Jun, P.; Kiyoung, S. Life Cycle Costing: Maintenance and Repair Costs of Hospital Facilities Using Monte Carlo Simulation. *J. Korea Inst. Build. Constr.* **2013**, *13*, 541–548.

43. Rahman, S.; Vanier, D. *Life Cycle Cost Analysis as a Decision Support Tool for Managing Municipal Infrastructure Building Regulation-Standards Processing View Project Municipal Infrastructure Investment Planning View Project*; In-House Publishing: Rotterdam, The Netherlands, 2004.

44. Park, M.; Kwon, N.; Lee, J.; Lee, S.; Ahn, Y. Probabilistic Maintenance Cost Analysis for Aged Multi-family Housing. *Sustainability* **2019**, *11*, 1843. [CrossRef]

45. Di Giuseppe, E.; Massi, A.; D'Orazio, M. Probabilistic Life Cycle Cost Analysis of Building Energy Efficiency Measures: Selection and Characterization of the Stochastic Inputs through a Case Study. *Procedia Eng.* **2017**, *180*, 491–501. [CrossRef]
46. Re Cecconi, F.; Moretti, N.; Maltese, S.; Tagliabue, L.C. A BIM-Based Decision Support System for Building Maintenance. In *Advances in Informatics and Computing in Civil and Construction Engineering*; Springer International Publishing: Cham, Germany, 2019; pp. 371–378.
47. Maltese, S.; Branca, G.; Re Cecconi, F.; Moretti, N. Ifc-based Maintenance Budget Allocation. *In Bo* **2018**, *9*, 44–51.
48. Ellul, C.; Stoter, J.; Harrie, L.; Shariat, M.; Behan, A.; Pla, M. Investigating the State of Play of Geobim Across Europe. *ISPRS-Int. Arch. Photogramm. Remote Sens. Spat. Inf. Sci.* **2018**, *XLII-4/W10*, 19–26. [CrossRef]

 *buildings*

Article

LCC Estimation Model: A Construction Material Perspective

Vojtěch Biolek * and Tomáš Hanák

Faculty of Civil Engineering, Brno University of Technology, Veveří 331/95, 602 00 Brno, Czech Republic
* Correspondence: biolek.v@fce.vutbr.cz

Received: 30 June 2019; Accepted: 6 August 2019; Published: 8 August 2019

Abstract: The growing pressure to ensure sustainable construction is also associated with stricter demands on the cost-effectiveness of construction and operation of buildings and reduction of their environmental impact. This paper presents a methodology for building life cycle cost estimation that enables investors to identify the optimum material solution for their buildings on the level of functional parts. The functionality of a comprehensive model that takes into account investor requirements and links them to a construction cost estimation database and a facility management database is verified through a case study of a "façade composition" functional part, with sublevel "external thermal insulation composite system (ETICS) with thin plaster". The results show that there is no generally applicable optimum ETICS material solution, which is caused by differing investor requirements, as well as the unique circumstances of each building and its user. The solution presented in this paper aims to aid investor decision-making regarding the choice of the building materials while taking the Life Cycle Cost (LCC) into account.

Keywords: building; construction material; life cycle costs; thermal insulation system

1. Introduction

Sustainable efforts are generally discussed from an environmental as well as economic perspective. On the one hand, there is a need to seek environmentally friendly solutions with minimum energy consumption and waste generation; on the other hand, there is the investor's intention to pursue cost-effective projects. Building projects especially are marked by the fact that they are complex, are carried out over a long period of time, and face a high level of uncertainty and several risks affecting the final project outcome [1].

The building project should thus be considered in terms of its entire life cycle. In this relation, the BLCC approach (Building Life Cycle Costs) plays an important role as it focuses on cost optimisation throughout the entire life cycle [2] of a building. Zabielski and Zabielska [3] formulate LCC as a sum of the cost of purchase (project execution costs), cost of ownership (maintenance) and the cost of disposal decreased by the residual value of the property. This kind of planning of the building life cycle is crucial for informed decision-making [4], since operational costs usually significantly exceed construction costs [5]. For instance, it is estimated that about 80% of the energy use relates to the operational stage of a building's life cycle [6].

A fundamental issue is to determine the lifespan of the building/building elements. In this regard, there is a lack of consensus in the relevant literature. Some authors consider the lifespan of 50 years [7–9], others use the value of 60 [10,11] years, while others even compare different service lifespans (30, 50 and 100 years) [12]. Generally speaking, the lifespan should correspond to the expected period of use, which may depend on the building's technical parameters (wood/concrete structure) or expected time of operation (from the investor's point of view), while it is also necessary to consider the lifespan of individual building elements. For example, Robati [13] uses 25 years as the period for replacement of glazed windows, so it is obvious that this particular element will be replaced several times during the lifespan of the building as a whole.

As a building's lifespan ranges across decades, the prediction becomes progressively less accurate with increasing prediction time. This inaccuracy and uncertainty of costs within the operational stage is associated with several factors, e.g., predicted inflation rates (energy prices), availability of new technologies, changes in applicable legislation, inspection costs, insurance, and local tax or labour costs [14,15]. The accuracy of cost prediction depends on various aspects involving the level of information detail on the building [16] (materials, conditions under which certain activities can be carried out, e.g., the cleaning service [17]) and information about materials and related data on deterioration behaviour [18].

Within the building life cycle, a major part of the costs will be incurred at the later stages, i.e., especially in the operational stage. For future costs such as maintenance and repairs, appropriate discount rate should be applied [19]. The value of the discount rate is important [20], and the influence is more significant with lower discount rates [21], and vice versa. As a result, building investments should be evaluated in terms of the NPV (Net Present Value) indicator [9].

One of the most crucial investment decision-making issues consists of striking a balance between construction and operation costs [22]. This problem is complex, and it also involves the effect of energy prices (growing energy prices result in a more significant role played by operation energy costs in the early years of the lifespan) [10] and many other factors mentioned above. That is why many researchers apply various optimisation techniques, e.g., mixed integer linear programming [23], genetic algorithms [24], hybrid algorithms [25], and regression models [26,27]. Many researchers also apply LCC minimisation with regards to a specific natural hazard. For instance, this refers to buildings threatened by earthquakes and wind damage [7], flooding [28] or seismic risks [29]. In this regard, the LCC approach also differs in that it takes into consideration the vulnerability of buildings to a particular risk, the risk exposure in a given location, as well as refurbishment costs incurred on account of the damage. There are numerous studies providing methodologies for estimating loss caused by natural hazards (see e.g., [30] for flood risk); however, distributing these losses over the lifespan of the building is subject to uncertainty from the NPV perspective.

Recently constructed buildings have considerably improved thermal characteristics compared to older buildings. Incidentally, older buildings are often renovated with the aim of reducing energy consumption. In this regard, it should be noted that reduced consumption of energy during the operational stage of the building life-cycle usually comes with increased use of materials and the related environmental costs that may counteract its financial benefits [31]. Therefore, sustainable material cycles, recycling options and disposal costs [32,33] should be considered during the preparation of building projects.

In the area of public works, procurement is governed by applicable national legislation, which in the case of the European Union (EU) is based on Directive 2014/24/EU of the European Parliament and of the Council. According to the directive [34], "life-cycle costing shall to the extent relevant cover parts or all of the following costs over the life cycle of a product, service or works: (a) Costs, born by the contracting authority or other user (acquisition, use, maintenance and end life costs); (b) Costs imputed to environmental externalities linked to the product, service or works during its life cycle, provided their monetary value can be determined and verified".

Unfortunately, no relevant databases of information on the expected lifetime of products, the time and extent to which they require repairs and the costs of maintenance of given structures are not available. That is why the LCC approach is rarely used in procurement practice. Nevertheless, such data should be processed in future BIM (Building Information Modelling) systems. In the future, the BIM model should serve as a source of information informing the work of the individual participants of the construction process. An approach that includes information with the BIM model will benefit from the data repository of transfer formats, allowing quick editing of the information and updating of the LCC value [35]. The integration of BIM and LCC serves to ensure better maintenance accessibility and enhanced collaboration between asset and maintenance management [36,37].

This paper therefore reflects the growing pressure to ensure sustainable construction, which is also associated with stricter demands on the cost-effectiveness of the construction and operation of buildings and reducing their environmental impact. The objective of the research is to propose a methodology enabling the selection of an optimum building material solution for the individual functional parts of a building in terms of life cycle costs. This case study involves the proposed and applied methodology for a selected functional part: "Façade—external thermal insulation composite system with thin plaster" in the context of the current state of the Czech construction sector.

The article is structured as follows: Firstly, the current state of knowledge is presented, followed by materials and methods and a description of the proposed methodology, where the methodology is then applied to the selected functional part in variant solutions and discussed. The final chapters summarise the research findings and limitations and outline future research directions.

2. Materials and Methods

A proposal for a methodology for LCC calculation with respect to construction materials requires several steps. With regard to LCC calculation standards, it is first necessary to define the required input data (see Section 2.1); specify the lifetime, the cycle and frequency of repairs, and the maintenance of the functional parts of buildings (Section 2.2); and then to propose a system for data exchange (Section 2.3).

The LCC indicator is calculated based on the formula indicated in the European ISO 15686-5:2017 [38] standard, which is based on the discounting of future costs during the examined period. Discounting means adjusting future costs (costs of reconstruction, utilities, maintenance, etc.) with respect to their present value. LCC are calculated according to following formula:

$$LCC = \sum_{t=0}^{T} \frac{C_t}{(1+r)^t} \tag{1}$$

where:

C_t denotes all costs as equivalent cash flows in year t;
r is the discount rate;
t is the analysed year ($t = 0, 1, 2 \ldots, T$);
T is the length of the life cycle in years.

Other models, such as those published by Bromilow and Pawsey [39] or Sobanjo [40], are based on the principle of different discounting of regular and irregular costs.

2.1. Input Data Requirements for LCC Calculation

For the purposes of LCC calculation, input data are required in three main areas. Specifically, this includes determination of the length of the examined period, the value of the discount rate, and the identification of the individual types of costs arising throughout the life cycle.

The length of the examined period either corresponds to the expected lifetime of the building, or can be set as a specific period corresponding to the investor's requirements. The expected lifetime of buildings in the Czech Republic is indicated in [41], where, e.g., a masonry building is expected to have a lifetime of 100 years.

The discount rate used when modelling LCC is up to the individual investors, but it should correspond to the rate of return of other similar projects or requirements for specific types of public projects. A 5% rate is commonly used [9,42]. Costs arising throughout the life cycle of a building then constitute the acquisition costs, operational costs (maintenance, repairs, replacement) and disposal costs. For the purposes of the present research activity, the discount rate was set at 5%.

The most complex part of the calculation consists in defining and quantifying the costs incurred in the operational stage of the building's life cycle. It is necessary to define the scope and frequency in which the individual parts of a building have to be maintained, repaired or replaced. Each structure has different requirements when it comes to maintenance and repair, including a different expected lifetime. As a general rule, however, one of the main factors that has an impact on the expected lifespan is the material used to build a given structure. Other factors include the quality of production, quality of construction work, and maintenance frequency.

The calculation of LCC makes it desirable to divide the analysed building into functional parts (FP). It is then necessary to establish the relevant repair, maintenance and replacement cycles, and costs; this study uses the data provided in [43]. For instance, "Exterior plasters, insulation" have their FP lifetime set to 30–60 years, with a repair cycle of 30 years and the scope of repairs of 20%.

2.2. Establishing the Lifetime, Cycle and Frequency of Repairs and Maintenance of Functional Parts

The process of establishing the aforementioned values is based on a survey of already built and operated buildings included in a facility management (FM) system. In this regard, it is vital that the FM system contain data on the individual costs of repairs, maintenance and replacement (R/M/R), including the time when the given intervention took place. Using the aforementioned recorded data, it is possible to calculate the average R/M/R costs and the average length of the cycle between individual R/M/R interventions, where the average costs are calculated using the following formula:

$$Cycle\_A = \frac{(DA_1 - DC) + \sum_{i=2}^{n}(DA_i - DA_{i-1})}{n} \tag{2}$$

where

$Cycle\_A$ is the average length of the cycle between two individual activities (A), provided separately for each R/M/R component;
DC is the date of construction;
DA_1 is the date of the 1st R/M/R activity;
DA_i is the date of the i th R/M/R activity;
n is the number of R/M/R activities.

The concept of collecting R/M/R information and its transformation into a database of lifespans, frequencies and costs of repairs and maintenance of functional parts is introduced in [44]. Calculating LCC requires effective communication between three systems: (1) The LCC calculation system; (2) the FM system; and (3) the building cost estimation system, which is necessary to establish unit prices and other information from the price database that influence LCC calculation over the entire life cycle of a building. Another system that can be incorporated is the system for creating BIM models. A BIM model is essentially a database of all of the information on the building, which can thus serve as a source of input data for LCC calculation (e.g., surface area, materials used, dimensions, characteristics, etc.); conversely, information obtained through the LCC calculation can be transferred back into the BIM model for future use.

Each of the above systems usually operates with a different data structure and a different classification of individual building structures. Accordingly, it was necessary to find a suitable structural division of a building that would be compatible with all the component systems. The proposed connecting database is based on a division of a building into four functional units—load-bearing structures, roof structures, façade and surface treatment of interior spaces. The individual functional units are divided into functional parts (FP); a detailed division is provided in Table 1. Functional parts are further divided according to the implementation possibilities or other distinct features of the given structural part. This brings multiple benefits over the building's life cycle, both in terms of managing its construction and cost management in the operational stage. This makes it possible to consider functional units as actual parts incorporated in a building that are supplied as a whole. Since there is no clearly available database for use in LCC calculation and the facility management system, the proposed connecting database appears to be the default option for both systems.

Table 1. Proposed connecting database with division to functional units and functional parts [45].

Functional unit	LOAD-BEARING STRUCTURES	ROOF STRUCTURES	FAÇADE	SURFACE TREATMENT OF INTERIOR SPACES
Functional part	foundations walls columns ceilings girders, (main) beams staircase load-bearing part of chimney	wooden roof frame roof covering metal sheeting of roof elements other roof elements—roof windows, skylights, antennas etc.	windows entrance door, gate façade composition exterior window sills other façade elements—covers, railing, blinds, etc.	wall plastering facings ceiling plastering suspended ceilings floor

2.3. General Description of the System

The entire process begins with the design of a building. If the design is made using a BIM tool, it is important for the maker of the BIM model to supply information to classify all building structures into functional units and functional parts. The BIM model's level of detail is high (LOD 200 to LOD 300) in order to include the selected construction solution, materials and dimensions, general information on the building and the size of the individual structures. If the design is made using traditional tools (2D design), the designer must input all the necessary information into the LCC calculation system manually. The required information (parameters) are dependent on the type of functional part, but generally speaking, this means its material characteristics and size. Among other information, functional parts also require information generally related to the whole building—its height, location etc.

Once a building is classified into distinctly defined functional parts, the system for LCC calculation will communicate with the building cost estimation system. Individual items of the price database carry an information on classification into functional parts, i.e., the items from which the information necessary for LCC calculation will be retrieved (e.g., unit prices, unit weight, rubble, time needed for (dis)assembly in standard hours).

The next stage consists in obtaining information for the calculation of costs in the operational stage of the building, i.e., the costs of repairs, maintenance and replacement (R/M/R). R/M/R costs information is transferred from the price framework of the building cost estimation system. As described above, there is a challenge consisting of the availability of information on the scope and frequency of R/M/R. The information can be entered into the LCC system in two ways:

- Information on the R/M/R scopes and frequencies of the individual functional parts is based on observation of already built buildings. The information is managed in the facility management system and can be transferred to the LCC calculation system via the connecting database (see Section 2.2).
- Repairs are simulated by the designer (LCC system user) based on experience or assumptions regarding the orientation or use of the building. The lifetime and maintenance can potentially be indicated by the manufacturer of the materials or can be simulated by the user as in the case of repairs. This possibility of recording can be used by users who lack data from the facility management system.

The final stage consists in entering the calculation conditions. The conditions are based on the LCC calculation formula itself—discount rate and the examined period. These parameters are dependent on the customs of the investor and the building's character.

Once all information necessary for the LCC calculation has been entered or transferred from the individual systems, the calculation itself takes place. The entire calculation can be divided into five stages:

- calculation of acquisition costs;

- calculation of replacement costs;
- calculation of maintenance costs;
- calculation of repair costs;
- calculation of disposal costs at the end of the building's lifetime.

The individual stages of the calculation, i.e., the structure of the calculation, differ for each functional part in terms of the manner of costing, e.g., the costing of a reinforced concrete wall differs from that of façade composition. The manner of costing is based on custom and the principles of building cost estimation.

After the calculation has been completed, an important added value of the proposed system consists in finding the most suitable solution that meets the same or better characteristics as the original solution. These properties depend on the type of the functional part and its typical characteristics, e.g., thermal insulation properties, dimensions, strength, etc. The proposed system will then calculate LCC for all alternatives and order them on the basis of various characteristics: acquisition costs, LCC, LCC minus acquisition costs, and so on. The user thus obtains a basis that will enable them to make a design based on the best possible solution.

If a BIM model is available, direct transfer of information from the LCC calculation system into the BIM model is possible. In particular, this refers to information on the acquisition price, frequency and cost of R/M/R, which benefits the planning of funding sources in the upcoming operational stage of the newly constructed building. This makes the BIM model multi-dimensional, since it contains information on costs, facility management and partially also information on time (time of assembly of the individual building structures).

The whole process of data exchange among the individual systems is shown in Figure 1. The diagram also shows other possibilities for linking the individual systems—LCA and energy consumption—which, however, are not part of the proposed methodology. Similarly, the methodology does not address costs associated with the entire project, i.e., costs associated with project documentation and other costs borne by the investor or the contractor. The proposed methodology covers only the costs associated with the structures incorporated in the building. The methodology addresses the structure of information and its processing in the individual systems, but not its mutual systemic interconnection at the data transfer level.

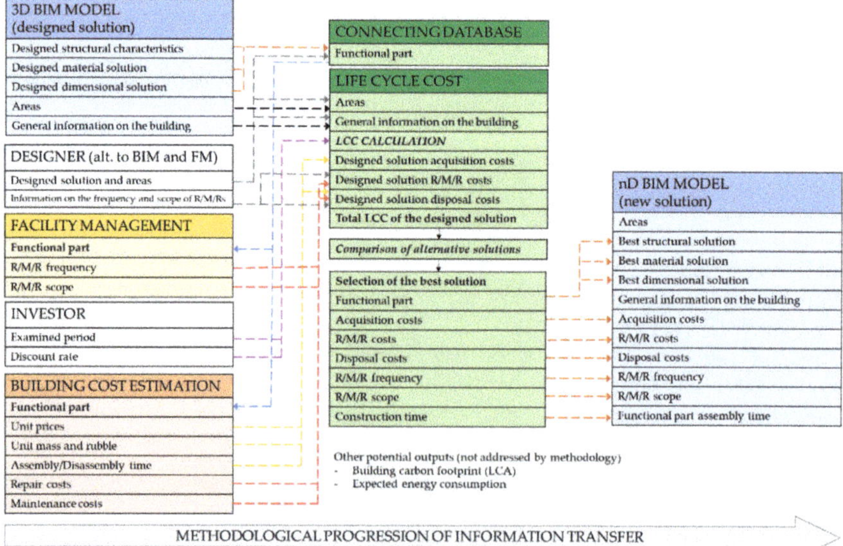

Figure 1. The process of exchanging information among the individual systems to ensure the best material solution of the building.

3. Applying the Model: Case Study of the "Façade Composition" Functional Part

To demonstrate the functionality of the system, the chapter below presents a proposed solution for calculating LCC for the functional part designated "façade composition". In the first stage of the process, the creator of the BIM model must define the structural, material and dimensional characteristics of the façade. The façade composition functional part can be divided into four sublevels depending on the construction solution:

- external thermal insulation composite system (ETICS) with thin plaster;
- external thermal insulation composite system (ETICS) with facing;
- plaster only;
- facing only.

The "external thermal insulation composite system (ETICS) with thin plaster" sublevel of the "façade composition" functional part, was selected for the case study. Viable types of external thermal insulation composite systems (hereinafter ETICS) and thin plasters are selected from the price database [46], which essentially contains all material and dimensional possibilities and is key for determining the costs. A total of 11 ETICS types and 13 thin plaster types were selected; a list of these, together with further details, is presented in Tables 2 and 3. There are 143 potential combinations. The thickness of the material depends on the type of ETICS and thin plaster used.

Table 2. The list of individual types of ETICS with their possible thickness values, thermal conductivity coefficient and indication of special properties (none = no special properties; SO = meant for the socle area; PO = fire resistant) [46–50].

Type of ETICS and Reference Products	Thickness [mm]	Thermal Conductivity Coefficient λ [W/(m·K)]	Property
EPS 70 façade board (Isover EPS 70F)	10; 20; 30; 40; 50; 60; 80; 100; 120; 140; 150; 160; 180; 200	0.039	none
EPS 100 façade board (Isover EPS 100F)	30; 50; 60; 80; 100; 120; 140; 160; 180; 200	0.037	none
EPS graphite façade board (Isover EPS GreyWall)	20; 30; 40; 50; 60; 80; 100; 120; 140; 160; 180; 200; 220; 240; 260; 280; 300	0.032	none
polystyrene socle board, façade (Isover EPS SOKL 3000)	20; 30; 40; 50; 60; 80; 100; 120	0.035	SO
polystyrene façade board for thermal insulation of bottom parts of buildings (Isover EPS PERIMETR)	40; 50; 60; 80; 100; 120; 140	0.034	SO
board with longitudinal mineral fibre (Isover TF PROFI)	40; 50; 60; 70; 80; 100; 120; 140; 160; 180; 200; 220; 240; 260; 280; 300	0.036	PO
board with perpendicular mineral fibre (Isover NF 333)	20; 30; 40; 50; 60; 70; 80; 100; 120; 140; 160; 180; 200; 220; 240; 260; 280; 300	0.041	PO
foamglass board, no surface treatment (FOAMGLAS®W+F)	50; 60; 80; 100; 120; 140	0.038	PO
XPS polystyrene board (BACHL XPS 300 G)	30; 40; 50; 60; 80; 100; 120	0.036 (up to 60 mm) 0.033 (over 60 mm)	SO
sandwich insulation board (polystyrene+wool) (Isover TWINNER)	100; 120; 140; 150; 160; 180; 200; 220; 240; 260; 280; 300	0.033 (up to 200 mm) 0.032 (over 200 mm)	PO
wood fibre board (STEICO Flex-wood fibre insulation)	40; 60; 80; 100; 120; 140; 160; 180; 200	0.036	none

Table 3. List of thin plasters with potential thickness [46].

Type of Thin Plaster	Thickness [mm]
mineral granular plaster	1.0; 1.5; 2.0
mineral grooved plaster	2.0
acrylic granular plaster	1.0; 1.5; 2.0
acrylic grooved plaster	2.0; 3.0
acrylic mosaic plaster	1.0; 2.0; 3.0
silicate granular plaster	1.0; 1.5; 2.0; 3.0
silicate grooved plaster	2.0
silicone granular plaster	1.0; 1.5; 2.0; 3.0
silicone grooved plaster	2.0; 3.0
silicone hydrophilic granular plaster	1.0; 1.5; 2.0; 3.0
silicone hydrophilic grooved plaster	2.0; 3.0
silicate-silicone granular plaster	1.0; 1.5; 2.0; 3.0
silicate-silicone grooved plaster	2.0

The above material and dimensional characteristics (Tables 2 and 3) are key to the proper classification of the proposed solution and must be inputted to the BIM model by its creator. Figure 2 shows the list of input information necessary for performing the LCC calculation. Aside from the material and dimensional characteristics of the façade layers, it also includes information that must be entered for the purposes of identification of all costs associated with implementing the façade layers. For the LCC calculation itself, it is also necessary to input the investor's requirements for LCC modelling—the discount rate and the length of the examined period based on the DCF model.

Area [m²]	100.00	
FD parameters – Façade – external thermal insulation composite system with thin plaster		
ETICS type:	EPS 100 façade board	
ETICS thickness [mm]:	120	
Use of thermally insulating plugs [yes/no]:	yes	
Use of dispersion (organic) reinforced plaster compound [yes/no]:	no	
Type of thin plaster:	mineral granular plaster	
Plaster thickness [mm]:	2	
ETICS characteristics:	no special characteristics	
General parameters		
Building height [m]:	10	
Landfill distance [km]:	30	
Years in which no maintenance is to be performed prior ETICS replacement [years]:	2	
Building lifetime [years]:	100	
Parameters for calculating economic efficiency:	Variant 1 and 3	Variant 2
Discount rate:	5%	5%
Examined period [years]:	100	30

Figure 2. Overview of input parameters for calculating LCC for the individual variants.

For a specific demonstration of how the proposed system works, the chosen default material solution consists of ETICS EPS 70 façade board (120 mm thick) with mineral granular plaster (2 mm thick)—see Figure 2. The input parameters of the discount rate and the examined period are modelled in three variants:

- Variants 1 and 3—The examined period corresponds to the entire lifetime of the building (100 years) and the discount rate is set to 5%; the variants differ in the R/M/R database (for more details see 3.6) [9,42,43].

- Variant 2—The examined period is set to 30 years, i.e., the minimum lifetime of "Exterior plaster, insulation" according to [43], while the discount rate remains identical to Variant 1, i.e., 5%.

3.1. Calculation of Acquisition Costs

The calculation of acquisition costs for implementation of the façade layers consists of the items of the price database [46]. The use of the individual items of the price database depends on the type of ETICS or thin plaster used, respectively. The external thermal insulation composite system is costed separately for assembly and supply of material. Assembly of ETICS is differentiated according to the system's type and thickness:

- assembly of polystyrene external thermal insulation boards—thickness under 40 mm, under 80 mm, under 120 mm, under 160 mm, under 200 mm, under 240 mm, over 240 mm;
- assembly of external thermal insulation mineral wool with longitudinal fibre—thickness under 40 mm, under 80 mm, under 120 mm, under 160 mm, over 160 mm;
- assembly of external thermal insulation mineral wool with perpendicular fibre—thickness under 40 mm, under 80 mm, under 120 mm, under 160 mm, under 200 mm, over 200 mm.

The price of the "ETICS assembly" item also includes the costs of assembly and supply of levelling compounds and fiberglass mesh. Each assembly item includes information not only on the unit acquisition price, but also information on the mass, which is essential for material transport calculations, and on the time demands of the work in standard hours, which is a required figure for calculating the assembly time.

The material corresponds to various types of ETICS indicated in Table 2. The thickness of the insulating material indicated by the manufacturer is distinguished. Each item of material also includes an information on the unit price and mass, where the thermal resistance is calculated according to the thermal conductivity coefficient and the insulating material thickness. Thermal resistance is important for finding variants from among the individual ETICS types with the same or improved characteristics.

ETICS costing also includes potential extra costs. The use of these extra costs is conditional on entering information into the BIM model in the form of an associated parameter. The following extra costs are included:

- for anchoring boards 22.5 m and higher above ground—determined according to the insulating material thickness;
- for use of thermally insulating plugs—determined according to ETICS type;
- for use of dispersion (organic) reinforced plaster.

The supply and assembly of thin plaster are indicated as one item in the price database. The price of the item also includes the costs of priming the substrate. As in the case of the ETICS, each item of the thin plaster includes information on the acquisition price, mass, and assembly time in standard hours. The thermal resistance of thin plaster is negligible and is disregarded.

Complete supply and assembly of the "external thermal insulation composite system with thin plaster" also carries some associated costs such as material transport, where the total mass of all the items used is added up. The items of material transport depend on the height, type and construction solution of the buildings, as well as on whether mechanisation is used fully, partially or not at all. The case study assumes full use of mechanisation in the construction process. Another cost involves the assembly, lease and removal of scaffolding and the possible use of safety nets. The price database distinguishes multiple types of scaffolding—light tubular scaffolding, heavy tubular scaffolding, light frame scaffolding, heavy frame scaffolding. Based on the calculation needs, the most commonly used type of scaffolding will be considered—light frame scaffolding with decking size of up to 1.2 m. The cost of scaffolding lease corresponds to the ETICS assembly time converted to working days. One of the user-entered parameters is the height of the building, which affects the use of items for extra costs associated with the assembly of ETICS, scaffolding and material transport.

3.2. Calculation of Replacement Costs

The replacement of a structure is assumed to occur at the end of the given functional part's lifetime, where the replacement costs are determined by the sum of disposal costs and the acquisition costs. The proposed system assumes that the lifetime of thin plaster is the same as the ETICS lifetime. The ETICS lifetime is based on the R/M/R database and is either transferred from the facility management system or can be entered by the designer based on his experience (see Section 2.3).

The price database [46] distinguishes between disposal of polystyrene boards and mineral wool boards, where it only takes into account the thickness of the disposed insulating material. The ETICS disposal item also includes the thin plaster disposal costs. Costs associated with the transport of rubble and the disposal of waste constitute an important part of the calculation. The result depends on the mass of the disposed material (rubble). Each disposal item includes information on the mass of the building structure being disposed; however, for an exact value, the mass of items used for the calculation of acquisition costs will be used. Costs associated with rubble transport are divided into two parts—vertical transport, which depends on the height of the building and relates to transport within the building, and horizontal rubble transport associated with moving the rubble from the construction site to a landfill. The distance of the landfill from the construction site is one of the parameters that must be entered by the user. The rubble dumping fee constitutes another cost, where its amount depends on classification of the waste according to waste catalogue decree.

3.3. Calculation of Maintenance Costs

Each structure has to be maintained in some way during its operational stage. In the case of the façade functional part, maintenance concerns the layer that is in contact with the air, i.e., thin plaster. As mentioned above, it is necessary to establish what needs to be carried out and how often. Two possibilities for creating the R/M/R database include creating it based on personal observations or according to the manufacturer's guidelines or own discretion (see Section 2.3).

The database determines the scope and frequency of maintenance, where it is not possible to alter the data as they are the result of long-term observations. The other option consists in simulating maintenance based on the manufacturer's recommendations and own experience. The main advantage of a simulation is the possibility to alter the inputs, since each building is exposed to different factors. Cities with high amounts of dust in the air (and thus higher dirt deposition on the façade) require a different level of maintenance than buildings in the countryside with lower traffic pollution levels. The user of the system has the manufacturer's recommendations available and can modify them based on the user's own experience. Based on the technical guideline of the plaster system manufacturer, it is necessary:

- to clean the façade with pressurised water every 3 to 5 years;
- to apply façade coating to plastered surfaces every 10 to 15 years.

The user can only influence the frequency of the individual stages of maintenance. The costs of the individual stages of maintenance are determined by the price database. Pressurised water cleaning has the same cost for all types of plasters. Only the coating type is determined by the plaster type. The price database [46] distinguishes 4 types of coatings for thin plasters, where the unit price is determined by the façade topography ranked 1 to 5. Topography levels 3 to 5 are practically absent in contemporary buildings because they are very costly due to the inclusion of ledges, window framing, pilasters, embedded columns, etc. Topography levels 1 and 2 are associated with the same unit price and the calculation thus need not be adjusted in any way. Table 4 shows the association of the individual types of coatings to the individual types of plasters.

The system gives the user an option to influence the calculation of the maintenance costs by enabling the user to determine when the LCC should include maintenance of thin plaster due to upcoming ETICS replacement.

Table 4. Coating types associated with the individual plaster types (texture not taken into consideration).

Type of Thin Plaster	Type of Coating
mineral plaster	lime
acrylic plaster	acrylic
silicate plaster	silicate
silicone and silicate-silicone plaster	silicone

3.4. Calculation of Repair Costs

The need to repair functional parts can, but does not have to, appear during their lifetime. This is the part of the LCC calculation that most relies on an own database of R/M/R and has appeared in comparable buildings that are already in operation. Nevertheless, even here, the user can set the expected scope of potential repairs based on his own experience. The only eventuality in which the ETICS with thin plaster functional part would have to be repaired during its lifetime consists of physical damage, such as perforation. From the point of view of the price database, a repair of the insulation system is determined by the ETICS type, thickness and the size of the part needing repair. Specific division is shown in Table 5.

Table 5. Division of ETICS repairs according to price database [46].

Polystyrene Thickness	Mineral Wool Thickness	Scope of Repair for Polystyrene and Mineral Wool
under 40 mm	under 40 mm	under 0.1 m^2
40 mm to 80 mm	40 mm to 80 mm	0.1 to 0.25 m^2
80 mm to 120 mm	80 mm to 120 mm	0.25 to 0.5 m^2
120 mm to 160 mm	120 mm to 160 mm	0.5 to 1.0 m^2
160 mm to 200 mm	over 160 mm	-
200 mm to 240 mm	-	-
over 240 mm	-	-

The unit price of replacement assumes cutting out the existing insulating material, applying binding compound and inserting fiberglass mesh. Unfortunately, the price system [46] does not include the possibility of choosing different insulation types as in newly constructed structures. To facilitate LCC calculation, items from the price database were adjusted so that they correspond to newly constructed ETICS, i.e., to include the assembly and material separately.

In the event of damage to the ETICS, thin plaster has to be repaired in the same scope as the ETICS. However, the plaster itself can be repaired separately, e.g., if it crumbles away or is worn by pressurised water cleaning or otherwise. The price database determines the scope of the repaired area, either by directly inputting the repaired area or using a percentage of the repaired area. Specific division of thin plaster is shown in Table 6. Types of plaster are identical to the division for newly applied plaster.

Table 6. Division of repairs of thin plaster and types of plaster.

Repair Scope by Area	Repair Scope as a Percentage
under 0.1 m^2	under 10%
0.1 to 0.25 m^2	10 to 30%
0.25 to 0.5 m^2	30 to 50%
0.5 to 1.0 m^2	-
1.0 to 4.0 m^2	-

The scope of repairs (see Tables 5 and 6) of ETICS and thin plaster indicates the size of the damage that needs to be repaired. The scope is indicated either as a percentage of the total area or the actual area to be repaired. For the purposes of the proposed system, the scope is defined as actual area to be repaired per each 10 m^2.

3.5. Calculation of Disposal Costs at the End of the Building's Lifetime

Each building, as well as its individual parts, has a lifetime. The lifetime of the entire building is determined by the lifetime of its load-bearing parts—foundations, walls, ceilings, etc. At the end of its lifetime, the building and its individual functional parts must be disposed of (demolished). The costs of disposal are detailed in the chapter of this paper dealing with the calculation of replacement costs. Calculating this cost is dependent on making the examined period correspond to the entire lifetime of the building. When entering an examined period for the LCC calculation shorter than the lifespan of the entire building, this cost will not be counted in.

3.6. Proposed Structure of the R/M/R Database

The above LCC calculation process increases requirements for the structure of the R/M/R database. The R/M/R database, i.e., its structure for the functional part designated "façade composition—external thermal insulation composite system with thin plaster", must indicate the ETICS and thin plaster separately. The reason for the division is the fact that it is not possible in terms of facility management to monitor all façade layers as a whole; for this functional part alone, this would be 143 possible combinations and, therefore, obtaining a relevant R/M/R information sample from facility management would not be realistic.

In the case of ETICS, the facility management system must record information on the lifetimes of the individual types of ETICS, the scope of repairs per 10 m^2 of area and the frequency of the repairs. The possibility for recording the scope of repairs is based on Table 5. In case the R/M/R database from the facility management system is not available, the designer must input the information manually.

As mentioned above, the lifetime of thin plaster corresponds to the lifetime of ETICS and this is taken into account in the LCC calculation; therefore, it is not necessary for this to be included in the R/M/R database. Thin plaster chiefly requires recording and determination of the frequency of maintenance, i.e., façade washing and re-coating. As with ETICS, it is necessary to monitor, i.e., record the scopes under Table 6, and the frequency of repairs for the individual types of thin plaster (aside from ETICS repairs where plaster repairs are automatically included).

Since the Czech Republic lacks a suitable R/M/R database, the functioning of the system is showcased by manual input of information based on the manufacturers' data or the expected lifetime and repair frequency of the individual ETICS types. The R/M/R database is modelled in three variants (see Figures 3 and 4):

- Variant 1—lifetime and scope and frequency of maintenance entered based on the manufacturer of ETICS [47]. The scope and frequency of repairs corresponds to the data included in [43] (with regard to repairs of ETICS, the proposed system does not offer the possibility of repair in the scope of 20%—this is replaced by the highest possible value of the scope of repairs, i.e., from 0.5 m^2 to 1.0 m^2, and an increased frequency, i.e., 15 years).
- Variant 2 and 3—the R/M/R database is filled in based on the possible assumed development of the lifetime and repairs of the ETICS. In comparison to Variant 1, the lifetime of the individual types of ETICS is adjusted according to the lifetime intervals indicated in [43]. Further adjusted is the frequency of repairs which is based on the expected resistance to mechanical damage on the part of the individual ETICS types.

R/M/R DATABASE FOR EXTERNAL THERMAL INSULATION COMPOSITE SYSTEM (ETICS)	LIFETIME [years]		REPAIRS		
				Frequency [years]	
	Variant 1	Variant 2 and 3	Scope per 10 m²	Variant 1	Variant 2 and 3
EPS 70 façade board	50	30	0.5 to 1.0 m²	15	5
EPS 100 façade board	50	30	0.5 to 1.0 m²	15	10
EPS graphite façade board	50	30	0.5 to 1.0 m²	15	10
polystyrene socle board, façade	50	40	0.5 to 1.0 m²	15	15
polystyrene façade board for thermal insulation of bottom parts of buildings	50	40	0.5 to 1.0 m²	15	15
board with longitudinal mineral fibre	50	60	0.5 to 1.0 m²	15	5
board with perpendicular mineral fibre	50	60	0.5 to 1.0 m²	15	5
foamglass board, no surface treatment	50	60	0.5 to 1.0 m²	15	10
XPS polystyrene board	50	50	0.5 to 1.0 m²	15	15
sandwich insulation board (polystyrene+wool)	50	45	0.5 to 1.0 m²	15	5
wood fibre board	50	40	0.5 to 1.0 m²	15	5

Figure 3. R/M/R database for the LCC calculation of the individual variants (ETICS).

R/M/R DATABASE FOR THIN PLASTER	MAINTENANCE		REPAIRS	
	Washing frequency [years]	Re-coating frequency [years]	Scope per 10 m2 or per % of total area	Frequency [years]
mineral granular plaster	3	9	10 to 30 %	20
mineral grooved plaster	3	9	10 to 30 %	20
acrylic granular plaster	3	9	10 to 30 %	20
acrylic grooved plaster	3	9	10 to 30 %	20
acrylic mosaic plaster	3	9	10 to 30 %	20
silicate granular plaster	4	12	10 to 30 %	20
silicate grooved plaster	4	12	10 to 30 %	20
silicone granular plaster	4	12	10 to 30 %	20
silicone grooved plaster	4	12	10 to 30 %	20
silicone hydrophilic granular plaster	4	12	10 to 30 %	20
silicone hydrophilic grooved plaster	4	12	10 to 30 %	20
silicate-silicone granular plaster	5	15	10 to 30 %	20
silicate-silicone grooved plaster	5	15	10 to 30 %	20

Figure 4. R/M/R database for the LCC calculation of the individual variants (thin plaster).

3.7. LCC Calculation and the Selection of the Best Variant

Previous chapters described the process of calculation and defining costs arising over the course of a building's life cycle depending on information transferred from the BIM model and R/M/R database. An important input for the calculation consists in setting parameters for the actual LCC calculation from the point of view of the investor—the examined period and the discount rate. Individual costs of repairs, replacement, maintenance and disposal of the given FP are subsequently adequately discounted so that the investor obtains a net current value of the investment for the examined period.

The LCC calculation system simultaneously identifies alternatives to the input ETICS layers. The alternatives to the ETICS must meet the condition of identical or better thermal conductivity in order to preserve the key parameter of designing the ETICS; in thin plaster, the alternatives must be of identical or higher thickness to maintain the building's aesthetic properties. Alternatives are subjected to the same LCC calculation as the designed solution. Subsequently, the LCC calculation system ranks the best variants of the ETICS and thin plaster in terms of acquisition costs and discounted LCC. The designer can then choose the best solution for his design.

The results of the calculation for all three modelled variants of input parameters and R/M/R database are shown in Figures 5–7. The first line includes the initial design variant, with the alternatives ranked below according to the lowest LCC value.

Variant 1	total LCC	thermal resistance (ETICS) [m²·K/W]	difference from the designed solution
EPS 70 façade board (120 mm) + mineral granular plaster (2 mm)	8,282.78 €	3.08	
EPS 70 façade board (120 mm) + silicate-silicone grooved plaster (2 mm)	7,298.39 €	3.08	-12%
EPS graphite façade board (100 mm) + silicate-silicone grooved plaster (2 mm)	7,455.22 €	3.13	-10%
EPS 100 façade board (120 mm) + silicate-silicone grooved plaster (2 mm)	7,491.10 €	3.24	-10%
polystyrene façade socle board (120 mm) + silicate-silicone grooved plaster (2 mm)	8,024.78 €	3.43	-3%
sandwich insulation board (polystyrene+wool) (120 mm) + silicate-silicone grooved plaster (2 mm)	8,417.64 €	3.64	2%
wood fibre board (120 mm) + silicate-silicone grooved plaster (2 mm)	8,576.81 €	3.33	4%
polystyrene façade board for thermal insulation of bottom parts of buildings (120 mm) + silicate-silicone grooved plaster (2 mm)	8,913.44 €	3.53	8%
XPS polystyrene board (120 mm) + silicate-silicone grooved plaster (2 mm)	9,514.89 €	3.64	15%
board with longitudinal mineral fibre (120 mm) + silicate-silicone grooved plaster (2 mm)	9,815.55 €	3.33	19%
board with perpendicular mineral fibre (140 mm) + silicate-silicone grooved plaster (2 mm)	9,870.73 €	3.41	19%
foamglass board, no surface treatment (120 mm) + silicate-silicone grooved plaster (2 mm)	14,017.24 €	3.16	69%

Figure 5. Output of the LCC calculation for Variant 1.

Variant 2	total LCC	thermal resistance (ETICS) [m²·K/W]	difference from the designed solution
EPS 70 façade board (120 mm) + mineral granular plaster (2 mm)	8,775.01 €	3.08	
polystyrene façade socle board (120 mm) + silicate-silicone grooved plaster (2 mm)	6,564.22 €	3.43	-25%
polystyrene façade board for thermal insulation of bottom parts of buildings (120 mm) + silicate-silicone grooved plaster (2 mm)	7,354.39 €	3.53	-16%
EPS graphite façade board (100 mm) + silicate-silicone grooved plaster (2 mm)	7,472.85 €	3.13	-15%
EPS 100 façade board (120 mm) + silicate-silicone grooved plaster (2 mm)	7,515.39 €	3.24	-14%
XPS polystyrene board (120 mm) + silicate-silicone grooved plaster (2 mm)	7,889.91 €	3.64	-10%
EPS 70 façade board (120 mm) + silicate-silicone grooved plaster (2 mm)	8,037.33 €	3.08	-8%
sandwich insulation board (polystyrene+wool) (120 mm) + silicate-silicone grooved plaster (2 mm)	8,073.72 €	3.64	-8%
wood fibre board (120 mm) + silicate-silicone grooved plaster (2 mm)	8,210.16 €	3.33	-6%
board with longitudinal mineral fibre (120 mm) + silicate-silicone grooved plaster (2 mm)	9,512.23 €	3.33	8%
board with perpendicular mineral fibre (140 mm) + silicate-silicone grooved plaster (2 mm)	9,626.03 €	3.41	10%
foamglass board, no surface treatment (120 mm) + silicate-silicone grooved plaster (2 mm)	12,418.65 €	3.16	42%

Figure 6. Output of the LCC calculation for Variant 2.

Variant 3	total LCC	thermal resistance (ETICS) [m²·K/W]	difference from the designed solution
EPS 70 façade board (120 mm) + mineral granular plaster (2 mm)	10,062.17 €	3.08	
EPS graphite façade board (100 mm) + silicate-silicone grooved plaster (2 mm)	8,288.88 €	3.13	-18%
polystyrene façade socle board (120 mm) + silicate-silicone grooved plaster (2 mm)	8,324.67 €	3.43	-17%
EPS 100 façade board (120 mm) + silicate-silicone grooved plaster (2 mm)	8,335.47 €	3.24	-17%
EPS 70 façade board (120 mm) + silicate-silicone grooved plaster (2 mm)	8,981.23 €	3.08	-11%
polystyrene façade board for thermal insulation of bottom parts of buildings (120 mm) + silicate-silicone grooved plaster (2 mm)	9,269.89 €	3.53	-8%
XPS polystyrene board (120 mm) + silicate-silicone grooved plaster (2 mm)	9,514.89 €	3.64	-5%
sandwich insulation board (polystyrene+wool) (120 mm) + silicate-silicone grooved plaster (2 mm)	9,925.14 €	3.64	-1%
wood fibre board (120 mm) + silicate-silicone grooved plaster (2 mm)	10,298.31 €	3.33	2%
board with longitudinal mineral fibre (120 mm) + silicate-silicone grooved plaster (2 mm)	11,257.97 €	3.33	12%
board with perpendicular mineral fibre (140 mm) + silicate-silicone grooved plaster (2 mm)	11,406.44 €	3.41	13%
foamglass board, no surface treatment (120 mm) + silicate-silicone grooved plaster (2 mm)	14,244.52 €	3.16	42%

Figure 7. Output of the LCC calculation for Variant 3.

Once the designer chooses the best solution, it is possible to transfer into the BIM model the information necessary for construction—acquisition price, assembly time—as well as information for the future operational stage of the building's life cycle—frequencies and costs of R/M/R and the disposal costs. The LCC value itself depends on the calculation parameters (the length of the examined period and the discount rate) and serves essentially as an evaluation criterion for choosing the best solution. An example of transferring selected parameters into the BIM model is shown in Figure 8.

BIM parameters	Variant 1	Variant 2	Variant 3
ETICS type:	EPS 70 façade board	polystyrene façade socle board	EPS graphite façade board
ETICS thickness [mm]:	120	120	100
Use of thermally insulating plugs [yes/no]:	yes	yes	yes
Use of dispersion (organic) reinforced plaster compound [yes/no]:	no	no	no
Type of thin plaster:	silicate-silicone grooved plaster	silicate-silicone grooved plaster	silicate-silicone grooved plaster
Plaster thickness [mm]:	2	2	2
ETICS characteristics:	no special characteristics	socle area	no special characteristics
Thermal resistance:	3.08 $m^2 \cdot$K/W	3.43 $m^2 \cdot$K/W	3.13 $m^2 \cdot$K/W
Acquisition price:	115,744.43 €	131,206.21 €	118,928.98 €
Assembly time of façade layers:	160.83 standard hours	160.9 standard hours	160.81 standard hours
Functional part lifetime [years]:	50	40	30
Cost of replacement of the FP:	4,928.76 €	5,561.08 €	5,046.08 €
Cost of 1 repair of the ETICS:	466.81 €	528.00 €	479.05 €
Frequency of repairs of the ETICS [years]:	20	15	10
Costs of 1 repair of thin plaster:	62.63 €	62.63 €	62.63 €
Frequency of repairs of thin plaster [years]:	20	20	20
Cost of 1 washing of thin plaster:	218.36 €	218.36 €	218.36 €
Frequency of washing of thin plaster [years]:	5	5	5
Cost of 1 re-coating of the façade:	1,256.58 €	1,256.58 €	1,256.58 €
Frequency of re-coating of the façade [years]:	15	15	15
Disposal (demolition) costs:	660.40 €	688.47 €	653.38 €

Figure 8. A showcase of information output transferrable to the BIM model.

4. Discussion

Section 3 showcased the proposed information exchange system and calculation of a building's LCC on the example of the functional part designated "façade composition". Multiple parameters and types of information enter the LCC calculation (see Figure 1) and influence the resulting value. As a case study, an LCC calculation was performed for an external thermal insulation composite system (ETICS) with thin plaster in three variants in terms of parameters and characteristics. The variants included different inputs of the length of the examined period (100 years in the first and third variants and 30 years in the second variant) and different input of the R/M/R database. The R/M/R database differed in the individual variants in terms of the entered lifetime and frequency of repairs of the ETICS, where in the first case, all ETICS types had the same lifetime and frequency of repairs, while different values were modelled in the second and third cases. The proposed design of façade layers included ETICS EPS 70 façade board (120 mm thick) with mineral granular plaster (2 mm thick). The system then calculated a discounted LCC design solution for the individual variants, where the result is always indicated in the first line of the resulting table (see Figure 5). The system then evaluated the other best permissible ETICS variants based on the requirement that the ETICS have the same or better thermal resistance, and thin plaster have the same or greater thickness.

The result for the ETICS type is clear in the first variant. Since all ETICS types have the same lifetime and repair frequency, the best type is also the cheapest solution (EPS 70 F), despite the fact that, e.g., the EPS graphite façade board has a higher thermal resistance (3.13 compared to 3.08 $m^2 \cdot$K/W) while being thinner (100 mm compared to 120 mm). The differences between the best three variants of the ETICS are very small; the difference of the overall LCC is approx. 2 percentage points. The choice of thin plaster in the first variant is influenced only by the frequency of maintenance, where the silicate-silicone grooved plaster is the best option in the long term, where over the course of 100-year lifetime it will have to be washed with water 12 times and recoated 6 times. This is in contrast to the originally designed mineral plaster, which would have to be washed 22 times and recoated 10 times.

In the second variant, the input parameters and modified R/M/R database yielded the polystyrene socle façade board as the best ETICS type. This result was achieved even though the acquisition price including plaster is approx. 20% higher than the originally designed solution (Figure 5), and the thermal resistance is more than 10% higher. The lifetime of the originally designed solution was set to correspond to the examined period, i.e., 30 years, so it will have to be replaced once. By contrast, the most favourable polystyrene socle façade board has a set lifetime longer than the examined period and additionally, in contrast to other ETICS types, it has a lower set frequency of repairs as it needs to be repairs twice during the examined period. The originally designed solution would have to be repaired 5 times during the examined period.

In the third variant, where the R/M/R was identical to Variant 2, the best solution was EPS graphite façade board (100 mm thick). Considering; however, that the best solution in Variant 2 (polystyrene socle façade board) where the examined period was shorter, has almost the same resulting LCC price, despite the fact that the best solution in Variant 3 (EPS graphite façade board) will have to be replaced 3 times during the examined period, which is one replacement more than in the case of the best solution in Variant 2. The highest lifespan—60 years—was set for ETICS featuring mineral wool and foamglass. While this system would only have to be replaced once during the building lifetime, it ranked near the bottom in the comparison of the total LCC. This was because of the higher maintenance frequency of mineral fibre boards. Foamglass is also affected by its very high acquisition price.

It is clear from the overview of the results of the individual variants that the R/M/R database and the length of the examined period play a very important role. The examined period and the R/M/R database affect the results in such a way that it is impossible to determine which façade layer arrangement is generally the best, which is documented by the variable LCC efficiency of the individual construction-material solutions in the three modelled variant solutions.

5. Conclusions

This paper presented a methodology for building LCC estimation that enables investors to identify the optimum material solution for their buildings on the level of individual functional parts. From a theoretical perspective, this paper contributes to the current body of knowledge with the proposed LCC system, which takes into consideration various data inputs and interconnects the construction cost estimation database, facility management database and investors' requirements into a comprehensive solution. Regarding managerial implications, the proposed LCC estimation system demonstrates the absence of a generally applicable optimum material solution for ETICS. Different investor requirements as well as the unique circumstances of each building and its user are a fact that underlines the need to apply comprehensive approaches to finding the best solutions. Differences between individual buildings lead to the fact that the results achieved (e.g., in the context of LCC calculations) will always be unique and, therefore, no ETICS or thin plaster type should be favoured in advance.

There are three important research limitations that should be mentioned. Firstly, the model's division in terms of materials and structures depends on the available price database, where the proposed system presented in this paper is based on databases used in the Czech Republic. Building structures and materials not indicated in the relevant price database cannot be assigned with costs, which means no LCC value can be determined. Nevertheless, if the methodology is applied generally, it could be used—with adequate modifications—also in other regions and with different price databases. Secondly, the calculation is unique for each functional part, because it is based on its own cost estimation principles and the LCC calculation thus has to be modified for each individual functional part separately. Thirdly, this system is limited only to those life cycle costs that are related to the building's structures and materials and omits future energy costs in the operational stage (heating, air conditioning, etc.).

Several future research directions can be outlined. It should be possible to follow up on the proposed system and the information necessary for LCA calculation in order to be able to select the best material and structural solution based on its carbon footprint as well, i.e., certainly in combination of LCC and LCA. This step would dramatically increase the potential for using the proposed system in the context of adhering to the principles of sustainable construction.

Another step could consist of incorporating utilities and energy consumption (e.g., heating, water and electricity) based on information obtained from already operated buildings and BIM model information. This would enable a more comprehensive evaluation of buildings in terms of their LCC. Finally, the information on the construction time (see Figure 5) has the potential for a broader use in creating construction time schedules or in identifying the optimum solution for a building that takes into account the construction time as one of the evaluation criteria (which can be significant in commercial development projects).

Author Contributions: Conceptualization, V.B. and T.H.; methodology, V.B. and T.H.; software, V.B.; calculation V.B. and T.H.; discussion V.B. and T.H.; resources, V.B. and T.H.; writing—original draft preparation, V.B. and T.H.; writing—review and editing, V.B. and T.H.; visualization, V.B.; supervision, T.H.; project administration, V.B.; funding acquisition, V.B.

Funding: This research was funded by Brno University of Technology, project Management of Enterprise and Investment Projects in Construction, grant number FAST-J-19-6052 and the APC was funded by FAST-J-19-6052.

Conflicts of Interest: The authors declare no conflict of interest.

References

1. Burcar Dunovic, I.; Radujkovic, M.; Vukomanovic, M. Internal and external risk based assessment and evaluation for the large infrastructure projects. *J. Civ. Eng. Manag.* **2016**, *22*, 673–682. [CrossRef]
2. Korytarova, J.; Hromadka, V. Building life cycle economic impacts. In Proceedings of the International Conference on Management and Service Science, Wuhan, China, 24–26 October 2010.
3. Zabielski, J.; Zabielska, I. Life Cycle of a Building (LCC) in the Investment Process-Case Study. In Proceedings of the Baltic Geodetic Congress, BGC-Geomatics, Olsztyn, Poland, 21–23 June 2018; pp. 254–259.
4. Fantozzi, F.; Gargari, C.; Rovai, M.; Salvadori, G. Energy upgrading of residential building stock: Use of life cycle cost analysis to assess interventions on social housing in Italy. *Sustainability* **2019**, *11*, 1452. [CrossRef]
5. Kovacic, I.; Zoller, V. Building life cycle optimization tools for early design phases. *Energy* **2015**, *92*, 409–419. [CrossRef]
6. REEB Consortium. ICT Supported Energy ICT Supported Energy Efficiency in Construction. 2010. Available online: https://ec.europa.eu/information_society/activities/sustainable_growth/docs/sb_publications/reeb_ee_construction.pdf (accessed on 24 April 2019).
7. Wen, Y.K.; Kang, Y.J. Minimum building life-cycle cost design criteria. II: Applications. *J. Struct. Eng.* **2001**, *127*, 338–346. [CrossRef]
8. Stephan, A.; Stephan, L. Life cycle energy and cost analysis of embodied, operational and user-transport energy reduction measures for residential buildings. *Appl. Energy* **2016**, *161*, 445–464. [CrossRef]
9. Lazzarin, R.M.; Busato, F.; Castelloti, F. Life cycle assessment and life cycle cost of buildings' insulation materials in Italy. *Int. J. Low Carbon Technol.* **2008**, *3*, 44–58. [CrossRef]
10. Han, G.; Srebric, J.; Enache-Pommer, E. Variability of optimal solutions for building components based on comprehensive life cycle cost analysis. *Energy Build.* **2014**, *79*, 223–231. [CrossRef]
11. Anuradha, I.G.N.; Perera, B.A.K.S.; Mallawarachchi, H. Embodied carbon and cost analysis to identify the most appropriate wall materials for buildings: Whole life cycle approach. In Proceedings of the MERCon 2018 4th International Multidisciplinary Moratuwa Engineering Research Conference, Moratuwa, Sri Lanka, 30 May–1 June 2018; pp. 43–48.
12. Juan, Y.; Hsing, N. BIM-based approach to simulate building adaptive performance and life cycle costs for an open building design. *Appl. Sci.* **2017**, *7*, 837. [CrossRef]
13. Robati, M.; McCarthy, T.J.; Kokogiannakis, G. Integrated life cycle cost method for sustainable structural design by focusing on a benchmark office building in Australia. *Energy Build.* **2018**, *166*, 525–537. [CrossRef]
14. Salvado, F.; Almeida, N.M.; Vale e Azevedo, A. Toward improved LCC-informed decisions in building management. *Built Environ. Proj. Asset Manag.* **2018**, *8*, 114–133. [CrossRef]
15. Che-Ghani, N.Z.; Myeda, N.E.; Ali, A.S. Operations and maintenance cost for stratified buildings: A critical review. *Proc. MATEC Web Conf.* **2016**, *66*, 41. [CrossRef]
16. Mong, S.G.; Mohamed, S.F.; Misnan, M.S. Key strategies to overcome cost overruns issues in building maintenance management. *Int. J. Eng. Technol.* **2018**, *7*, 269–273. [CrossRef]
17. Haugbølle, K.; Raffnsøe, L.M. Rethinking life cycle cost drivers for sustainable office buildings in Denmark. *Facilities* **2019**. [CrossRef]
18. Farahani, A.; Wallbaum, H.; Dalenbäck, J. Optimized maintenance and renovation scheduling in multifamily buildings—A systematic approach based on condition state and life cycle cost of building components. *Constr. Manag. Econ.* **2019**, *37*, 139–155. [CrossRef]

19. Buyle, M.; Audenaert, A.; Braet, J.; Debacker, W. Towards a more sustainable building stock: Optimizing a flemish dwelling using a life cycle approach. *Buildings* **2015**, *5*, 424–448. [CrossRef]

20. Liu, L.; Rohdin, P.; Moshfegh, B. LCC assessments and environmental impacts on the energy renovation of a multi-family building from the 1890s. *Energy Build.* **2016**, *133*, 823–833. [CrossRef]

21. Silvestre, J.D.; Castelo, A.M.P.; Silva, J.J.B.C.; Brito, J.M.C.L.; Pinheiro, M.D. Retrofitting a building's envelope: Sustainability performance of ETICS with ICB or EPS. *Appl. Sci.* **2019**, *9*, 1285. [CrossRef]

22. Bottinelli, A.; Louf, R.; Gherardi, M. Balancing building and maintenance costs in growing transport networks. *Phys. Rev. E* **2017**, *96*, 032316. [CrossRef] [PubMed]

23. Milić, V.; Ekelöw, K.; Moshfegh, B. On the performance of LCC optimization software OPERA-MILP by comparison with building energy simulation software IDA ICE. *Build. Environ.* **2018**, *128*, 305–319. [CrossRef]

24. Konstantinidou, C.A.; Lang, W.; Papadopoulos, A.M.; Santamouris, M. Life cycle and life cycle cost implications of integrated phase change materials in office buildings. *Int. J. Energy Res.* **2019**, *43*, 150–166. [CrossRef]

25. Bandara, R.M.P.S.; Fernando, W.C.D.K.; Attalage, R.A. Optimizing life cycle cost of buildings through simulation-based optimization: A case study. In Proceedings of the Annual International Conference on Architecture and Civil Engineering, Singapore, 14–15 May 2018; pp. 363–370.

26. Kim, J.; Kim, T.; Yu, Y.; Son, K. Development of a maintenance and repair cost estimation model for educational buildings using regression analysis. *J. Asian Archit. Build. Eng.* **2018**, *17*, 307–312. [CrossRef]

27. Au-Yong, C.P.; Ali, A.S.; Ahmad, F. Enhancing building maintenance cost performance with proper management of spare parts. *J. Qual. Maint. Eng.* **2016**, *22*, 51–61. [CrossRef]

28. Balasbaneh, A.T.; Bin Marsono, A.K.; Gohari, A. Sustainable materials selection based on flood damage assessment for a building using LCA and LCC. *J. Clean. Prod.* **2019**, *222*, 844–855. [CrossRef]

29. Cheng, M.; Wei, H.; Wu, Y.; Chen, H.; Wu, C. Optimization of life-cycle cost of retrofitting school buildings under seismic risk using evolutionary support vector machine. *Technol. Econ. Dev. Econ.* **2018**, *24*, 812–824. [CrossRef]

30. Zeleňáková, M.; Gaňová, L.; Purcz, P.; Horský, M.; Satrapa, L. Determination of the potential economic flood damages in Medzev, Slovakia. *J. Flood Risk Manag.* **2018**, *11*, S1090–S1099. [CrossRef]

31. Blengini, G.A.; Di Carlo, T. The changing role of life cycle phases, subsystems and materials in the LCA of low energy buildings. *Energy Build.* **2009**, *42*, 869–880. [CrossRef]

32. Brown, M.T.; Buranakarn, V. Emergy indices and ratios for sustainable material cycles and recycle options. *Resour. Conserv. Recycl.* **2003**, *38*, 1–22. [CrossRef]

33. Fregonara, E.; Giordano, R.; Ferrando, D.G.; Pattono, S. Economic-environmental indicators to support investment decisions: A focus on the buildings' end-of-life stage. *Buildings* **2017**, *7*, 65. [CrossRef]

34. European Union. *Directive 2014/24/EU of the European Parliament and of the Council of 26 February 2014 on Public Procurement and Repealing Directive 2004/18/EC*; European Union: Brussels, Belgium, 2014.

35. Santos, R.; Costa, A.A.; Silvestre, J.D.; Pyl, L. Integration of LCA and LCC analysis within a BIM-based environment. *Autom. Constr.* **2019**, *103*, 127–149. [CrossRef]

36. Saridaki, M.; Psarra, M.; Haugbølle, K. Implementing life-cycle costing: Data integration between design models and cost calculations. *J. Inf. Technol. Constr.* **2019**, *24*, 14–32.

37. Spagnolo, S.L. Information integration for asset and maintenance management. In *Integrating Information in Build Environments: From Concept to Practice*; Sanchez, X.A., Hampson, D.K., London, G., Eds.; Routledge: London, UK, 2018; pp. 133–149.

38. ISO. *Buildings and Constructed Assets-Service Life Planning*; ISO 15686-5:2017; ISO: Geneva, Switzerland, 2017.

39. Bromilow, F.J.; Pawsey, M.R. Life cycle cost of university buildings. *Constr. Manag. Econ.* **1987**, *5*, S3–S22. [CrossRef]

40. Sobanjo, J.O. *Facility Life-Cycle Cost Analysis on Fuzzy Sets Theory. Durability of Building Materials and Components 8*; Institute for Research in Construction: Ottawa, ON, Canada, 1999.

41. Kupilík, V. *Závady a Životnost Staveb (Defects and Lifetime of Buildings)*; Grada: Praha, Czech Republic, 1999; p. 288.

42. Korytárová, J.; Papežíková, P. Assessment of Large-Scale Projects Based on CBA. *Procedia Comput. Sci.* **2015**, *64*, 736–743. [CrossRef]

43. Marková, L. *Náklady Životního Cyklu Stavby: Náklady Investora, Celospolečenské Dopady (Building Life Cycle Costs: Investor Costs, Societal Implications)*; Akademické nakladatelství CERM: Brno, Czech Republic, 2011; p. 125.

44. Biolek, V.; Hanak, T.; Marovic, I. Data flow in Relation to Life-Cycle Costing of Construction Projects in the Czech Republic. *IOP Conf. Ser. Mater. Sci. Eng.* **2017**, *245*, 072032. [CrossRef]
45. Biolek, V.; Domansky, V.; Vyskala, M. Interconnection of construction-economic systems with BIM in the Czech environment. *IOP Conf. Ser. Earth Environ. Sci.* **2019**, *222*, 012022. [CrossRef]
46. ÚRS. *Katalog Stavebních Konstrukcí a Prací ÚRS, Cenová Úroveň 2019/1 (ÚRS Catalogue of Building Structures and Works; Price Level as of 2019/1)*; URS: San Francisco, CA, USA, 2019.
47. Isover Saint-Gobaint. *Katalog výrobků Isover (Isover Product Catalogue)*; Isover Saint-Gobaint: Ludwigshafen, Germany, 2019.
48. Bachl. *Fyzikální Vlastnosti Extrudovaného Polystyrenu XPS (Physical Properties of XPS Extruded Polystyrene)*; Karl Bachl GmbH & Co. KG: Röhrnbach, Germany, 2015.
49. Foamglas. *Technický List FOAMGLAS® W+F (FOAMGLAS® W+F Technical Sheet)*; Foamglas, Institut Bauen und Umwelt e.V.: Berlin, Germany, 2008.
50. STEICO. *Technický list STEICOFlex 036 (STEICOFlex 036 Technical Sheet)*; STEICO: Feldkirchen, Germany; Berlin, Germany, 2016.

Article

The Influence of 30 Years Outdoor Weathering on the Durability of Hydrophobic Agents Applied on Obernkirchener Sandstones

Jeanette Orlowsky *, Franziska Braun and Melanie Groh

Department of Building Materials, Technische Universität Dortmund, August-Schmidt-Str. 8, 44227 Dortmund, Germany; franziska.braun@tu-dortmund.de (F.B.); Melanie.groh@tu-dortmund.de (M.G.)
* Correspondence: Jeanette.orlowsky@tu-dortmund.de; Tel.: +49-231-755-4840

Received: 29 November 2019; Accepted: 13 January 2020; Published: 20 January 2020

Abstract: The durability of eleven different water repellents applied on one sandstone type was studied after a long-term weathering at seven different locations in Germany. By measuring colour changes, it could be shown that the formation of black crusts, the deposition of particles and biogenic growth caused a gradual darkening as well as significant changes in total colour over time. Additionally, the water absorption behaviour was investigated with two different methods: applying a low pressure using the pipe method and capillary water absorption measurements from a wet underlay. Afterwards, the test results were analysed with four different evaluation methods: calculation of the protection degree from pipe method and capillary water absorption, determination of the velocity of water uptake during capillary water absorption and calculation of the damaged depth of the stone surface using single-sided NMR technique. The growing damaged depth leads to an increase of the water uptake velocity and to a decrease of the protection degree of the applied hydrophobing agents. Three protective agents based on isobutyltrimethoxysilane showed already after two years of outdoor weathering a clear loss of performance, which significantly increased after 30 years of exposure.

Keywords: conservation; natural stone; long-term weathered; water repellents; durability; single-sided NMR

---

1. Introduction

Most of our historical stone buildings and monuments are made of sedimentary rocks. Their mineral composition and physical structure, in combination with varying pore space characteristics, limit their durability in terms of weathering influences. In order to reduce the susceptibility to weathering, water repellents based on organosilicon compounds can be used. The long-term impact of these stone treatments is a controversial issue [1–3]: on the one hand, the capillary water ingress is significantly reduced, which prevents stone deterioration processes based on the influence of water. On the other hand, the possible decrease of water vapour diffusion through the pore system can lead to hygric expansion inside the stone, which causes in combination with freeze-thaw cycles spalling of the treated stone area. Furthermore, the long-term behaviour and efficiency of these stone treatments depends on diverse aspects and is still a scientific issue [4–7]. The type of stone (e.g., sandstone, limestone, igneous or metamorphic rocks) as well as the chemical composition of the used conservative (e.g., acrylic polymers or organosilicon compounds) and also its application have a significant influence on the success of a conservation action, e.g., an incorrect performance or a poor adhesion of the water repellent with the stone substrate can lead to severe damage over time [8–10].

Finding answers concerning the above-mentioned issues, the Zollern-Institute (Deutsches Bergbau Museum, Bochum) performed between 1986 and 2017 a field study in Germany, where various types of natural stones were exposed to different weathering conditions. In addition, reference samples were also deposited in an archive over the entire period. For each type of stone and each location, 11 specimens were applicated with different hydrophobing agents based on organosilicon compounds, while 2 specimens stayed untreated [11,12].

This study focuses on the evaluation of these long-term weathered samples treated with 11 different water repellents, as shown on one sandstone type. The aim of this research is the performance analysis of the applied hydrophobic agents. The influence of the composition of the agents, leading among other things to various penetration depths, is determined with three different test methods. Based on the work done in [13], a damaged depth inside the weathered stone is calculated for the first time. The term damaged depth describes the extent of damage in the uppermost area of the natural stone surface. A comparison between damaged depth of the stone surfaces and penetration depth of the agents allows statements concerning the performance and explains partly unsolved results concerning the here evaluated protection degrees.

2. Experimental Matrix

Investigations were carried out on a frequently used building stone in Germany, namely the Obernkirchener Sandstone (OKS). The mineral composition as well as the geotechnical properties of this sandstone type are listed in Table 1.

Table 1. Petrographic and petrophysical properties of Obernkirchener Sandstone.

Stone Type	Obernkirchener Sandstone OKS
colour	beige to yellowish-grey, 5 Y 8/1–5 Y 7/2
mineral content [14]	quartz 81%, rock fragments 17%, muscovite (subordinated)
matrix	quartzitic, kaolinitic
classification	fine-grained quartzitic sandstone
total porosity [%]	20
bulk density [g/cm^3]	2.16
apparent density [g/cm^3]	2.71
average pore radius [µm]	3.4

The samples have a triangle-prismatic-shaped structure with a length of 300 mm. Figure 1 shows the geometry of the samples as well as the further preparation for the investigations. To assess the weathering effects and the degradation of the hydrophobic agents on a time-related basis, specimens were taken from these stone samples after an exposure time of 2, 24 or 30 years. At each time step, a 30 mm thick slice was cut out from the lateral surfaces (marked green) of each sample (Figure 1). All investigations described in this study were carried out on these segments at the TU Dortmund. The triangle-shaped slices, which were cut out after 2 years of natural weathering, were stored in an archive until the examinations were carried out in the years 2018 and 2019.

In 1986, the dry samples were applicated with 11 different water repellents based on organosilicon compounds. In order to generate a uniformly wetted surface everywhere, the samples were completely immersed in a container filled with the respective hydrophobic agents and treated by flood-process with an application time of one minute. Table 2 provides an overview of the chemical composition and active ingredient content of the used products as well as characteristic values of OKS treated with these agents. The penetration depth of the hydrophobic agents was measured with a calliper on the triangle-shaped slices, which were moistened for this purpose [1,2,13]. The measurement was carried out on the cut side. The mean values and the standard deviations in Table 2 were calculated in each case from 42 measuring points.

Figure 1. Left: The geometry of the samples and the sampling of the sample slices after 2 and 30 years, here marked green (years = a). Right: Outdoor weathering of the samples in Nuremberg after an exposure time of 30 years.

Table 2. Overview of the applied protective agents and characteristic values of Obernkirchener Sandstone treated with these agents (±standard deviation).

No.	Protective Agent and Content [M.-%]	Penetration Depth [mm]	Water Absorption after 1 h [kg/m^2]	Increase of Water Vapour Diffusion Resistance Value [%]
0	untreated Obernkirchener Sandstone, 2a indoor (reference)	-	1.38	
1	34% propyl-/ 5% octyltrimethoxysilane	7.0 ± 1.8	0.03	16
2	35% isobutyltrimethoxysilane	3.3 ± 3.1	0.04	5
3	20% isobutyltrimethoxysilane + 20% tetraethoxysilanehydrolysat	3.3 ± 2.1	0.03	15
4	20% isobutyltrimethoxysilane + 20% tetraethoxysilane	2.6 ± 2.5	0.04	12
5	low-molecular methylethoxysiloxane + tetraethoxysilane (Σ75%)	6.7 ± 1.5	0.03	11
6	7.5% low-molecular methylethoxysiloxane	4.0 ± 0.6	0.02	18
7	6.7% oligomer methylethoxysiloxane	3.3 ± 0.5	0.03	17
8	5% methyl-/isooctyl silicone resin	3.4 ± 1.4		14
9	6.7% oligomer methyl-/isooctylmethoxysiloxane	2.9 ± 1.1	0.03	13
10	oligomer methyl-/isooctylmethoxy-siloxane + tetraethoxysilane (Σ8.3%)	3.6 ± 0.8	0.03	8
11	8% polymeric methylmethoxy-siloxane (silicone resin)	3.4 ± 0.6	0.03	17

The hydrophobic agents with the identification numbers 1, 2, 3 and 4 are based on silane whereby agents 1 and 2 include the highest silane contents and agents 3 and 4 additionally contain silicic acid ester (tetraethoxysilane). The impregnation depth of the water repellents containing isobutyltrimethoxysilane (agents 2, 3 and 4) is more than twice lower compared to agent 1 and has a high standard deviation. These high standard deviations suggest that the hydrophobic agents, based on silane, penetrate uneven into the pore system of the sandstones. Furthermore, these materials showed no clear line between hydrophilic and hydrophobic zone [13]. The other materials, especially based on siloxane, showed after wetting the surface a clear boundary between dry (hydrophobic) and wet (hydrophilic). Figure 2 visualizes exemplary two samples with high penetration depth and compares the boundaries of agent No. 1 and 5. Agents 5, 6, 7, 9 and 10 are based on siloxane and allow the comparison of low-molecular and oligomer methylethoxysiloxane as well as methylethoxy- and methyl-/isooctylmethoxysiloxane. The third group is based on silicone resin (agents 8 and 11).

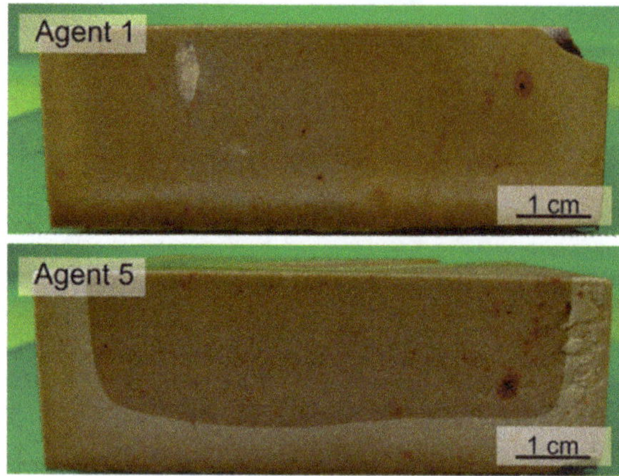

Figure 2. Visual penetration depth of agent No. and 5 after wetting the samples.

The capillary water absorption measured according to DIN EN 15801 [15] demonstrates the effectiveness of all hydrophobing agents applied on OKS. The increase in water vapour diffusion resistance was determined in relation to the untreated samples. For this reference value, the water vapour diffusion resistance of the two untreated samples was averaged from each location (mean value of 24 samples and standard deviation: 37.8 ± 4). The values were measured by wet cup test according to DIN EN 15803 [16,17]. Due to application of the hydrophobic agents, the water vapour diffusion resistance increased up to 18% but stayed still under the threshold value of 50% required by [2]. Obviously, the addition of silicic acid ester has no influence on the water vapour diffusion resistance.

All samples of the triangle-prismatic-shaped Obernkirchener Sandstones were exposed at seven different locations in Germany in 1986/1987. Three locations were situated in North Rhine-Westphalia, Dortmund, Duisburg and Eifel, and three in Bavaria, Nuremberg, Munich and Kempten. Table 3 lists a detailed description of the exposure sites and their surrounding environments. The seventh site was indoor and can be used as reference.

Table 3. Description of the exposure sites and their setting.

Location	Exposure Site	Setting
Dortmund	located in the west of the city, on an old colliery site (former heavy industrial site) with traffic around, exposure stations situated under trees	residential area/industrial area
Duisburg	located in the north of the city, exposure stations situated on the green area of a schoolyard under trees	residential area/heavy industrial area
Eifel	located on agricultural land, exposure stations situated on the edge of a forest	urban area and rural space in the countryside
Nuremberg	located in the north-west of the city, near a water treatment plant, on a busy street behind bushes	urban area
Munich	located in the south-west of downtown with a high amount of traffic around the site, exposure stations situated partly under trees	residential area/business area
Kempten	exposure stations situated on the roof of a tree nursery	residential area

For the above-stated weathering locations, their mean climatic data is shown in Table 4. These are averages of the last 10 years of exposure (Dortmund, Duisburg, Eifel: 2000–2009; Nuremberg, Munich, Kempten: 2007–2016) from measuring stations near the locations and with similar environmental conditions. An extensive evaluation of the climatic data for temperature, relative humidity and precipitation on the basis of hourly values has been carried out, in order to obtain differentiated information with regard to the duration of certain temperature levels, the number of freeze-thaw changes, annual precipitation times and precipitation frequencies. In addition, the averaged pollutant values for nitrogen- and sulphur-dioxide are given for this period.

In the considered period, the temperatures below the freezing point were more present at the Bavarian locations, especially in Kempten. In addition, the number of hours with a relative humidity above 80% was higher for the locations Eifel and Kempten compared to the other sites. A high precipitation rate was recorded in Kempten, while it did not rain many times. The pollution data showed a low NO_2 rate in Eifel, whose exposure site is situated in an agricultural landscape, while in the urban location Munich the values are significant higher. Concerning SO_2, the heavy industrial setting in Duisburg had the highest pollution rate.

Table 4. Climatic data for the weathering locations, mean values over the last 10 years of exposure (a = year, h = hours).

Title	Exposure Site above Mean Sea Level	Temperature		Freeze-Thaw-Changes	Relative Humidity		Precipitation		Pollution	
		<0 °C	>25 °C		<50%	>80%			NO_2	SO_2
	[m a.s.l.]	[times in h/a]		[1/a]	[times in h/a]		[times/a]	[mm/a]	[µg/m^3]	[µg/m^3]
Dortmund	126	587	274	97	935	5014	527	947	30.4	5.7
Duisburg	18	441	352	83	1050	4775			31.1	11.1
Eifel	579	990	64	91	320	6129	521	820	9.2	5.4
Nuremberg	299	968	309	130	975	4753	447	631	28.2	x *
Munich	552	914	333	97	1126	3929	408	938	67.9	4.0
Kempten	661	1343	206	189	751	5166	381	1177	22.0	x *

Sources: https://www.lanuv.nrw.de/umwelt/luft/immissionen/messorte-und-werte, * no values available
https://opendata.dwd.de/climate_environment/CDC/; https://www.umweltbundesamt.de/themen/luft/luftschadstoffe/stickstoffoxide; https://www.umweltbundesamt.de/themen/luft/luftschadstoffe/schwefeldioxid.

3. Experimental Methods

3.1. Optical Surface Changes and Colorimetry

Optical changes caused by the influence of outdoor weathering or by the conservation action itself were evaluated by colour measurements performed according to DIN EN 16851 [18] and DIN EN 15886 [19]. Measurements were carried out using the spectrophotometer spectro-guide sphere gloss from BYK-Gardner with a standardised light of D65, a 10° image field size and a measuring geometry of d/8°. The examinations were carried out on the triangle-shaped slices (Figure 1). On both the treated weathered side and the cut side (unweathered-untreated $\triangleq$ reference), 5 measuring points were determined, and the mean values and standard deviations were calculated. The results were presented in the CIE L*a*b* colour system, where colour is expressed with the following parameters: L* is the lightness (black (0) to white (100)), a* the red (a*)–green (-a*) coordinate and b* the yellow (b*)–blue (-b*) coordinate in a Cartesian coordinate system.

An ideal hydrophobing agent does not significantly influence the visual appearance of the natural stone surfaces, and despite of natural weathering processes, the treated surfaces should also be stain-repellent, which is why these coordinate values should not change noticeable over time. To evaluate variations in the optical appearance of the samples over time, with ΔL*, Δa* and Δb*, the total colour difference ΔE* was calculated according to Equation (1):

$$\Delta E^* = \sqrt{(\Delta L^*)^2 + (\Delta a^*)^2 + (\Delta b^*)^2} \tag{1}$$

A $\Delta E^* > 5$ units is rated as a different colour and represents in this paper the limit value of noticeable changes (dashed lines in the diagrams) [20,21].

3.2. Measurement of Water Absorption by Pipe Method (Karsten Tube)

According to DIN EN 16302 [22] an open glass pipe was mounted with a fixing compound on the stone surface and the graded tube was filled with water. The amount of water absorbed over an area of 5.7 cm^2 has been determined in defined time steps up to one hour. The tests were carried out using a tube for horizontal surfaces. The water pressure of 10 mbar was kept constant during the measurements. The investigations were done on untreated and treated stones after 0, 2 and 24 or 30 years of natural weathering. Afterwards the protection degree PD_{LP} was calculated with Equation (2) according to DIN EN 16581 [18]. While the standard just mentions that the value should be near 100% for a hydrophobic function, this paper proposes the following three classes:

- Protection degree 100% to 95%: hydrophobic;
- Protection degree 94% to 90%: influenced by a hydrophobic effect;
- Protection degree ≤ 89%: hydrophilic.

$$PD_{LP}(\%) = \frac{(W_f)_B - (W_f)_A}{(W_f)_B} \times 100 \tag{2}$$

with

PD_{LP}—protection degree under low water pressure for 1 h; in%
$(W_f)_B$—absorbed amount of water of untreated sandstone after 2 years indoor storage in mL/cm^2;
$(W_f)_A$—absorbed amount of water of treated sandstone in mL/cm^2.

3.3. Measurement of Capillary Water Absorption

According to DIN EN 15801 [15] the capillary water absorption was determined by placing the samples with their weathered side on a wet underlay and measuring the weight changes after different time periods up to 72 h. The test was done on untreated and treated OKS after 2 and 24 or 30 years of natural weathering. Afterwards, the mass changes related to the contact area can be drawn in a diagram as function of the square root of time. Using the square root of time, nearly linear gradients for the amount of water uptake occur (compare [3,15,23]). As described in Figure 3 and Equation (3), the velocity of water uptake between 5 and 120 min was calculated.

$$\vartheta_{Q_{0.08-2}} = \frac{Q_{A,2} - Q_{A,0.08}}{\sqrt{2} - \sqrt{0.08}} \ \text{in} \left[\frac{g}{m^2 \cdot \sqrt{h}} \right]. \tag{3}$$

with

$\vartheta_{Q0.08\text{-}2}$—velocity of water uptake between $\sqrt{0.08}$ and $\sqrt{2}$ h in g/m^2 $\sqrt{h}$;
$Q_{A,2}$—absorbed amount of water after 2 h in g/m^2;
$Q_{A,0.08}$—absorbed amount of water after 0.08 h in g/m^2.

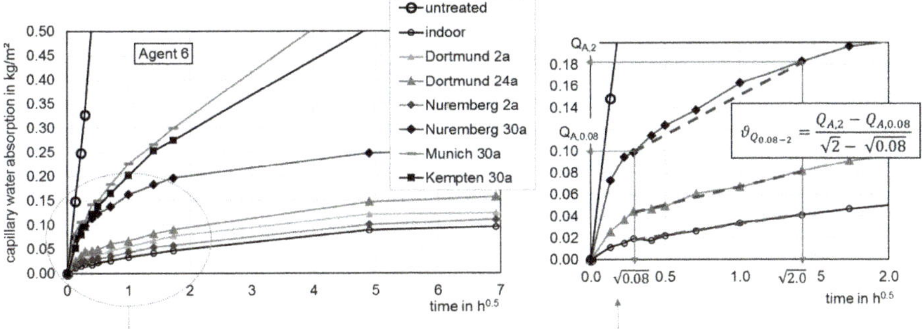

Figure 3. Evaluation of the capillary water absorption with regard to the velocity of water uptake during a period of 5 to 120 min, exemplary shown for agent 6 (a = year).

Furthermore, the protection degree PD_{Ci} was calculated with Equation (4):

$$PD_{Ci}(\%) = \frac{(Q_{Bi} - Q_{Ai})}{Q_{Bi}} \times 100 \tag{4}$$

with

Q_{Bi}—absorbed amount of water of untreated sandstone after 2 years indoor storage at the time t_i;
Q_{Ai}—absorbed amount of water of treated sandstone at the time t_i.

3.4. Moisture Distribution Inside the Stone—Degradation Depth

The NMR MOUSE® PM 25 (Mobile Universal Surface Explorer, registered trademark of RWTH Aachen University, Aachen, Germany) allows the non-destructive detection of moisture distribution inside porous materials via nuclear magnetic resonance. The measurement technique is described amongst others in [13,24,25].

In situ capillary water absorption tests were carried out on the top Table above the NMR sensor while the NMR measuring volume (40 × 40 × 0.2 mm) was moved through the stone, beginning at the stone surface up to a final depth of 4.4 mm. Figure 4 shows the test set up and an exemplary measuring curve. The amplitude (signal of proton density) depends on the amount of water in the porous stone.

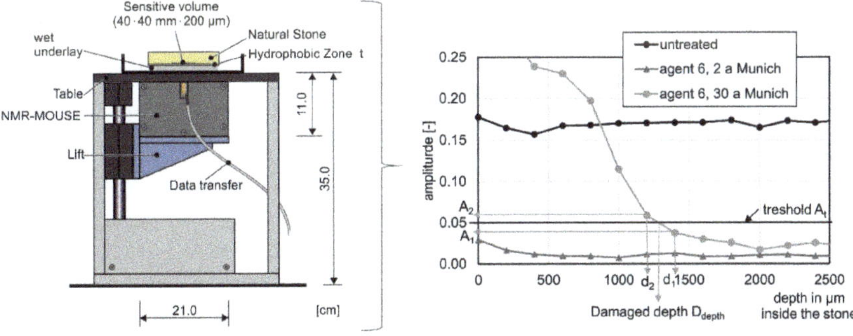

Figure 4. Test set up of in situ NMR measurements during capillary water absorption and determination of the depth of damage after 30 years of outdoor weathering in Munich, shown as an example with protective agent 6.

Based on 51 measurements, a correlation between amplitude and amount of absorbed water was determined for untreated and treated OKS [13]. A threshold for the amplitude was defined: if the amplitude is ≤0.05 the water content is ≤0.1 kg/m^2 and the protection degree is ≥95%. According to the critical values mentioned in [2] and our own experience, this threshold allows the classification between hydrophobic or just an influenced zone. With this threshold and assuming a linear gradient of the amplitude between the two measuring points nearest to the threshold, the damaged depth can be calculated with Equation (5):

$$D_{depth} = \frac{(A_t - A_1) \cdot (d_2 - d_1)}{(A_2 - A_1)} + d_2 \tag{5}$$

with

D_{depth}—damaged depth inside the natural stone in [μm];
A_t—threshold amplitude, for Obernkirchener Sandstone 0.05 in [–];
A_1—measured amplitude directly below the threshold amplitude in [–];
A_2—measured amplitude directly above the threshold amplitude in [–];
d_1—corresponding depth in which amplitude A_1 is measured in [μm];
d_2—corresponding depth in which amplitude A_2 is measured in [μm].

Based on water absorption measurements on treated OKS samples (Table 2) it could be shown that all applied hydrophobing agents are initially effective. As a result of a natural weathering time of 30 years, changes occur in the hydrophobized zone. This is caused by soiling, biogenic growth, a reduction of the water-repellent effect and loosening of the stone structure. These changes are associated with an increase in the amount of absorbed water in the uppermost area of the stone surface (Figure 4). With the determination of a limit value for the loss of hydrophobicity, this damaged depth can now be determined with Equation (5).

4. Results and Discussion

4.1. Optical Surface Changes and Colorimetry

Figure 5 presents the results of the total colour difference (ΔE*) and changes in lightness (ΔL*) for different treated OKS samples. It can be seen that the initial treatment (Figure 5, indoor 2a) did not cause significant changes of ΔE* as the values are below the limit value of 5 (except agent 11). However, due to the application of these different hydrophobing agents the surface darkens, leading from the beginning on to optical aesthetic changes of the stone surfaces (see Figure 6, indoor 2a). After a storage time of 30 years (indoor 30a), the surface discoloured, probably because of the dusty environment in the archive.

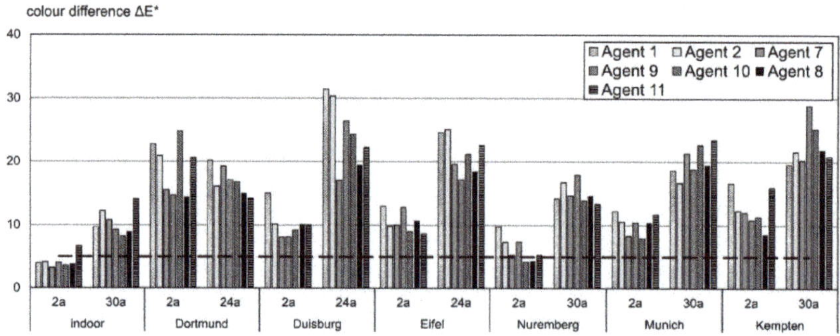

Figure 5. Total colour difference ΔE* of Obernkirchener Sandstone (OKS) samples, treated with silanes (agent 1, 2), siloxanes (agent 7, 9, 10) and silicone resins (agent 8, 11) from different exposition sites.

Except for treated OKS samples exposed in Dortmund, ΔE^* and ΔL^* increased after 24 years of outdoor exposure on all exposition sites by a factor of 2. A maximum ΔE^* has been measured on silane treated samples, which were exposed for 24 years in Duisburg (agent 1 = 31; agent 2 = 30). On the exposition sites in North Rhine-Westphalia, the colour differences and differences in ΔL^* are higher for samples impregnated with silane-based products than for siloxane or silicone resin impregnated stone surfaces. In contrast, ΔE^* increased for siloxane treated samples from the locations in southern Germany. In addition, these samples show a significant darkening over time (Figure 6).

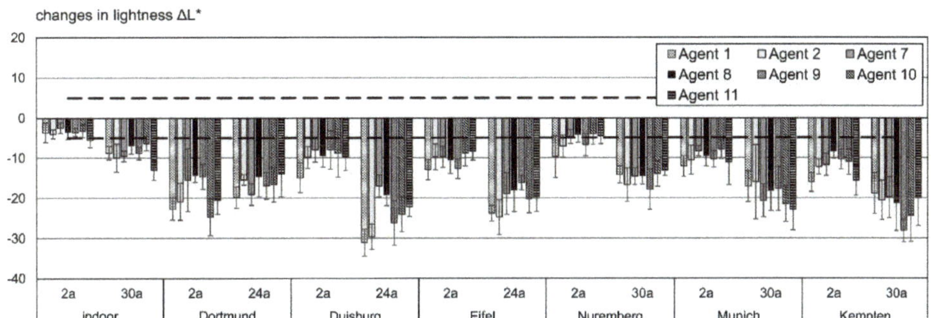

Figure 6. ΔL^* values of OKS samples, treated with silanes (agent 1, 2), siloxanes (agent 7, 9, 10) and silicone resins (agent 8, 11) from different exposition sites.

Slight changes in total colour and lightness could be detected for the treated samples exposed in Dortmund, where no distinct variations appeared during a time period of 24 years. The visible and measured colour changes are caused by biological growth, the deposition of particles and the formation of black crusts on the OKS surfaces, which lead to a gradual darkening as well as a significant change of the optical appearance (compare also with [26,27]).

4.2. Comparison of the Protection Degree Calculated from Water Absorption by Pipe Method and Capillary Water Absorption

Table 5 summarises the protection degree PD_{LP} calculated from water absorption measurements by pipe method after 1 h (Equation (2)) for the 11 protective agents applied on OKS. In general, the hydrophobic surfaces withstand a water pressure of 10 mbar for one hour after a long-term weathering at the different outdoor locations. After 24 as well as 30 years of outdoor weathering, only protective agent 9 (6.7% oligomer methyl-/isooctylmethoxysiloxane) has a consistent loss of its hydrophobic effect. After 2 years of outdoor weathering, the silane-based agents 2, 3 and 4 show partly a reduction in performance. The increase in the water-repellent effect after 30 years compared to 2 years of outdoor weathering with agents 2, 3 and 4 seems somewhat surprising. Two different reasons could cause this effect: (a) the influence of biogenic growth and (b) the very uneven penetration of the isobutyltrimethoxysilanes into the stone. The following results (Table 6 and Figures 7 and 8) show that these agents are no longer effective even after 30 years.

Table 6 summarises the protection degree PD_{Ci} calculated from capillary water absorption tests after 1 h measuring time (Equation (4)) for the 11 protective agents applied on OKS after outdoor weathering. In contrast to Table 5, all protective agents show after a long-term outdoor weathering a loss of effectiveness. Especially at the southern locations in Germany, a decrease of the hydrophobic effect can be seen on the treated samples after an exposure time of 30 years. The silane-based agents 2, 3 and 4 show already after 2 years of outdoor weathering a performance loss.

Comparing Tables 5 and 6, it is obvious that the two unequal measuring methods lead to different results concerning the performance of hydrophobic agents. This fact is well known in literature (compare for example [28]). To sum up, the application of 10 mbar water pressure does not affect the hydrophobic function while a superficial water contact with the total area of the treated stone with the wet underlay leads after more than 24 years outdoor weathering to a loss of the hydrophobic effect for all protective agents. A further evaluation of the capillary water absorption for a more detailed discussion concerning the effectiveness of the different protective agents is necessary.

Table 5. Protection degree PD_{LP} calculated from water absorption by pipe method after a weathering time of 2, 24 and 30 years, respectively, at the 7 locations (PD classes: 100% to 95%: hydrophobic (white cell), 94% to 90%: influenced by a hydrophobic effect (light grey cell), ≤89%: hydrophilic (grey cell)).

Agent	Indoor		Dortmund		Duisburg		Eifel		Nuremberg		Munich		Kempten	
	2a	30a	2a	24a	2a	24a	2a	24a	2a	30a	2a	30a	2a	30a
1	100	100	100	95	98	100	98	96	100	97	100	92	98	97
2	100	100	80	91	91	97	50	99	100	95	100	95	100	91
3	99	99	89	95	100	96	57	96	78	95	55	96	100	95
4	100	100	64	95	23	96	86	95	100	96	95	96	100	95
5	97	100	95	99	98	98	98	97	99	95	100	96	100	95
6	99	100	100	100	95	97	100	96	99	98	100	96	100	95
7	100	100	95	95	100	96	100	95	100	95	100	96	100	91
8	96	100	100	97	100	96	100	98	100	96	100	92	100	95
9	100	100	100	91	100	91	100	95	100	92	100	95	99	91
10	98	100	95	98	100	96	100	95	100	93	99	95	100	94
11	100	100	95	96	100	96	100	95	100	95	100	96	100	96

Table 6. Protection degree PD_{Ci} calculated from capillary water absorption tests after a weathering time of 2, 24 and 30 years, respectively, at the 7 locations (PD classes: 100% to 95%: hydrophobic (white cell), 94% to 90%: influenced by a hydrophobic effect (light grey cell), ≤89%: hydrophilic (grey cell)).

Agent	Indoor		Dortmund		Duisburg		Eifel		Nuremberg		Munich		Kempten	
	2a	30a	2a	24a	2a	24a	2a	24a	2a	30a	2a	30a	2a	30a
1	98	97	95	93	95	88	94	86	96	90	95	81	96	87
2	97	98	85	88	85	81	42	81	95	85	93	80	95	76
3	98	97	82	88		90	66	89	69	87	71	85	95	81
4	97	96	65	90	83	91	76	87	93	89	87	85	83	80
5	97	96	95	90	93	92	95	91	95	89	95	87	95	87
6	98	98	96	95	97	88	95	92	97	89	96	84	96	85
7	98	97	95	86	97	92	96	89	97	84	95	83	95	79
8	98	97	96	93		93	97	92	98	88	95	81	96	87
9	98	97	94	89	97	86	95	85	96	83	95	78	96	82
10	97	98	90	90	95	90	96	90	96	87	95	84	96	83
11	98	97	96	94	97	93	96	90	97	90	96	88	96	87

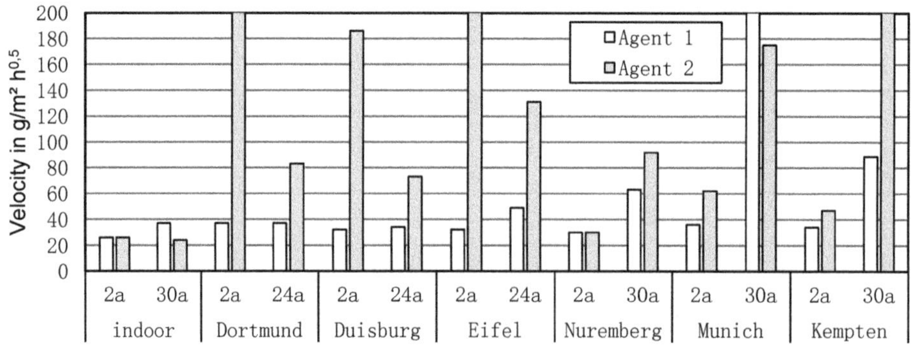

Figure 7. Velocity of water uptake between 5 and 120 min for OKS samples treated with agents 1 and 2 and weathered at the 7 different locations.

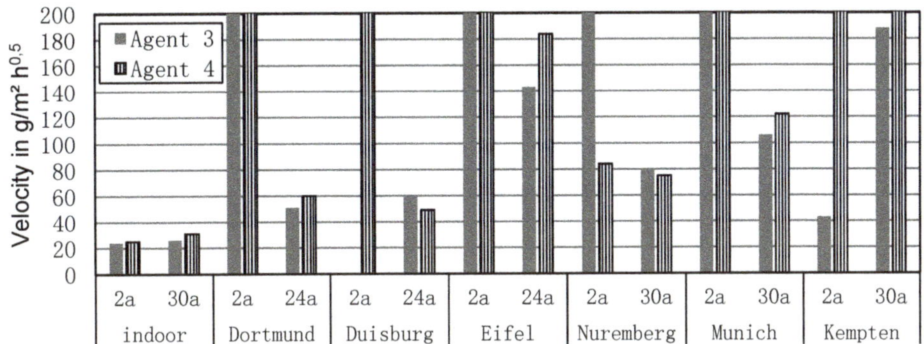

Figure 8. Velocity of water uptake between 5 and 120 min for OKS samples treated with agents 3 and 4 and weathered at the 7 different locations.

4.3. Evaluation of the Protective Agents by the Velocity of Water Uptake

The velocity of water uptake between 5 and 120 min during the capillary absorption test was calculated with 3. While the untreated OKS has a velocity of water uptake $\vartheta_{Q0.08\text{-}2}$ of nearly 1400 g/m² h$^{0.5}$, the treated samples stored indoor have a value of about 25 g/m² h$^{0.5}$. Figure 7 shows the differences in water uptake velocities of agent 1 and 2 applied on OKS. After 2 years of natural weathering of OKS samples treated with agent 1, the rate of water uptake increased only negligibly from 26 g/m² h$^{0.5}$ (indoor) to maximum 37 g/m² h$^{0.5}$ (Dortmund). The extension of the weathering time to 30 years led, at Nuremberg, Munich and Kempten, to more than a doubling of $\vartheta_{Q0.08\text{-}2}$. The highest value was calculated for Munich with 214 g/m² h$^{0.5}$. In contrast, the OKS samples treated with agent 2 showed already after 2 years of exposure more than a doubling of $\vartheta_{Q0.08\text{-}2}$ (except Nuremberg and Kempten). After 24 or 30 years, the values reached 73 to 238 g/m² h$^{0.5}$ (Kempten). These results explain the low protection degree PD_{Ci} of agent 2 (Table 6). An obvious reason for the insufficient effectiveness of agent 2 (35% isobutyltrimethoxysilane) is the uneven distribution of the agent inside the stone matrix of OKS, which varies between 0.8 and 8.3 mm. This irregular distribution was determined for all agents with isobutyl-structures. As a consequence, the velocity of water uptake also increased significantly for OKS samples treated with agents 3 and 4, already after a weathering time of 2 years. Due to the lower ingredient content of 20% isobutyltrimethoxysilane, the effectiveness of these agents is less pronounced than for agent 2 (Figure 8). The decrease of the velocity of water uptake after 24 years weathering especially in Dortmund and Duisburg compared to 2 years could be caused due to the uneven distribution of the agent inside the stone matrix and/or the bionic growth on the stone surface.

Figure 9 shows the velocity of water uptake for OKS samples treated with agents 5 and 6. The methylethoxysiloxane-agents are more effective than the silanes. The results show an increase in the velocity of water absorption after 24 years. This effect is more pronounced at locations in southern Germany.

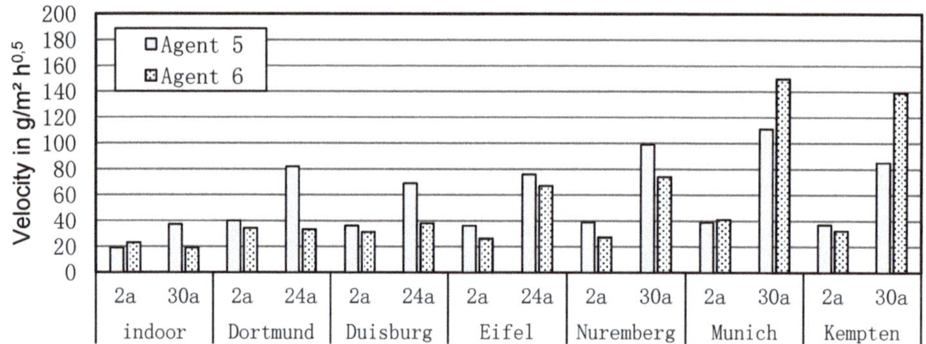

Figure 9. Velocity of water uptake between 5 and 120 min for OKS samples treated with agents 5 and 6 and weathered at the 7 different locations.

A comparison of the water uptake velocities of OKS samples treated with agents 7, 9 and 10 is shown in Figure 10. Two years of weathering at the different locations does not significantly influence $\vartheta_{Q0.08\text{-}2}$ compared to the reference (2a indoor). After 24 and 30 years, respectively, outdoor weathering the OKS samples treated with agent 9 shows the highest increase of $\vartheta_{Q0.08\text{-}2}$ up to 249 g/m^2 h$^{0.5}$ in Munich. At that time at each location, a loss of functionality can be stated for agent 9, which fits to the results shown in Tables 5 and 6. In contrast, the OKS samples treated with agent 7 show, after 24 years natural weathering in Duisburg, a lower increase of $\vartheta_{Q0.08\text{-}2}$. Agents 7 and 9 have the same oligomer siloxane content of 6.7%, but agent 9 has a methyl-/isooctylmethoxy-structure while agent 7 consist just of a methylethoxy-structure. The addition of tetraethoxysilane to the oligomer siloxane with methyl-/isooctylmethoxy-structure (agent 10) does not noticeably influence the results of $\vartheta_{Q0.08\text{-}2}$.

Figure 11 shows the velocity of water uptake for OKS samples treated with silicone resin. The lower agent content and/or the methyl-/isooctyl-structure of agent 8, compared to agent 11, which contains a methylmethoxy-structure (Table 2), leads to a higher water uptake rate after a long-term outdoor weathering. Again, the sample exposition at the southern locations of Germany led, after 30 years, to a significant increase of $\vartheta_{Q0.08\text{-}2}$. Samples from the rural region Eifel show, also after 24 years outdoor weathering, a clearer increase of $\vartheta_{Q0.08\text{-}2}$ compared to samples exposed to the city regions Dortmund and Duisburg.

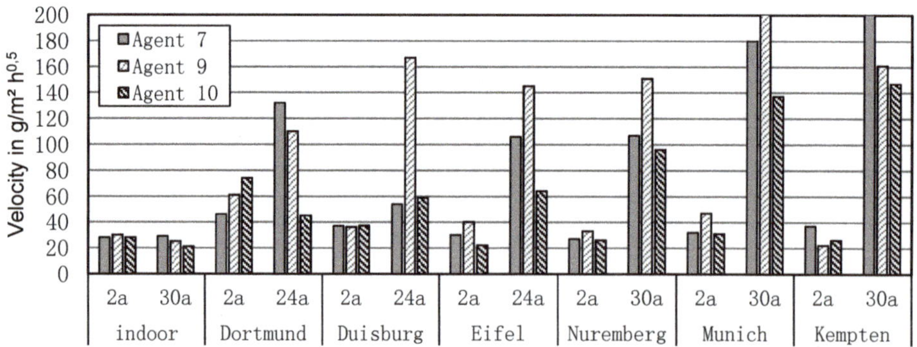

Figure 10. Velocity of water uptake between 5 and 120 min for OKS treated with agents 7, 9 and 10 and weathered at the 7 different locations.

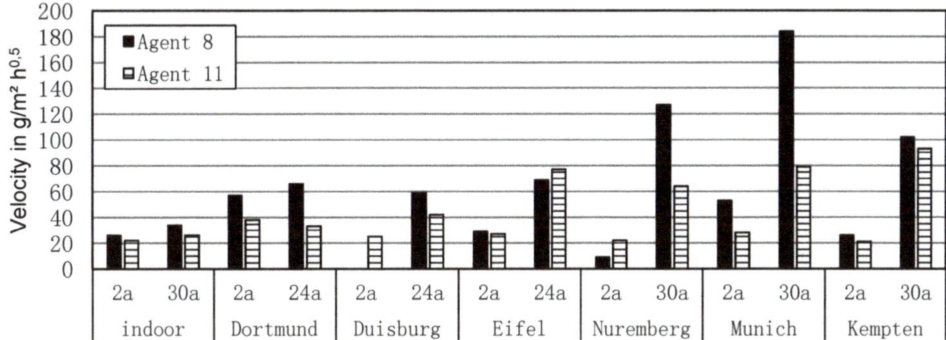

Figure 11. Velocity of water uptake between 5 and 120 min for OKS treated with agent 8 and 11 and weathered at the 7 different locations.

4.4. Evaluation of the Protective Agents by the Degradation Depth

Due to the non-destructive single-sided NMR technique described in Section 3.4, it is possible to detect indirectly the depth of damage (insufficient effectiveness) inside the treated Obernkirchener Sandstone. In Figure 12, the calculated damaged depth inside the OKS (Equation (5)) is compared with the measured penetration depth of the hydrophobic agents (measured visually with a calliper by wetting a cut edge). The high variation of the penetration depth of agent 2 is obvious. At the Eifel location, the damaged depth is higher than the penetration depth, which with 0.14 mm has the lowest value. The samples treated with agent 2 show a higher damaged depth compared to the samples treated with agent 1, which corresponds to the results discussed before (Figure 7).

Figure 13 compares the penetration depth of agents 3 and 4 with their damaged depths. The acquisition of the damaged depths with the NMR technique is only possible to a depth of 4400 µm. If the damaged depth reaches this value, it is marked with an arrow (Duisburg, Eifel, Munich). If the depth of damage exceeds the penetration depth, the speed of water absorption is over 180 g/m² h$^{0.5}$ (Figure 8), and the protection degree PD_{Ci} calculated from capillary water absorption is insufficient (Table 6). Once again, the depth of damage of these agents is often higher after 2 years than after 24 or 30 years, which in addition to the scattering of the penetration depth, suggests a reduction in water absorption due to pollution and bionic growth. This behaviour leads to the assumption that after an early damage of the hydrophobicity within 2 years, a stronger or different increase of bionic growth and pollution causes a compaction of the stone surface in such a way that the water absorption is reduced. This assumption will be considered in further investigations.

Figure 14 confirms the functionality of agents 5 and 6 even after 30 years of outdoor weathering. The depth of damage reaches a maximum of 1280 µm in Munich. Similar results were determined for agents 8 and 11 (Figure 15). In general, the damaged depth increases with increasing weathering time.

Figure 16 compares the penetration depths of agents 7, 9 and 10 with their damaged depths. Again, the damaged depth increased after 24 or 30 years of outdoor weathering, especially for samples exposed to the southern locations, which are characterised by a longer weathering time and a rougher climate. Agents 7 and 9 tend to show a higher depth of damage after 24 or 30 years, respectively, compared to agent 10.

For OKS samples treated with agents 1, 5, 6, 8, 10 and 11, a maximum damaged depth of about 1.5 mm was found, which is in general more than a third lower than the penetration depth. These results explain the increase of capillary water uptake which resulted in a decrease of the protection degree PD_{Ci} and an increase of the rate of water uptake during the first 120 min measuring time. This fact might also explain the slight changes of the protection degree PD_{LP}. The performance of the hydrophobing agents in deeper stone depths prevent the ingress of water under low pressure.

A correlation between the determined damaged depth and the penetration depth of the hydrophobic agents could not be established from the data. This result can be stated for all samples treated with the different agents. This indicates that a high penetration depth of the active substance does not necessarily mean that the durability of the treatment is increased. If the damaged depth reaches nearly the penetration depth, a loss of performance can be found for all investigated samples, for example, agent 9 after 30 years in Kempten.

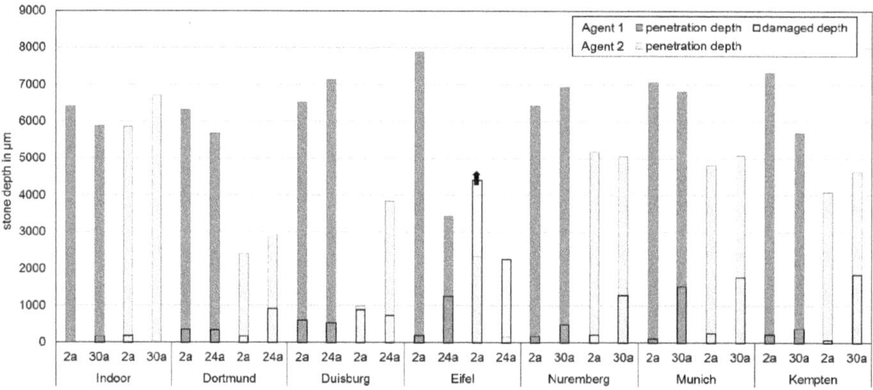

Figure 12. Comparison of the penetration depth of agents 1 or 2 applied on OKS with the damaged depth calculated with Equation (5) (depth 0 = surface of the stone).

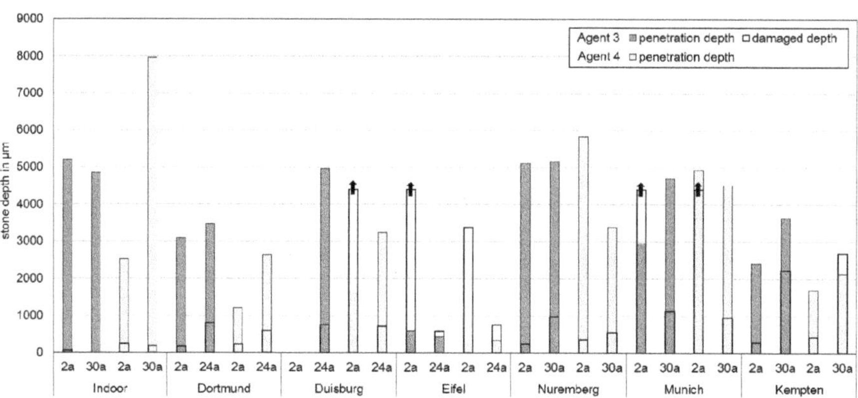

Figure 13. Comparison of the penetration depth of agents 3 or 4 applied on OKS with the damaged depth calculated with Equation (5) (depth 0 = surface of the stone).

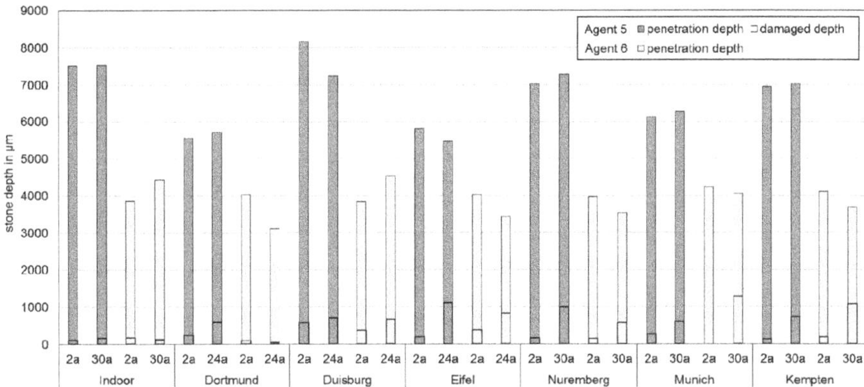

Figure 14. Comparison of the penetration depth of agents 5 or 6 applied on OKS with the damaged depth calculated with Equation (5) (depth 0 = surface of the stone).

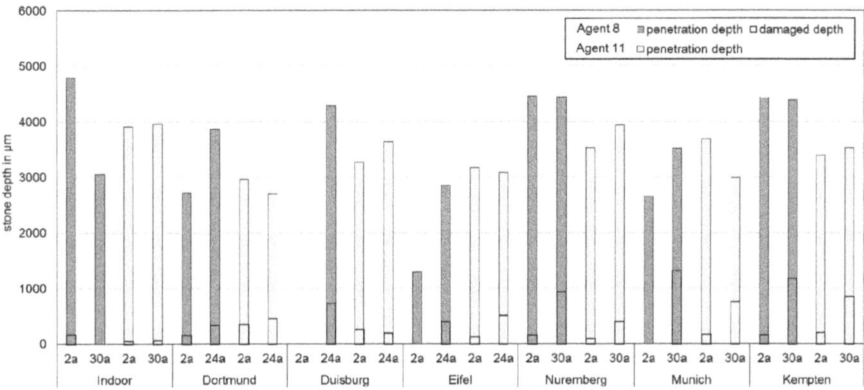

Figure 15. Comparison of the penetration depth of agents 8 and 11 applied on OKS with the damaged depth calculated with Equation (5) (depth 0 = surface of the stone).

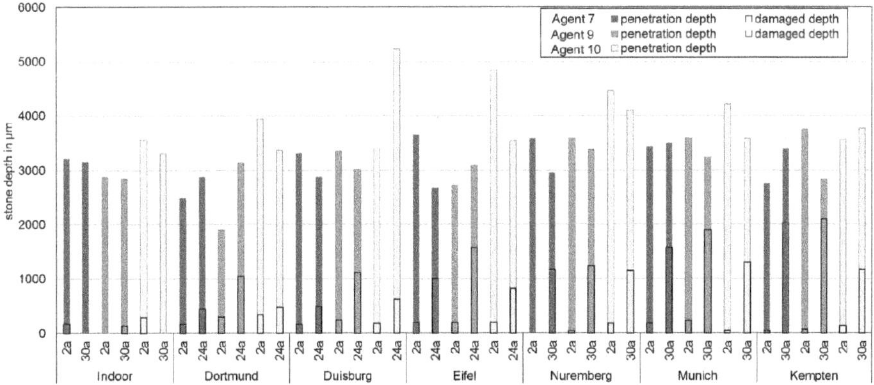

Figure 16. Comparison of the penetration depth of agents 7, 9 and 10 applied on OKS with the damaged depth calculated with Equation (5) (depth 0 = surface of the stone).

5. Conclusions

The influence of up to 30 years of outdoor weathering at six different locations in Germany on the effectiveness of 11 different hydrophobic agents applied on Obernkirchener Sandstones is analysed in this paper. Periods of 24 and 30 years of outdoor weathering led to an ageing of the treated stone surfaces, expressed by discolouration and staining. By measuring colour changes, it could be shown that the formation of black crusts, the deposition of particles and biogenic growth caused a gradual darkening as well as significant changes in total colour over time. Beside the optical changes and colorimetry, the water absorption behaviour was investigated with two different methods: applying low pressure with the pipe method and capillary water absorption from a wet underlay. Afterwards, the test results were analysed with four different evaluation methods: calculation of the protection degrees PD_{LP} (from pipe method) and PD_{Ci} (from capillary water absorption) as well as of the velocity of water uptake during capillary water absorption measurements between 5 and 120 min and calculation of the damaged depth inside the stone using single-sided NMR technique.

It can be concluded that the degradation of the treated stones has been increased due to a longer weathering period of up to 30 years. This degradation can be illustrated with the damaged depth, which has been calculated from NMR measurements during capillary water absorption. The increase of the damaged depth leads to a higher velocity of water uptake and to a decrease of the protection degree PD_{Ci}. As long as the effective hydrophobic zone is three times the depth of the damaged area, the functionality of the hydrophobing agent is still given.

Only agents 2, 3 and 4 which are based on isobutyltrimethoxysilane show already after 2 years of outdoor weathering a clear loss of performance. The irregular distribution of these agents inside the matrix of OKS caused a high variation of the penetration depth, which partly led to a noticeable deterioration of the hydrophobic layer.

Comparing the protective agents containing siloxane, the low-molecular methylethoxysiloxanes (agent 5 and 6) show even with a low content of 7.5% (agent 6) a good performance, which is similar, partly better than the oligomer methylethoxysiloxane (agent 7). Especially after 30 years of outdoor weathering, the agents with oligomer siloxane based on an isooctylmethoxy-structure (agent 9 and 10) have a higher performance loss than the agents 5 and 6.

The effectiveness after 30 years outdoor weathering of the protective agents based on silicone resin is comparable to that of low-molecular siloxanes. Agent 8 shows a more reduced hydrophobic effect due to the low ingredient content and possibly to the isooctyl-structure.

Comparing the six different locations, it can be concluded that the degradation of the treated stones is higher in southern Germany than in North Rhine-Westphalia. This fact can be explained by a longer weathering time of 6 years as well as the rougher environment. In North Rhine-Westphalia, the rural location Eifel, which has a high amount of temperatures under 0 °C and relative humidity above 80% within one year, leads in general to a higher degradation compared to Duisburg and Dortmund.

To sum up, with the help of complementary evaluation methods, the durability of different hydrophobic agents applied on Obernkirchener Sandstone and under the influence of outdoor weathering can be analysed in detail. After 30 years, all agents show a decrease in performance, but some protective agents still provide an effective hydrophobic layer.

Author Contributions: Conceptualization, J.O. and F.B.; methodology, J.O. and F.B.; validation, F.B. and M.G.; formal analysis, F.B. and M.G.; investigation, F.B. and M.G.; resources, F.B. and M.G.; data curation, F.B. and M.G.; writing—original draft preparation, J.O; writing—review and editing, M.G., J.O. and F.B.; visualization, G.O., F.B. and M.G.; supervision, J.O.; project administration, J.O. and F.B.; funding acquisition, J.O. and F.B. All authors have read and agreed to the published version of the manuscript.

Funding: This work was performed as part of a project funded by the Deutsche Forschungsgemeinschaft (DFG, German Research Foundation)—project number 342881843.

Acknowledgments: We would like to thank Stefan Brüggerhoff, for providing the stone samples for our investigations.

Conflicts of Interest: The authors declare no conflict of interest.

References

1. Charola, A.E. Water-repellent treatments for building stones: A practical overview. *APT Bull. J. Preserv. Technol.* **1995**, *26*, 10–17. [CrossRef]
2. *WTA Merkblatt 3–17: Hydrophobierende Imprägnierung von mineralischen Baustoffen*; WTA Merkblatt 3-17-10/D:2010-06; Fraunhofer IRB Verlag: Stuttgart, Germany, 2010.
3. De Ferri, L.; Lottici, P.P.; Lorenzi, A.; Montenero, A.; Salvioli-Mariani, E. Study of silica nanoparticles–polysiloxane hydrophobic treatments for stone-based monument protection. *J. Cult. Herit.* **2011**, *12*, 356–363. [CrossRef]
4. Tsakalof, A.; Manoudis, P.; Karapanagiotis, I.; Chryssoulakis, I.; Panayiotou, C. Assessment of synthetic polymeric coatings for the protection and preservation of stone monuments. *J. Cult. Herit.* **2007**, *8*, 69–72. [CrossRef]
5. Siegesmund, S.; Snethlage, R. (Eds.) *Stone in Architecture: Properties, Durability*; Springer Science & Business Media: Berlin, Germany, 2011. [CrossRef]
6. Di Tullio, V.; Cocca, M.; Avolio, R.; Gentile, G.; Proietti, N.; Ragni, P.; Errico, M.E.; Capitani, D.; Avella, M. Unilateral NMR investigation of multifunctional treatments on stones based on colloidal inorganic and organic nanoparticles. *Magn. Reson. Chem.* **2015**, *53*, 64–77. [CrossRef] [PubMed]
7. Braun, F.; Orlowsky, J. Steter Tropfen höhlt (nicht) den Stein—Zur Dauerhaftigkeit von Hydrophobierungsmaßnahmen am Baumberger Sandstein. *Bauen im Bestand* **2018**, *4*, 8–12.
8. Moropoulou, A.; Kouloumbi, N.; Haralampopoulos, G.; Konstanti, A.; Michailidis, P. Criteria and methodology for the evaluation of conservation interventions on treated porous stone susceptible to salt decay. *Prog. Org. Coat.* **2003**, *48*, 259–270. [CrossRef]
9. La Russa, M.F.; Barone, G.; Belfiore, C.M.; Mazzoleni, P.; Pezzino, A. Application of protective products to "Noto" calcarenite (south-eastern Sicily): A case study for the conservation of stone materials. *Environ. Earth Sci.* **2011**, *62*, 1263–1272. [CrossRef]
10. La Russa, M.F.; Ruffolo, S.A.; Rovella, N.; Belfiore, C.M.; Pogliani, P.; Pelosi, C.; Andaloro, M.; Crisci, G.M. Cappadocian ignimbrite cave churches: Stone degradation and conservation strategies. *Period. Mineral.* **2014**, *83*, 187–206.
11. Mirwald, P.W. Umweltbedingte Gesteinszerstörung untersucht mittels Freilandverwitterungsexperimenten. *Sonderheft Bausubstanzerhaltung in der Denkmalpflege, Bautenschutz + Bausanierung S.* **1986**, 24–27.
12. Brüggerhoff, S.; Wagener-Lohse, C. Gesteinsverwitterung in Freilandversuchsfeldern—Erfahrungen mit ihrer Errichtung und Nutzung. *Sonderheft Bausubstanzerhaltung in der Denkmalpflege S.* **1989**, 76–80.
13. Braun, F.; Orlowsky, J. Non-destructive detection of the efficiency of long-term weathered hydrophobic natural stones using single-sided NMR. *J. Cult. Herit.* **2019**. [CrossRef]
14. Grimm, W.-D. *Bildatlas Wichtiger Denkmalgesteine der Bundesrepublik Deutschland*; Teil II Bildband; 2. erweiterte Auflage; Ebner Verlag: Ulm, Germany, 2018; ISBN 978-3-87188-247-0.
15. DE. *Conservation of Cultural Property—Test Methods—Determination of Water Absorption by Capillarity*; DIN EN 15801:2010-04; DIN German Institute for Standardization: Berlin, Germany, 2010.
16. DE. *Conservation of Cultural Property—Test Methods—Determination of Water Vapour Permeability (δp)*; DIN EN 15803:2010-04; DIN German Institute for Standardization: Berlin, Germany, 2010.
17. Braun, F.; Orlowsky, J. Effect of Different Silicic Acid Ester on the Properties of Sandstones with Varying Binders. *Restor. Build. Monum.* **2018**, *23*, 1–13. [CrossRef]
18. DE. *Conservation of Cultural Heritage—Surface Protection for Porous Inorganic Materials—Laboratory Test Methods for the Evaluation of the Performance of Water Repellent Products*; DIN EN 16581:2015-03; DIN German Institute for Standardization: Berlin, Germany, 2010.
19. DE. *Conservation of Cultural Property—Test Methods—Colour Measurement of Surfaces*; DIN EN 15886:2010-12; DIN German Institute for Standardization: Berlin, Germany, 2010.
20. Malaga, K.; Müller, U. Validation of improvement of procedures for performance testing of anti-graffiti agents on concrete surfaces. In Proceedings of the Sixth International Conference on Concrete under Severe Conditions, Environment and Loading, Mérida, Mexico, 7–9 June 2010.
21. García, O.; Malaga, K. Definition of the procedure to determine the suitability and durability of an anti-graffiti product for application on cultural heritage porous materials. *J. Cult. Herit.* **2012**, *13*, 77–82. [CrossRef]

22. DE. *Conservation of Cultural Heritage—Test Methods—Measurement of Water Absorption by Pipe Method*; DIN EN 16302:2013-04; DIN German Institute for Standardization: Berlin, Germany, 2010.

23. Peruzzi, R.; Poli, T.; Toniolo, L. The experimental test for the evaluation of protective treatments: A critical survey of the "capillary absorption index". *J. Cult. Herit.* **2003**, *4*, 251–254. [CrossRef]

24. Orlowsky, J. Analyzing of coatings on steel-reinforced concrete elements by mobile NMR. *Arch. Civ. Eng.* **2016**, *62*, 65–82. [CrossRef]

25. Baias, M.; Blümich, B. Nondestructive Testing of Objects from Cultural Heritage with NMR. In *Modern Magnetic Resonance*; Springer Publishing: New York, NY, USA, 2018; Volume 12, pp. 293–304. [CrossRef]

26. Gibeaux, S.; Thomachot-Schneider, C.; Eyssautier-Chuine, S.; Marin, B.; Vazquez, P. Simulation of acid weathering on natural and artificial building stones according to the current atmospheric SO_2/NO_x rate. *Environ. Earth Sci.* **2018**, *77*, 327. [CrossRef]

27. Grossi, C.M.; Brimblecombe, P.; Esbert, R.M.; Alonso, F.J. Color changes in architectural limestones from pollution and cleaning. *Color Res. Appl.* **2007**, *32*, 320–331. [CrossRef]

28. Vandevoorde, D.; Cnudde, V.; Dewanckele, J.; Brabant, L.; de Bouw, M.; Meynen, V.; Verhaeven, E. Validation of in situ applicable measuring techniques for analysis of the water adsorption by stone. *Procedia Chem.* **2013**, *8*, 317–327. [CrossRef]

Article

Probabilistic Life-Cycle Assessment of Service Life Extension on Renovated Buildings under Seismic Hazard

Roberta Di Bari [1,*], **Andrea Belleri** [2], **Alessandra Marini** [2], **Rafael Horn** [1,3] **and Johannes Gantner** [3]

[1]   Institute for Acoustics and Building Physics (IABP), University of Stuttgart, 70174 Stuttgart, Germany; rafael.horn@iabp.uni-stuttgart.de

[2]   Department of Engineering and Applied Sciences, University of Bergamo, 24044 Dalmine, Italy; andrea.belleri@unibg.it (A.B.); alessandra.marini@unibg.it (A.M.)

[3]   Fraunhofer Institute for Building Physics (IBP), 70569 Stuttgart, Germany; johannes.gantner@pom.ch

*   Correspondence: roberta.di-bari@iabp.uni-stuttgart.de; Tel.: +49-7119-7031-8111

Received: 3 December 2019; Accepted: 25 February 2020; Published: 9 March 2020

Abstract: Existing buildings can reach a performance enhancement and extend their nominal service life through renovation measures such as seismic rehabilitation. In particular, when buildings have almost exhausted their service life, seeking an optimal solution should consider whether costs and environmental effects are worthwhile, or new construction is preferred. In this paper, a methodology to consider seismic hazard into probabilistic approaches for life-cycle analyses is presented considering the possibility of structural enhancement over an extended building lifespan. A life-cycle-based decision support tool for building renovation measures is developed and applied to a selected case study. Unlike standard "static" analyses, which in this work show shortcomings by underestimating impacts of vulnerable buildings, such an approach brings out environmental and economic advantages of retrofit measures designed to improve the structural performance.

Keywords: Life Cycle Assessment uncertainties; seismic hazard; building renovation; retrofit

---

1. Introduction

In a perspective of sustainable development, the Europe 2020 strategy aims for a control of environmental impacts as an instrument of European economic growth. On such an issue, the construction sector is subjected to a particular attention: with reference to recent information of International Energy Agency (IEA), the existing building stock is responsible for 50% of material depletion and 40% of energy consumption, generates 36% of greenhouse gases (GHG) emissions and a third of the total waste [1]. Moreover, people spend 90% of their time in buildings that should guarantee the highest level of safety, comfort and wellness [2].

As a consequence of the recent economic crisis of the construction sector in all European countries, the number of new buildings is significantly decreased, even if by 2050 2.6 billion people will require new housing, work places and infrastructure due to rapid urbanization and population growth [2]. With respect to this, the renovation of existing buildings through retrofit is an effective strategy: the overall building performance is improved and additions of new volumes and surfaces are allowed, without encountering any demolition and rebuilding, which has proved to have higher environmental impacts [3,4]. This challenge calls research for improvements to established design procedures and construction approaches, which could be more effective and support the achievement of a net carbon neutral society.

In the last few decades, methodologies have been developed to assess environmental and economic impacts, such as life-cycle approaches, used in the building sector especially in the context of building certification labelling systems. They support the decision-making process and represent an instrument for developing and applying new and more sustainable materials, technologies and measures. Recently, the results' transparency and the trustworthiness of such analyses have been discussed. One of the debated topics is the influence of uncertainties and the importance to include them in the analysis to improve the robustness of life-cycle assessment (LCA) and to allow a better interpretation and communication of the results [5]. Another important aspect is the inclusion of unexpected events such as earthquakes: in a context of buildings mostly outdated (almost 40% [2]) and with seismic vulnerabilities, such events could lead in the worst case scenario to irreversible consequences such as high reconstruction costs, environmental damage and lost human lives. On the other hand, users could implement preventive renovation measures that could minimize the environmental impacts and final costs, if such measures have been conceived following a life-cycle thinking perspective. Thanks to these, the building can extend or even restart its own service life [6].

With regard to the methodological point of view, standard life-cycle (or so called "static") approaches do neither consider dynamic effects, such as impact profile variation in time, service life variation, technological improvements and subject willingness, nor sources of uncertainty, such as unexpected events, that might lead to service-life reduction. Thus, stakeholders might be wrongly supported leading to miscalculation or missing information on the building life-cycle. Given the above, the focus of the paper is to address a methodology for the inclusion of seismic hazard and building structural quality enhancement: in life-cycle analyses carried out under consideration of probabilistic approaches in order to provide a tool for decision making on renovation measures.

2. Life-Cycle Assessment (LCA) State of the Art

LCA, defined as "an evaluation of the inputs and outputs of a product system", is a widely applied methodology whose framework is provided in standards ISO 14040 and ISO 14044. The environmental assessment of products is conducted thought 4 steps, namely goal and scope, life-cycle inventory (LCI), life-cycle impact assessment (LCIA) and a final results interpretation [7,8]. By contrast with industrial processes, buildings cannot be standardized as easily and several issues cause a particular complexity of the assessment. Among them, we mention:

- the long lifespan of entire building (50–100 years);
- the shorter, but varying lifespan of building parts and components;
- the complex interaction and interconnection of numerous materials and processes;
- the unique character of each building;
- the building evolution over time due to maintenance, refurbishment or renovation measures.

Most of them cannot be exactly predicted during the planning and design phase: the long lifespan and user behavior and choices can especially require more assumptions and, if neglected, may question the environmental performance and the results' credibility [5]. For such reasons, research on LCA has focused on uncertainties and their effects. The aforementioned standards address uncertainties analysis within the interpretation phase without any specifications or any systematic procedures and therefore many studies provide different approaches [9,10].

In literature there are several uncertainties classifications which consider common sources (i.e., data comprehensiveness, subjective factors, temporal or local conditions) which may all affect the results of LCA studies [11–13]. For the investigated case, uncertainty due to methodology and external factors, which are the focus of this work, have been prioritized depending on their influence on results coming from environmental assessment. Further sources are assumed negligible.

2.1. Uncertainties Due to Methodology

As for many mathematical models, LCA does not refrain from approximation: the basis of such approximation is related to a linearized calculation of the vector r of environmental impacts according to (1):

$$r = Q \times H \times G^{-1} \times u, \tag{1}$$

where Q is the matrix of characterization factors, H is the environmental intervention matrix of emissions per unit process of product systems, G is the technology matrix representing the inter-process flows needed for functioning of the product system, and u is the external supply vector, related to the functional unit [11,14].

Such a vector cannot always express the variability given, i.e., spatial and temporal factors, and consequently this can mislead final decisions [15]. In order to solve this issue and improve the reliability in LCA, the ISO norms recommend technical standards which lead to the implementation of input variability in form of simple probability distribution. In addition to this, a global sensitivity analysis is generally accepted for a comprehensive uncertainty analysis [14].

Beside such procedures, probabilistic and sampling-based methods have been developed and applied: while sensitivity analysis tries to characterize output uncertainty by apportioning it to its input constituents, probabilistic methods such as Bayesian inference, Gaussian process (GP), and polynomial chaos expansion methods characterize and propagate uncertainties throughout the system to outputs and quantify them [16].

Among them, the Monte Carlo sampling is mostly applied, where parameters are provided dependently from probability functions and the output vector is calculated. Simulations are repeated thousands of times and the results are provided in the form of probability distribution rather than a unique value. Especially in a LCA framework, with an adequate uncertainties quantification, efforts can be focused on collecting more (accurate) information on LCA phases that contribute the most to the LCA output. It has been shown that the trade-off between parameter uncertainty and model-form uncertainty can help in determining an optimal model to reduce the overall uncertainty [17].

2.2. Design Changes Due to External Factors

The complexity of a building design is due to a variety of factors which influence the final decision during the whole planning process. Some of them may induce design changes even during the operating phase of a building [18]: several studies tried to identify all of them and agree on a distinction between internal and external factors [18–21]. By contrast with internal factors, which could be included during the project management, external factors are parties not directly involved. In addition to this, none of all external factors can be foreseen and consequently a prediction may be possible only through statistical methodologies [22]. As reported by Yana et al., they can be distinguished in [18]:

- political and economic matters;
- third parties' requests;
- technological advancement;
- natural environment.

Political and economic matters can dictate policies and regulations; economic fluctuations (price increase, inflations etc.) can furthermore influence the decision making. Requests coming from third parties (i.e., neighborhood) concern mostly end users and regulatory bodies. A technological advancement can be helpful by promoting more performing and optimized products. Lastly, the natural environment involves a multitude of matters: weather and geological conditions or natural disasters that are hardly predictable. The latter, even without a high frequency, represent a still open issue for the engineering design criteria. A common example is earthquakes, which are included in the overall building design for seismic-prone regions [23].

By a LCA-methodological point of view, such issues have been handled so far in different ways.

The problem is usually solved a priori, by considering a seismic resistant building and consequently neglecting further vulnerabilities. The impact related to seismic hazards is limited to renovation measures, which are carried out in pre-specified periods of the building operating phase [24]. Even if capable of bringing an idea about the impact of the single intervention, such examples do not envisage unforeseen future losses and the random nature of seismic events.

In FEMA's Advanced Engineering Building Module (AEBM) [25] and in PEER-PBEE studies [26], probabilistic methodologies are investigated rather than the abovementioned deterministic one. On the basis of a probabilistic seismic hazard assessment (see Section 3.1), seismic losses are evaluated and the environmental impacts expressed by a repair or retrofit measure belonging to a predetermined set [25,26]. As already noticed by Calvi et al., such approaches do not take into account a possible building performance upgrade over time of the buildings lifecycle [27].

In general, all the mentioned applications do not include user subjectivity, which, depending on its own economic availability and aesthetical preference, represents an active actor within the choice of renovation measures.

2.3. Uncertainties in Service Life

In the field of life-cycle assessment, there is wide consensus on the consideration of a service lifespan coming from structural and seismic design requirements: building components are designed for a service life between 30–70 years and then refurbished or substituted, while the building is planned with a 50 years' service life (Design working life category 4: Building structures and other common structures [28]). The structure shall be designed such that deterioration over the considered building lifespan does not impair the structural performance, considering regular maintenance. After the working life, the structure needs to be checked for possible deterioration and retrofitted or dismantled. Such assumption differs however from information coming from building stock data. According to the latest report of Building Performance Institute Europe (BPIE), a substantial share of the existing building stock in Europe is older than 50 years and has not always been subjected to renovation measures to enhance the overall performance or many of the buildings still in use are hundreds of years old [29].

With regard to building renovation works, the cost of the intervention is a relevant factor which could compromise the definition of the plan for necessary measures. Therefore, social and technical benefits have to find a compromise with the national economic situation and the fund availability. A model for works planning can be based on predicting the time when building critical elements may reach degradation levels that exceed acceptable values [30]. Nevertheless, even a degradation level cannot be scientifically defined or a service life cannot be assessed by deterministic approaches. Therefore, in the literature, probabilistic or engineering methods which combine both deterministic and probabilistic ones, such as the factor method (ISO:15686-1) [31], have been developed but not frequently applied by architects and planners.

Eurocodes [28] state that to achieve an adequately durable structure, the designer should take into account various aspects such as the intended or foreseeable use of the building, the expected environmental conditions, the structural detailing, and the intended maintenance during the designed working life, among others. However, in some conditions, the knowledge of such factors has changed during the building life: for instance, different and not predictable use of the building may have been arisen since the initial design, the knowledge of the environmental factors, such as seismic hazard, has changed during time due to the increased availability of earthquake data, and an increased knowledge of the structural and material performance has been available due to the advancement of scientific research.

All these issues contribute to increasing uncertainties in the working life and, consequently, in the impacts assessment.

3. Method: Probabilistic life-cycle analyses with Inclusion of Seismic Hazard Model

The presented methodology consists in a probabilistic model enriched by factors related to technological, economic and environmental changes, ensuring its flexibility over an extended building lifespan. The service life is dictated by using stage and user aware choices, and, especially for construction with significant structural vulnerabilities, by earthquakes.

The main steps are:

1. **"As is" situation analysis**. Investigations about the reference building (when t = 0, today's situation), in order to verify its compliance with current building code, minimal requirements of design limit states and vulnerabilities.
2. **Events choice**. With particular respect to the geographical location, earthquakes with different magnitude and frequency are taken into account. The frequency is evaluated by Geophysics and Volcanology institutions on the basis of the classical probabilistic seismic hazard analysis (PSHA) model.
3. **Decision tree set up**. In the occurrence of earthquakes, different scenarios are considered. In the worst case, where the building will present a damage level greater than 40% of the building value, a demolition and a re-build will generally occur. Over time the reference building may be subjected to further improvements by user decision and under particular technological, environmental and economic conditions which influence intervention willingness.
4. **Future scenarios analysis**. Because of an eventual adjustment, the response to seismic loads (here called Building Quality—BQ) could be improved.
5. **Probabilistic Model setting**. A Monte Carlo sampling method is set up and all aforementioned considerations are included.

The inclusion of Probabilistic Seismic Hazard Analysis (PSHA) in life-cycle assessment (LCA) and life-cycle cost (LCC), while introducing more uncertainties in the analyses (e.g., aleatory and epistemic uncertainties related to the structural performance), provides a rational way to deal with them. In particular, accounting for unforeseen events such as earthquakes allows a more comprehensive LCA and LCC and to analyze the effects of events that could alter classic results.

3.1. Life-Cycle Impact Assessment (LCIA) and Cost (LCC) Method of Measurement

The total environmental impact and cost are calculated as the sum of impacts (2) and costs (3) recorded every year of the observation time of n years.

$$GWP_{tot} = \sum_{i=1}^{n} GWP_{prod,\, i} + \sum_{i=1}^{n} GWP_{B4-B5,\, i} + \sum_{i=1}^{n} GWP_{EoL,\, i} \tag{2}$$

$$\text{\euro}_{tot} = \sum_{i=1}^{n} \text{\euro}_{prod,\, i} + \sum_{i=1}^{n} \text{\euro}_{B4-B5,\, i} + \sum_{i=1}^{n} \text{\euro}_{EoL,\, i} \tag{3}$$

where

$GWP_{prod,\, i}/\text{\euro}_{prod,\, i}$, GWP (LCA) or costs (LCC) due to production on year i.

$GWP_{B4-B5,\, i}/\text{\euro}_{B4-B5,\, i}$, GWP (LCA) or costs (LCC) due to renovation measures on year i.

$GWP_{EoL,\, i}/\text{\euro}_{EoL,\, i}$, GWP (LCA) or costs (LCC) due to end of life on year i.

3.1.1. Construction Phase and Building Renovation Measures

Regarding the embodied carbon derived by the building production, replacement or refurbishment, a dynamic measurement has been considered, by adjusting the value at t = 1 year (i.e., at present) by using price increase rate and discount rate for cost analyses (5) and a *decarbonisation rate* for the environmental assessment (4), which expresses technological advancements suitable to reduce environmental impacts due to manufacturing. The discount rate has been neglected in environmental analyses [10].

$$GWP_{prod,\ i} = GWP_{prod,\ i-1} \cdot \frac{1 - r_d}{\left(1 + r_f\right)^i} \tag{4}$$

$$€_{prod,\ i} = €_{prod,\ i-1} \cdot \frac{1 - r_d}{\left(1 + r_f\right)^i} \tag{5}$$

where

i is a year belonging to the investigate period (100 years)
$GWP_{prod,\ i-1}$, is the embodied carbon in the $(i-1)$ year
$€_{prod,\ i-1}$, are cos ts in the $(i-1)$ year
r_d, is the considered decarbonisation rate (LCA)/price increase rate (LCC)
r_f, is the considered discount factor

3.1.2. End-of-Life

In this approach, the end-of-life (EoL) costs and environmental impacts are dictated by a final dismantling after 100 years' analysis. As well as for the production, the value is corrected by taking into account average decarbonisation, price increase and discount rates [10].

$$GWP_{EoL,\ i} = GWP_{EoL,\ 99y} \cdot \frac{1 - r_{d,avarage}}{\left(1 + r_{f,avarage}\right)^{100}} \tag{6}$$

$$€_{EoL,\ i} = €_{EoL,\ 99y} \cdot \frac{1 - r_{d,avarage}}{\left(1 + r_{f,avarage}\right)^{100}} \tag{7}$$

where

$$r_{d,avarage} = \frac{\sum_i^n r_{d,i}}{n} \text{ is the evaluated decarbonisation (LCA)/price increase (LCC) rate} \tag{8}$$

$$r_{f,avarage} = \frac{\sum_i^n r_{f,i}}{n} \text{ is the evaluated discount rate (LCA/LCC)} \tag{9}$$

3.2. Seismic Hazard

3.2.1. Probabilistic Seismic Hazard Analysis (PSHA)

By contrast with other cases such as weather conditions, seismic events cannot be *forecast*, but the current state of knowledge allows predictions: the available models are capable to address the likelihood of a seismic event [32].

In order to maintain conservative assumptions, the Poisson Model is generally used for seismic hazard assessment to represent the occurrence rate of an earthquake through time-independent models. The probability density function is expressed as following:

$$f_{Exp}(t) = \lambda\, e^{\lambda t} \tag{10}$$

where:

$$\lambda = N/\Delta T, \text{ mean rate of events per unit time} \tag{11}$$

$$T_R = \lambda^{-1} = \Delta T/N, \text{ is the average return period} \tag{12}$$

On the basis of this, Cornell developed the probabilistic seismic hazard analysis (PSHA), which is still nowadays used for deriving hazard curves and embedded into seismic design regulations [33]. In PSHA the annual probability of exceedance of a ground motion $Y \geq y$ is:

$$\gamma(y) = \sum_i v_i \iiint f_m(m) f_r(r) f_\varepsilon(\varepsilon) \, P[Y > y \mid m, r, \varepsilon] dm \, dr \, d\varepsilon \tag{13}$$

where:

v_i is the activity rate of the ith-source of earthquakes
$f_m(m)$ is the probability density function of earthquake magnitude
$f_r(r)$ is the probability density function of source-to-site distance
$f_\varepsilon(\varepsilon)$ is the ground motion shape uncertainty density function
$Y > y | m, r, \varepsilon$ is the conditional probability that Y exceeds y, given m, r, ε

For simplicity, hazard curve and annual frequencies can be provided by seismologists (such as the Institute for Geophysics and Volcanology, INGV, in Italy) [34]. For each geographical location different earthquake intensity, expressed for instance in terms of peak ground acceleration (PGA), are associated with their mean annual frequency of occurrence and, according to building design regulation, to a design limit state [32,35].

3.2.2. Inclusion of PSHA Method in Probabilistic LCA

In the present paper, the occurrence of a seismic event is taken into account by a Monte Carlo simulation. The building lifespan is subdivided into years and for each year the occurrence of an earthquake with a given intensity is considered by comparing a random value n_r with the earthquake occurrence probability P (λ) under the following conditions:

$$\text{random value } (n_r) \begin{cases} > P(\lambda_1, t), & \text{no earthquake} \\ \leq P(\lambda_1, t), & \text{light earthquake occurence} \\ \leq P(\lambda_2, t), & \text{medium earthquake occurence} \\ \leq P(\lambda_3, t), & \text{strong earthquake occurence} \end{cases} \tag{14}$$

where:

λ_j is the mean rate of earthquake occurrence.

The occurrence probability $P(\lambda)$ is the Probability Mass Function (PMF) for Poisson's distribution (see Equations (10)–(12)). In this application only one seismic event (n = 1) is allowed in the considered time step, t = 1 year. Consequently, different earthquakes cannot occur simultaneously, and, if that is the case, only the highest intensity earthquake is considered. The generalized equation can be assumed here as following:

$$P(\lambda, n) = \frac{e^{-\lambda} \lambda^n}{n!} \underset{n=t=1}{\longrightarrow} \lambda \, e^{-\lambda t} \tag{15}$$

This assumption allows the evaluation of a yearly increases of environmental impacts and costs overall the building lifespan. Due to the discrete and limited number of the earthquake intensity levels, it is expected to observe scattered results in the cumulative distribution functions.

3.3. Life Duration of Building and Building Parts

For this analysis, a wider time duration of 100 years has been considered, in order to assume an extended service life of the whole building, where the structure is not highly damaged. While structural service life is dictated more dynamically by earthquake occurrence, non-structural building parts present in this application a pre-defined lifespan: every 30 years' service life, their replacement takes place. The chosen building lifespan leads to the generation of 10,000 random value runs, i.e., 100 times the building lifespan.

4. Case Study

A case study has been selected as a proof of concept for the proposed methodology: an integrated structural and architectural renovation intervention, which can potentially improve the energy performance through further measures. An 8-storey apartment building is considered located in L'Aquila, a region of high seismicity in Italy, and designed for gravity loading, without accounting for seismic actions. Two different structural retrofit measures have been considered and analyzed. In this section, the results of a static and the proposed probabilistic LCA are reported and compared. Table 1 shows the specification of the LCA analysis.

Table 1. Specification of life-cycle assessment (LCA) analysis according to [7,8,36]

Specification of LCA Analysis	
Goal and Scope	Evaluation of Global Warming Potential (GWP) for the construction and demolition of a reinforced concrete (RC) residential building for social housing. Evaluation of potential scrap steel coming from RC reinforcements. The use stage is neglected because of missing information about energy performance of the building.
System Boundaries	B4: Building parts replacements (non-structural elements), which include end-of-life of substituted parts and production of new ones. B5: Refurbishment or addition of new elements which were not initially foreseen. C + D: End-of-life and credits. For the end-of-life of steel systems the pattern of re-use is modelled.
Functional Unit	[m^2 net surface] multi apartments building designed for 50 years service life
Impact Categories	GWP [kg CO_2 eq.] Characterization Method CML 2001–January 2016

4.1. Static LCA

The LCA analysis has been carried out for the reference building and for the selected retrofit measures, specifically a diagrid system and shear walls. For both examples the modelling and the final impact evaluation have been realized with help of GaBi ts pc-tool [37] and Generis software [38], on the basis of environmental databases updated on 2019.

4.1.1. Rebuilding

Data collection has been made for the whole considered building and for each part of it. The stratigraphic information has been collected and included in the final model in GaBi ts software. For the End-of-Life of the building, a whole demotion has been envisaged with a rebuilding process. The new building will be considered equivalent to the old one in terms of geometry but with structural elements fully complying with current seismic regulations.

Life-Cycle Impact Assessment: according to the results of the impact assessment for global warming potential, 490 kg CO_2 eq./m^2 are calculated. Most of the environmental impacts are due to the construction process during which structural reinforced concrete (RC) elements are the most relevant (see Figure 1). The calculated value does not differ too much from results of analyses carried out for new buildings in the Italian territory, which estimate 488.5 kg CO_2 eq./m^2 [39]. Hence, for the rebuilding process the same LCIA result can be claimed valid.

Reconstruction Costs Assessment: the lack of precise information about manufacturing processes and further costs related to reconstruction of damaged buildings does not allow a direct calculation of the costs. An indirect evaluation is possible consulting likely reconstruction costs per net surface of RC buildings after strong earthquakes which include, e.g., professional fees, geotechnical tests, dismantling, strengthening and repair costs.

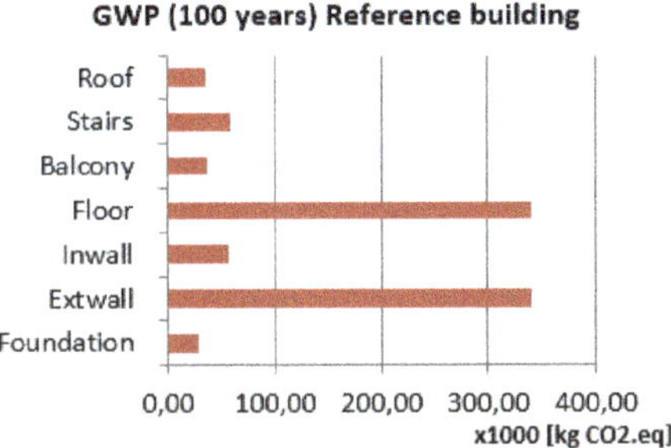

Figure 1. LCIA results of reference building (Source GaBi ts [37]).

For most of the highly damaged RC buildings (usability rating class E according to Italian practice) following the recent earthquakes in L'Aquila on 2009, a mean unit cost for re-construction of 1300 €/m^2 net surface has been evaluated [40,41]. Demolition activities are, meanwhile, evaluated for RC buildings with 152 €/m^2 net surface [42]. The discount due to the scrap steel of reinforcement bars has been considered negligible.

4.1.2. Replacement

The first renovation alternative consists in a substitution of the non-structural elements belonging to the external envelope and to the internal partition walls. The total emission derives by considering the production of the new elements and the dismantling of the old ones (see Table 2). With regard to costs, lump-sum costs are derived by information from local authorities [42] and evaluated as 245 €/m^2 net surface.

Table 2. Replacement measure specification (Source: Generis [38]).

Construction	GWP [kg CO$_2$ eq./m^2 Element]	Quantity [m^2]	Total GWP [kg CO$_2$ eq.]
External wall: bricks with composite thermal insulation	44.1	1942	85,642.2
Internal partition wall: double coating and acoustic insulation	27.9	2204	61,491.6
Sloped roof with insulation	98.6	188	18,536.8
-		Total	165,671
-		/m^2 Net Surface Area	90.0

4.1.3. Diagrid Retrofit System

A diagrid is a global intervention which provides an external reinforcement directly attached to the building. An additional exoskeleton reshapes the building façade, improves the structural performance by sustaining static and seismic loads that exceed the capacity of the existing structure. The global system is activated by transferring the seismic inertia loads through floor diaphragms. As a result, a box-structural behavior of the retrofitted building is obtained (see Figure 2) [43,44].

The envisaged system is made by cross bracing steel elements with variable circular hollow cross sections. At each floor, a floor diaphragm made by steel profiles, metal sheets and protected by lightweight concrete connects the grid to the existing floors. A concrete foundation system transfers loads from the structure to the soil.

Life-Cycle Impact Assessment: the evaluation of the potential environmental impacts is reported in Table 3. After the data collection, impacts are calculated for raw material supply and manufacturing (A1–A3), disposal (C3–C4) and credits due to recycling (D) modules, according to EN 15804 [36].

By contrast with other renovation measures, the main advantage of the diagrid is due to its dried technique: this allows an easier dismantling and more chances to the scrap steel elements to be recycled or, in the best case integrally reused.

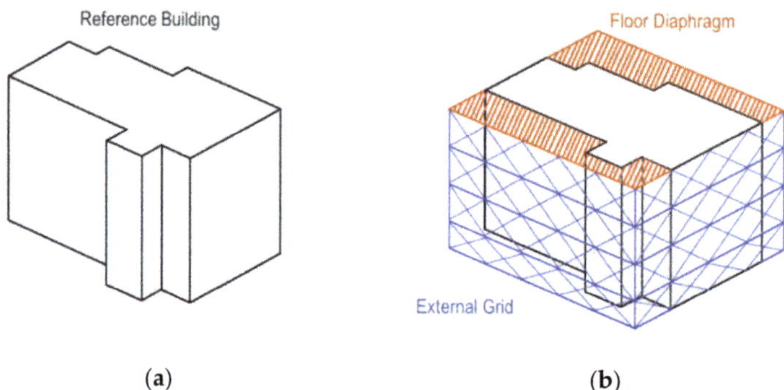

Figure 2. A schematic representation of (**a**) Reference building and (**b**) Diagrid retrofit system.

Table 3. Diagrid data collection and LCIA results (Source: Generis [38]).

Element [unit]	Quantity	Global Warming Potential [kg CO$_2$ eq./unit]			Global Warming Potential [kg CO$_2$ eq.] Total		
-		A1 – A3	C1 – C4	D	A1 – A3	C1 – C4	D
Steel profiles [kg]	105,221	0.994	8.87×10^{-4}	−0.223	104,631	93.35	−23,464.30
Metal sheet [kg]	1447	2.68	0.003	−1.58	3878	4.63	−2280
Lightweight concrete [m^3]	9.21	72.77	2.81	0	670.5	25.88	0
Concrete (C30/37) Foundation [m^3]	60.92	219	18	−21.4	13,340	1096	−1304
-			Total		122,520	1220	−27,048
			[/m^2 Net Surface]		66.05	0.66	−14.58

Costs Assessment: for a screening cost assessment of the diagrid, three groups of activities have been considered (see Table 4): (I) pouring of lightweight concrete per floor diaphragm, (II) steel components erection, (III) new foundations together with total required materials. Information is derived by local authorities and include raw materials, manufacturing and machine costs [42] and it provides a conservative estimation. Since they do not include value added tax (VAT) and by considering the lack of information, they are increased by 22%. Lastly, the recovery steel elements have been considered with reference to the current scrap European prices updated on the end of year 2019. Furthermore, the final unit price per square meter in Table 4 is calculated with reference to the net surface in the as-is layout, and reduces quite remarkably if one considers the extension of the net surface in the post-retrofit condition, thus including the red shaded area in Figure 2.

Table 4. Diagrid cost analysis [42].

Element	Unit Cost		Quantities	Total [€]
HEM steel profiles	3.09	€/kg	13,465	41,606.85
Metal sheet	3.65	€/kg	1447	5281.55
Lightweight concrete	240.74	€/m^3	4.81	1157.96
Grid–Circular hollow profiles	4.15	€/kg	91,756	380,787.10
Concrete Foundation	111.54	€/m^2	100	11,154
-	Total		439,987.45 €	
-	+VAT 22%		536,784.69 €	
-	-		253,20 €/m^2	
-	Scrap Steel Value		212.35 €/ton	−10.68 €/m^2

4.1.4. Shear Wall Retrofit System

The second retrofit system considered is composed by steel cross bracings located in specific parts of the building. The additional system is made by steel circular hollow profiles and it is assembled by employing X- or V-cross bracing. The V-Cross bracing elements are used in the neighborhood of openings for sake of usability (see Figure 3).

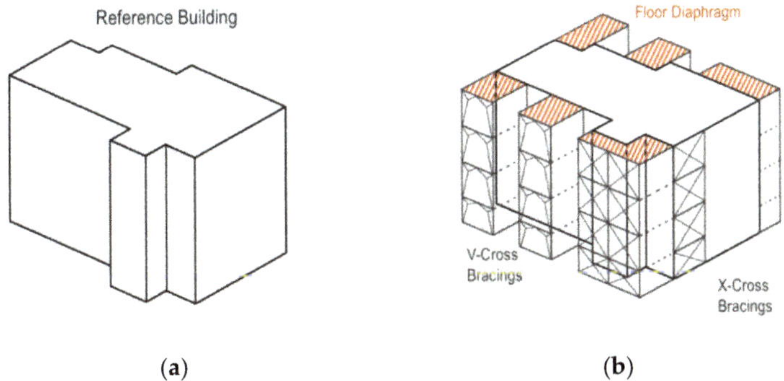

Figure 3. A schematic representation of (**a**) Reference building and (**b**) Shear walls retrofit system.

The location of such elements has been selected to re-establish the floor symmetry and regularity. An additional floor is realized where new surfaces are added. The connection between the new system and the building has been made in correspondence to the floor RC edge beams by steel elements and connectors. A concrete foundation system transfers loads from the structure to the soil. The unit price per square meter in Table 5 is calculated by spreading the total cost over the net surface in the as-is condition, and does not reflect the beneficial effect of the expansion of the net surface (red shaded area in Figure 3).

Table 5. Shear walls system data collection and LCIA results (Source: Generis [38]).

Element [unit]	Quantity	Global Warming Potential [kg CO$_2$ eq./kg]			Global Warming Potential [kg CO$_2$ eq.] total		
-	-	A1 − A3	C3 − C4	D	A1 − A3	C3 − C4	D
Steel hollow profile [kg]	138,099	0.99	8.87×10^{-4}	0.22	137,326	122.52	−30,796
Lightweight concrete [m^3]	4.78	72.77	2.81	0	347.84	13.42	0
Concrete (C30/37) Foundation [m^3]	30.46	219	18	−21.4	6670	548	−651.79
				Total	1,444,344	684.2	−31,448
				[/m^2 net surface]	77.81	0.37	−16.95

Life Cycle Impact Assessment: by collecting all materials, the total impact of the shear walls system has been evaluated through environmental databases according to EN 15804 (see Table 5) [36]. Due to the high quantity of steel and the large cross section of the chosen profiles, the total GWP is higher in the shear wall systems compared to the diagrid system.

Costs Assessment: the screening cost assessment considers steel components erection, additional floor and foundation system construction together with total required materials (see Table 6). Information is derived by local authorities and includes raw materials, manufacturing and machine costs [42]. Costs are increased by 22% VAT as well as for the diagrid system and the recovery steel elements refers to the current scrap European prices updated on the end of year 2019.

Table 6. Shear walls system cost analysis [42].

Element	Unit Cost		Quantities	Total [€]
Diaphragm—Lightweight concrete	240.74	€/m^3	4.78	1151.70
Grid–Circular hollow profiles	4.15	€/kg	138,099	573,112.75
Concrete Foundation	111.54	€/m^2	50	5577
	Total		**579,841 €**	
	+VAT 22%		**707,406 €**	
			333.68 €/m^2	
	Scrap Steel Value		212.35 €/ton	**−13.83 €/m^2**

4.1.5. Comparing Results

As shown in Figure 4, the impact and cost assessment carried out by static approaches does not encourage renovation measures associated with a structural enhancement, especially because of the evaluated total costs.

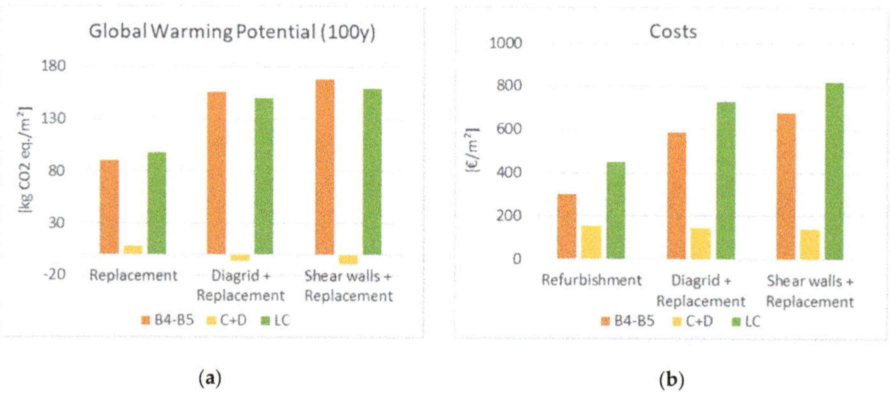

(a)   (b)

Figure 4. (**a**) LCIA and LCC (**b**) results, by considering B4–B5, C + D life-cycle modules.

This may be due to the selected functional unit, which, by referring on net surface of dwellings, can reflect the equal performances of renovation measures only in "standard" conditions, namely under static permanent or variable loads and wherever seismic events do not occur. In a context of seismic hazard, a building provided with diagrid system can be comparable to a new building in terms of the remaining structural service life, resistance and displacements due to seismic loads, performances which cannot be improved through replacement of non-structural elements.

Rather than reviewing and establishing a new functional unit, in the next section such a shortcoming is handled by revising the methodology and applying probabilistic approaches. As an advantage, the seismic hazard can be included as well as further sources of uncertainties.

4.2. Probabilistic LCA

This section shows the results of the probabilistic LCA under seismic hazard with Monte Carlo simulations. The environmental impact and costs at year-1 are given by the results of static LCA–LCC and processed according to the Section 3.

4.2.1. Seismic Event Choice

Earthquake magnitudes are derived for the geographical location of L'Aquila. The chosen events refer to the designed nominal service life of dwellings (50 years) and with higher return periods (300 and 600 years). For the considered case study, earthquakes between 50–300 years return period may lead to a mean damage level not higher than 25% of the building. Higher return periods, due to higher damages, lead to demolition and reconstruction. Table 7 reports the characteristics of the considered earthquakes [34].

Table 7. Earthquakes characteristics for geographical location of L'Aquila (Source INGV [34]). Note: PGA is the peak ground acceleration on stiff soil in terms of m/s^2.

Earthquake Return Period	Mean Annual Frequency of Earthquake	PGA 50th Percentile
50	0.02	0.1038
300	0.003333	0.2316
600	0.001667	0.3602

4.2.2. Fan Diagram—Today's Scenarios

Depending on human decisions, there are three main groups of possibilities:

- No retrofit. Service life of non-structural elements is extended. Nevertheless, the overall structural performance does not comply with current regulations (Building quality—BQ 1)
- Diagrid/shear wall intervention. The service life of the building can be extended: the intervention adjusts the building structural response under seismic loading according to current regulation (BQ 3).
- Demolition and re-building. This entails complications, such as the relocation of inhabitants, waste production and long-term works. For such reasons, demolition and rebuilding is not considered as an option to start the analysis with.

4.2.3. Fan Diagram—Future Scenarios

By considering the quality of the building at year-0, future scenarios are analyzed (Figure 5).

- Reference building without adequate adjustments may resist low-intensity earthquakes (λ_1). After repairs, some vulnerabilities can be solved (BQ 2) but the occurrence of medium- and high-intensity earthquakes (λ_2–λ_3) cause severe damage and eventually its collapse. This leads to a new construction: the new building is considered equivalent in shape and complying with seismic design requirements (BQ 3).
- The building with a new retrofit intervention enhances immediately its behavior (BQ 3). In the case of strong earthquake occurrence (λ_3) the building may be subjected to moderate damage evaluated for 25% of the building value in terms of costs and environmental impacts.

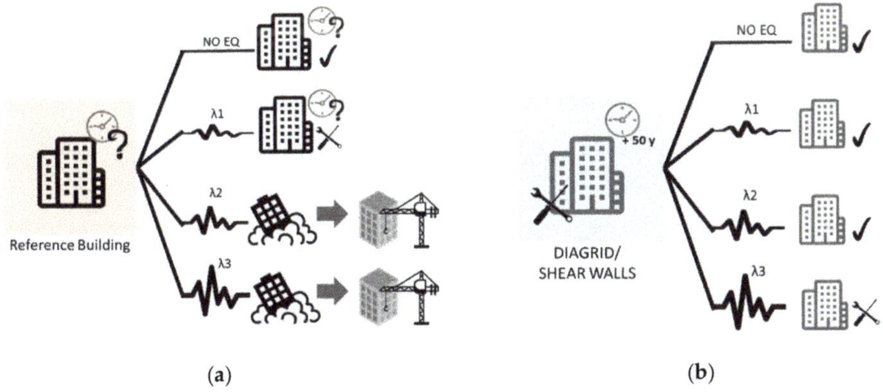

Figure 5. Future scenarios depending on the building quality: (**a**) not-retrofitted reference building; (**b**) retrofitted reference building.

Scenarios are converted into a matrix (see Table 8) in which interventions related to earthquake events and building quality (BQ) are converted in GWP and costs (Tables 9 and 10). Both matrices are processed in a Monte Carlo Simulation by means of the software MATLAB [45]. In Figure 6 a flow chart shows the building quality function variation due to earthquakes' occurrence.

Table 8. Interventions due to seismic event occurrence and building quality (BQ).

Events BQ	No Earthquake	EQ1	EQ2	EQ3
1: As is	0	1/4 * (Building)	Dem. + Building	Dem. + Building
2: Repaired	0	0	1/4 * (Building)	Dem + Building
3: Adjusted	0	0	0	1/4 * (Building)

Table 9. GWP in kg CO_2/m^2 due to seismic event occurrence and building quality.

Events BQ	No Earthquake	EQ1	EQ2	EQ3
1: As is	0	123	497	497
2: Repaired	0	0	123	497
3: Adjusted	0	0	0	123

Table 10. Costs expressed in €/m^2 due to seismic event occurrence and building quality.

Events BQ	No Earthquake	EQ1	EQ2	EQ3
1: As is	0	143	1252	1252
2: Repaired	0	0	143	1252
3: Adjusted	0	0	0	143

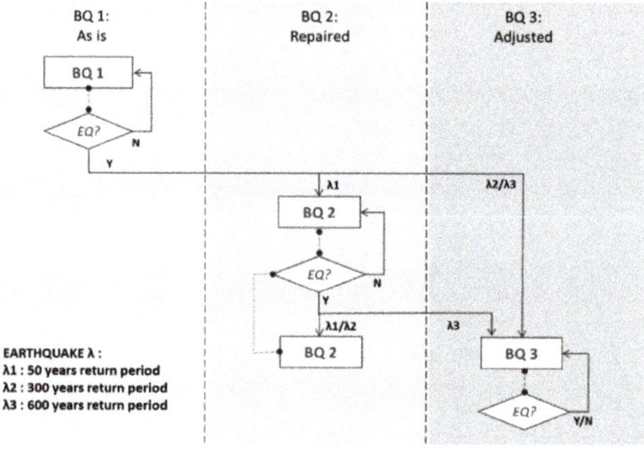

Figure 6. Building quality function variation depending on earthquake occurrence.

For impacts and costs discounting rates, the Italian price increase rate is derived from direct measurements or as the average value of the inflation rate (evaluated in 2019 as 0.82% [46]). The discount rate is derived from government bonds (expiring date at least of 10 years—2.8% [47]). LCA analyses consider a "business as usual" modelled decarbonisation rate of 1.6% [48].

4.2.4. Monte Carlo (MC) Simulation Results

After obtaining a series of possible CO_2 emissions and costs values, results are processed. Relevant statistical values are derived, e.g., minimum, maximum, mean and median values (see Table 11).

Table 11. Probabilistic LCA–LCC results (10,000 runs MC simulation).

GWP (100 Years), [kg CO_2 eq./m^2]	Min	Max	Median	Mean	St. dev.
Reference Building (RB)	299	2,221	458	542	234
RB+Retrofit Diagrid (DG)	253	566	253	268	39
RB+Shear Walls (SW)	262	576	262	277	39
Costs (100 Years), [€/m^2]	**Min**	**Max**	**Median**	**Mean**	**St. dev.**
Reference Building (RB)	639	3,704	845	938	326
RB+Retrofit Diagrid (DG)	445	881	446	458	40
RB+Shear Walls (SW)	534	969	534	546	40

It is worth to notice that, due to the recorded values distribution, the median and arithmetic mean are different, especially for RB case, this may be related to different reasons, such as a non-Gaussian distribution, and, more likely, to the discrete number of the considered earthquake intensity levels and to the given deterministic repair costs and environmental impacts for each earthquake scenario. In addition, RB cases are characterized by higher dispersion. Besides these limitations, the results obtained still allow general considerations on the effects of including the earthquake occurrence in LCA and LCC analyses.

The results are grouped in ranges of values and their occurrence within such ranges is counted. The following graphs represent the probability trends. Two groups of values are distinguished (see Figure 7): one in correspondence to low-medium values (200–500 kg CO_2 eq.) and another to high environmental impacts (500 to 1000 kg CO_2 eq.). The second group reveals the effects of the occurrence of medium and strong intensity earthquakes. For a not-retrofitted building, critical consequences are expected and consequently high economic and environmental impacts due to the demolition and re-building process are observed (Figure 7a). A building provided with structural retrofit systems (Figure 7b,c) does not reveal high impact occurrence because preventive initiatives have been undertaken. Few cases are recorded in the shear wall case (Figure 7c), due to higher environmental impacts of reinforcement production.

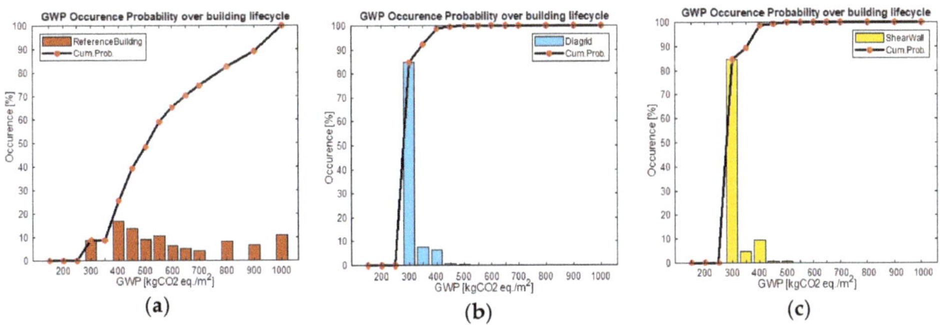

Figure 7. GWP occurrence probability for (**a**) not-retrofitted reference building, retrofitted building with (**b**) diagrid and (**c**) shear walls.

With regard to LCC results, a non-retrofitted building presents costs occurrence that is rather homogenous between 700–1100 €/m^2 while values higher than 1200 €/m^2 have more than 20% probability of occurrence. Retrofitted buildings (Figure 8b,c) do not record high costs.

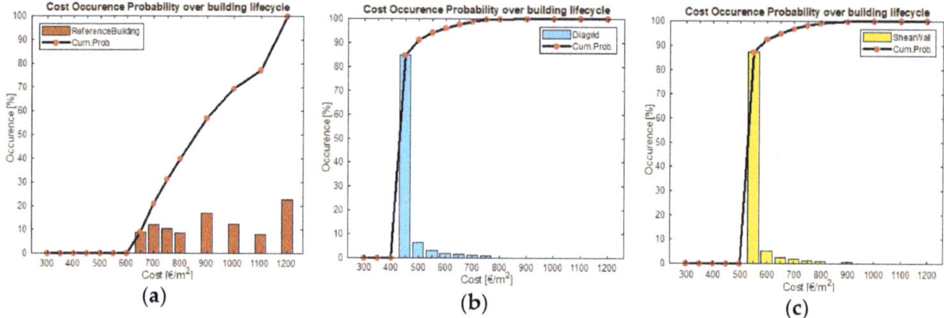

Figure 8. Cost occurrence probability for (**a**) non-retrofitted reference building, retrofitted building with (**b**) diagrid and (**c**) shear walls.

Lastly, probabilistic LCA and LCC results are shown over the 100 years' observation time of the building (Figure 9). For each year average GWP and costs are calculated and summed with previous values in order to analyze the yearly increase. This visualization of the results enables to evaluate the measure impact over time. As shown by the graphs, every 30 years a building refurbishment increases costs and impacts. GWP curves present a linear trend, while costs, because of discounting, are non-linear. This is especially noticeable for the reference building, where sensitivity to change is higher. In this regard, the building quality variation is visible through the curve gradient. When the building achieves the highest building quality, both GWP and costs curve trends become as flat as for buildings provided with structural retrofit. Nevertheless, this occurs for the reference building only beyond 75 years' analysis. The reference building seems to be advantageous in terms of GWP only in a short or medium period (20 years ca. for both diagrid and shear walls). In comparison with lifecycle costs of the reference building, the diagrid proved to be the most advantageous renovation with a payback time of 44 years. It is worth noting that if we would have considered standard profiles instead of circular ones for the diagrid, we would have had a payback time of 33 years. In any case, if we account for the extra floor surface provided by the new retrofit system ($92m^2$ per floor for the diagrid and 55 m^2 per floor for the shear walls system), the cost over the building lifecycle reduces (dashed lines in Figure 9) with a payback time of 27 years (22 years with standard profiles) for the diagrid and 47 years for the shear walls system.

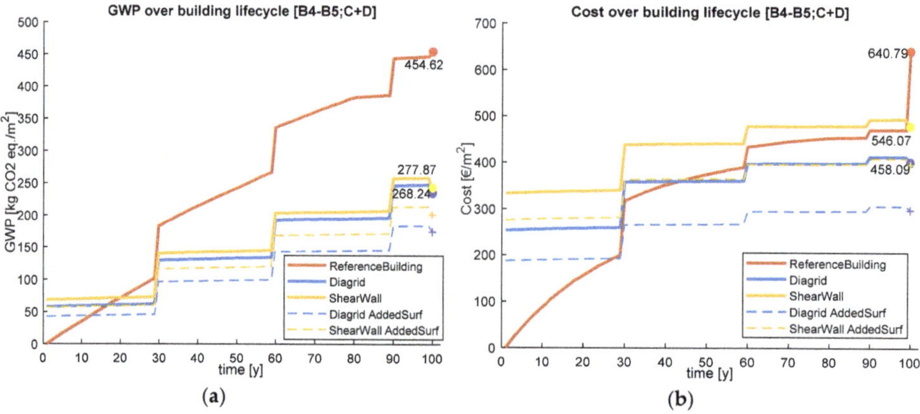

Figure 9. Future scenarios depending on the building quality: (**a**) GWP and (**b**) cost over building lifecycle.

5. Discussion

The comparison between the two methodologies highlights that probabilistic LCA and LCC can solve miscalculation due to uncertainties and thus prevent from misleading LCA-based decisions. The probabilistic approach enriches the analyses by adding additional functions and parameters related to environmental, economic contingencies and external factors, leading to a more complex model which can allow stakeholders to make more aware choices.

According to the results obtained, static analyses are not able to consider the advantage of a seismic retrofit, since it may be not capable through a single trustworthy value to reveal the performance of the construction under seismic actions and the risk of long-term losses due to the lack of a suitable anti-seismic structural system. On the contrary, the presented method performs both probabilistic LCA–LCC and includes a statistical assessment in order to provide impact occurrence probability for different ranges of values. Among the analyzed solutions, the retrofit system with a diagrid achieves more economic and environmental savings in shorter terms, while the analyzed shear wall system can be considered not efficient for the considered case study.

In comparison with the PEER-PBEE probabilistic model, the proposed approach introduces a function related to the building quality. Thank to this, the structural enhancement due to repairs of seismic adjustments has been included and consequently a reduced likelihood of damage in the case of subsequent earthquakes. As a disadvantage, the missing variation of the building quality function may lead to the considered case study overestimating results that would not always encourage anti-seismic reinforcements within the decision making process over the building life cycle (Appendix A).

To validate this study, in particular referring to the occurrence rate of the simulated earthquakes, a comparison between the results of the model and the target annual frequency of exceedance has been made: for 10,000 runs over 100 years' observation time, the recorded frequency can slightly diverge (see Table 12) but an accuracy of $10^{-2} \div 10^{-3}$ is achieved. More MC runs would improve the convergence between values model and empirical values, but, on the other hand, generate problems due to the used calculation software limits and the high quantity of values to be processed.

Table 12. Comparison with Mean Annual Frequency of Exceedance from probabilistic seismic hazard analysis model and frequency recorded by Monte Carlo (MC) Simulation.

Earthquake Return Period	Mean Annual Frequency of Exceedance	MC Sim Recorded Frequency
50	0.02	0.019142
300	0.033333	0.003337
600	0.001667	0.001657

It is necessary to underline the "Proof-of-Concept" character of the presented methodology, which still contains inaccuracies and approximations. A first approximation can be related to the probabilistic seismic hazard assessment (PSHA) model, which is currently used and it results advantageous because it describes earthquake occurrence and effects by requiring a limited number of parameters. Due to the complexity of introducing Monte Carlo simulations in the LCA evaluation, future research will focus on the development and investigation of simplified procedures to include earthquake effects in LCA, analogously to what has developed in earthquake engineering, such as software (as PACT—Performance Assessment Calculation Tool, FEMA P-58 [48]) or direct loss assessment procedures (such as the displacement-based loss assessment [49]).

More significant inaccuracies are related to the correct assessment of the building damage and to the cost of retrofit. In this work, three seismic events with increasing intensity have been considered, which lead to a certain building damage level. Actually, there is a variety of events with different consequences, depending on the vulnerability of the existing building: in this regard, the inclusion of a wider set of earthquakes would improve the technical aspect of the analyses, by adding more scenarios. Specific local damages and interventions may be included in order to derive more accurate predictions under seismic hazard on building components level.

The framework could be enriched through the inclusion of more sources of LCA uncertainties. On one hand, this enables more transparent results but on the other hand increases the complexity of the model. Hence, it is important to identify uncertainties that, individually or together with other factors, influence results' robustness. Lastly, a more detailed data collection or the extension of LCA-LCC system boundaries, through i.e., the inclusion of building energy demand information, would improve the analysis with more comprehensive results.

Author Contributions: Conceptualization and methodology, R.D.B., J.G.; Analysis R.D.B., A.B., A.M.; investigation, R.D.B.; data curation, R.D.B., A.B., A.M.; writing—original draft, R.D.B.; visualization R.D.B.; writing—review and editing, R.D.B., A.B., A.M. and R.H.; Supervision, A.B., A.M., R.H. and J.G. All authors have read and agreed to the published version of the manuscript.

Funding: This research and the APC was funded by the Deutsche Forschungsgemeinschaft (DFG, German Research Foundation) under Germany´s Excellence Strategy—EXC 2120/1. Grant number 390831618.

Acknowledgments: Supported by the Deutsche Forschungsgemeinschaft (DFG, German Research Foundation) under Germany´s Excellence Strategy—EXC 2120/1-390831618.

Conflicts of Interest: The authors declare no conflict of interest.

Appendix A

In Table A1 are the results of 10,000 runs of Monte Carlo simulation of LCA–LCC analyses by varying the starting building quality of the reference building on year y = 1. BQ variation due to earthquake occurrence is not considered.

Table A1. Probabilistic LCA–LCC results without BQ variation due to earthquake occurrence.

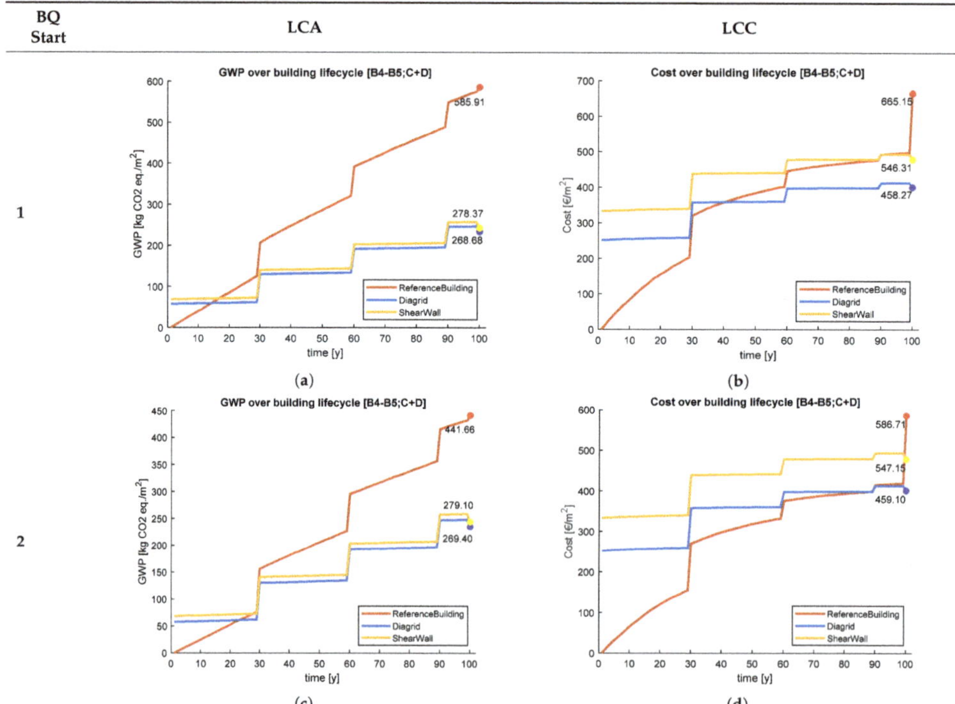

Table A1. *Cont.*

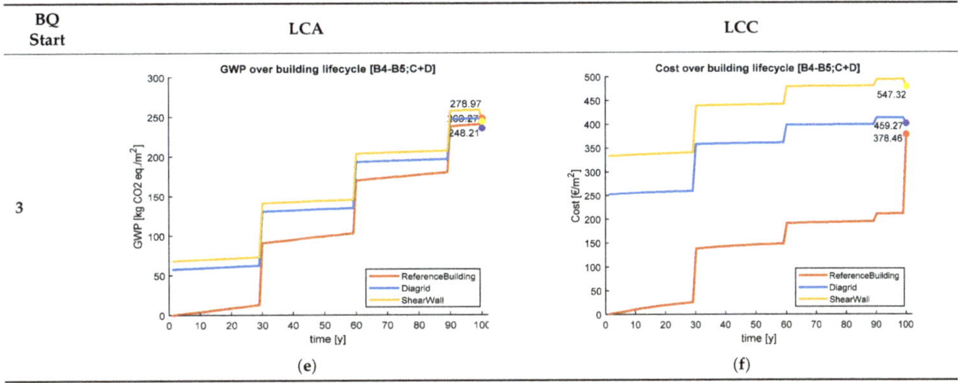

In Table A2 are the results of 10,000 runs Monte Carlo simulation of LCA–LCC analyses by varying the starting building quality of the reference building on year y = 1. BQ variation due to earthquake occurrence is considered.

Table A2. Probabilistic LCA–LCC results with BQ variation due to earthquake occurrence.

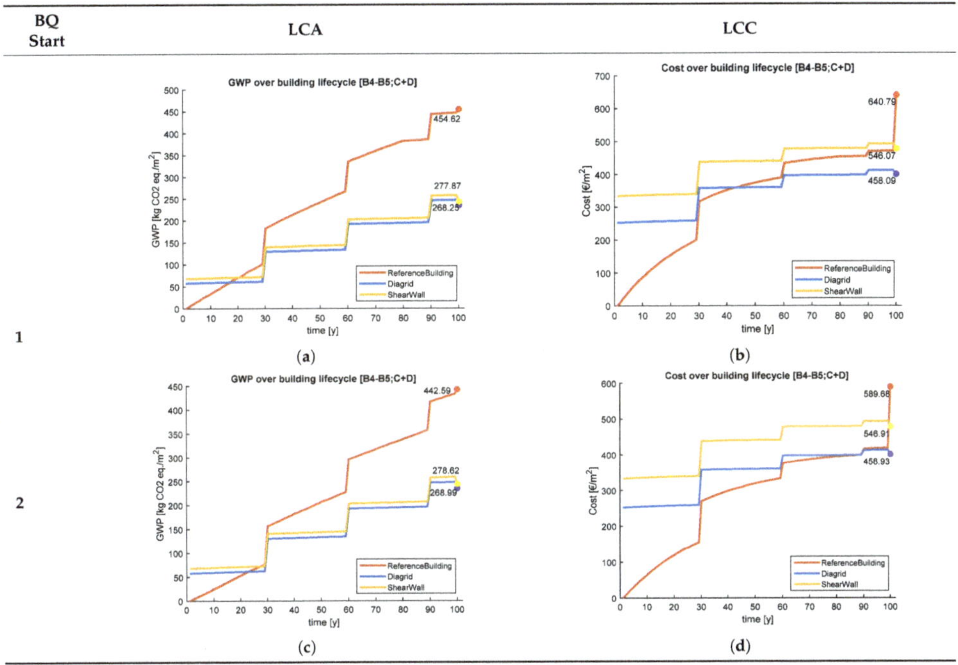

BQ Start	LCA	LCC
3	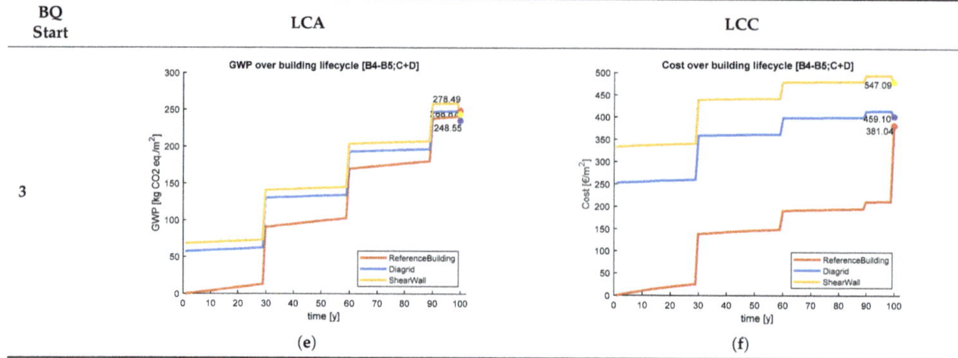	

(e)   (f)

References

1. Global Alliance for Buildings and Construction. Global Status Report 2018. Available online: https://www.unenvironment.org/resources/report/global-status-report-2018 (accessed on 30 November 2019).
2. Eurostat. People in the EU—statistics on housing conditions. Available online: https://ec.europa.eu/eurostat/statistics-explained/index.php/People_in_the_EU_-_statistics_on_housing_conditions (accessed on 30 November 2019).
3. Frey, P.; Dunn, L.; Cochran, R.; Spataro, K.; McLennan, J.F.; Di Nola, R.; Heider, B. The Greenest Building: Quantifying the Environmental Value of Building Reuse. Available online: http://www3.cec.org/islandora-gb/en/islandora/object/islandora%3A1018 (accessed on 30 November 2019).
4. Hasik, V.; Escott, E.; Bates, R.; Carlisle, S.; Faircloth, B.; Bilec, M.M. Comparative whole-building life cycle assessment of renovation and new construction. *Build. Environ.* **2019**, *161*, 106218. [CrossRef]
5. Buyle, M.; Braet, J.; Audenaert, A. Life cycle assessment in the construction sector: A review. *Renew. Sustain. Energy Rev.* **2013**, *26*, 379–388. [CrossRef]
6. Belleri, A.; Marini, A. Does seismic risk affect the environmental impact of existing buildings? *Energy Build.* **2016**, *110*, 149–158. [CrossRef]
7. Internaational Organization for Standardization. *Life Cycle Assessment—Principles and Framework, DIN EN ISO 14040:2006*; Internaational Organization for Standardization: Geneva, Switzerland, 2006.
8. Internaational Organization for Standardization. *Life cycle Assessment—Requirements and Guidelines, DIN EN ISO 14044*; International Organization for Standardization: Geneva, Switzerland, 2006.
9. Igos, E.; Benetto, E.; Meyer, R.; Baustert, P.; Othoniel, B. How to treat uncertainties in life cycle assessment studies? *Int. J. Life Cycle Assess.* **2019**, *24*, 794–807. [CrossRef]
10. Gantner, J.; Fawcett, W.; Ellingham, I. Probabilistic Approaches to the Measurement of Embodied Carbon in Buildings. In *Embodied Carbon in Buildings: Measurement, Management, and Mitigation*; Pomponi, F., de Wolf, C., Eds.; Springer: Cham, Switzerland, 2018; pp. 23–50. ISBN 978-3-319-72796-7.
11. Huijbregts, M.A.J. Application of uncertainty and variability in LCA. *Int. J. Life Cycle Assess.* **1998**, *3*, 273. [CrossRef]
12. Williams, E.D.; Weber, C.L.; Hawkins, T.R. Hybrid Framework for Managing Uncertainty in Life Cycle Inventories. *J. Ind. Ecol.* **2009**, *13*, 928–944. [CrossRef]
13. Björklund, A.E. Survey of approaches to improve reliability in lca. *Int. J. Life Cycle Assess.* **2002**, *7*, 64. [CrossRef]
14. Huijbregts, M.A.J. Part II: Dealing with parameter uncertainty and uncertainty due to choices in life cycle assessment. *Int. J. Life Cycle Assess.* **1998**, *3*, 343–351. [CrossRef]
15. Roy, P.-O.; Deschênes, L.; Margni, M. Uncertainty and spatial variability in characterization factors for aquatic acidification at the global scale. *Int. J. Life Cycle Assess.* **2014**, *19*, 882–890. [CrossRef]
16. Ziyadi, M.; Al-Qadi, I.L. Model uncertainty analysis using data analytics for life-cycle assessment (LCA) applications. *Int. J. Life Cycle Assess.* **2019**, *24*, 945–959. [CrossRef]
17. Van Zelm, R.; Huijbregts, M.A.J. Quantifying the trade-off between parameter and model structure uncertainty in life cycle impact assessment. *Environ. Sci. Technol.* **2013**, *47*, 9274–9280. [CrossRef] [PubMed]

18.	Yana, A.G.A.; Rusdhi, H.A.; Wibowo, M.A. Analysis of Factors Affecting Design Changes in Construction Project with Partial Least Square (PLS). *Procedia Eng.* **2015**, *125*, 40–45. [CrossRef]

19.	Alnuaimi, A.S.; Taha, R.A.; Al Mohsin, M.; Al-Harthi, A.S. Causes, Effects, Benefits, and Remedies of Change Orders on Public Construction Projects in Oman. *J. Constr. Eng. Manag.* **2010**, *136*, 615–622. [CrossRef]

20.	Hsieh, T.-Y.; Lu, S.-T.; Wu, C.-H. Statistical analysis of causes for change orders in metropolitan public works. *Int. J. Proj. Manag.* **2004**, *22*, 679–686. [CrossRef]

21.	Moayeri, V. Design Change Management in Construction Projects Using Building Information Modeling (BIM). Ph.D. Thesis, Concordia University, Montreal, QC, Canada, 2017.

22.	Duffey, R. Predicting Rare Events Risk Exposure Uncertainty and Unknown Unknowns. In Proceedings of the 19th Advances in Risk and Reliability Technology Symposium, Stratford-upon-Avon, UK, 12–14 April 2011.

23.	European Committee for Standardization. *Eurocode 8: Design of Structures for Earthquake Resistance. Part 1: General Rules, Seismic Actions and Rules for Buildings, EN 1998-1:200*; European Committee for Standardization: Brussels, Belgium, 2004.

24.	Vitiello, U.; Salzano, A.; Asprone, D.; Di Ludovico, M.; Prota, A. Life-Cycle Assessment of Seismic Retrofit Strategies Applied to Existing Building Structures. *Sustainability* **2016**, *8*, 1275. [CrossRef]

25.	Kircher, C.A. HAZUS®-MH: Advanced Engineering Building Module (AEBM)—Technical and User's Manual. Available online: https://www.fema.gov/media-library/assets/documents/16638 (accessed on 30 November 2019).

26.	Günay, M.S.; Mosalam, K.M. PEER Performance Based Earthquake Engineering Methodology, Revisited. *J. Earthq. Eng.* **2013**, *17*, 829–858. [CrossRef]

27.	Calvi, G.M.; Sousa, L.; Ruggeri, C. Energy Efficiency and Seismic Resilience: A Common Approach. In *Multi-Hazard Approaches to Civil Infrastructure Engineering*; Gardoni, P., LaFave, J.M., Eds.; Springer: Cham, Switzerland, 2016; pp. 165–208. ISBN 978-3-319-29713-2.

28.	European Commitee for Standardization. *Eurocode 0—Basis of Structural Design, EN 1990:2002*; European Commitee for Standardization: Brussels, Belgium, 2005.

29.	Buildings Performance Institute Europe (BPIE). Europe's buildings under the microscope: A country-by-country review of the energy performance of buildings. Available online: http://bpie.eu/publication/europes-buildings-under-the-microscope/ (accessed on 30 November 2019).

30.	Garrido, M.A.; Paulo, P.V.; Branco, F.A. Service life prediction of façade paint coatings in old buildings. *Constr. Build. Mater.* **2012**, *29*, 394–402. [CrossRef]

31.	Ortega, L.; Serano, B.; Fran, J. Proposed method of estimating the service life of building envelopes. *J. Constr.* **2015**, *14*, 60–68.

32.	Shi, B.; Wang, Z.; Woolery, E.W. Understanding Seismic Hazard and Risk Assessments: An Example in the New Madrid Seismic Zone of the Central United States. In Proceedings of the 8th U.S. National Conference on Earthquake Engineering, San Francisco, CA, USA, 18–22 April 2006.

33.	Cornel, C.A. Engineering Seismic Risk Analysis. *Bull. Seismol. Soc. Am.* **1968**, *58*, 1583–1606.

34.	Istituto Nazionale di Geofisica e Vulcanologia (INGV). Interactive Seismic Hazard Maps. Available online: http://esse1-gis.mi.ingv.it/s1_en.php (accessed on 30 November 2019).

35.	Ministero delle Infrastrutture e dei Trasporti. *Norme Tecniche per le Costruzioni, D.M. 17 Gennaio 2018.*; Gazzetta Ufficiale n. 42; Ministero delle Infrastrutture e dei Trasporti: Rome, Italy, 2018.

36.	European Committee for Standardization. *Sustainability of Construction Works—Environmental Product Declarations- Core Rules for the Product Category of Construction Products, EN 15804:2012 +A2:2019*; European Committee for Standardization: Brussels, Belgium, 2019.

37.	Thinkstep. *GaBi Software Version 9.2*; Thinkstep—A Sphera Company: Leinfelden-Echterdingen, Germany, 2019.

38.	Fraunhofer IBP. Generis Online Software. Available online: www.generis.live (accessed on 10 February 2020).

39.	Gervasio, H.; Dimova, S. Model for Life Cycle Assessment (LCA) of buildings: EFIResources: Resource Efficient Construction towards Sustainable Design. Available online: https://ec.europa.eu/jrc/en/publication/model-life-cycle-assessment-lca-buildings (accessed on 30 November 2019).

40.	Di Ludovico, M.; Prota, A.; Moroni, C.; Manfredi, G.; Dolce, M. Reconstruction process of damaged residential buildings outside historical centres after the L'Aquila earthquake: Part II—"heavy damage" reconstruction. *Bull. Earthq. Eng.* **2017**, *15*, 693–729. [CrossRef]

41. Risi, M.T.D.; Del Gaudio, C.; Verderame, G.M. Evaluation of Repair Costs for Masonry Infills in RC Buildings from Observed Damage Data: The Case-Study of the 2009 L'Aquila Earthquake. *Buildings* **2019**, *9*, 122. [CrossRef]

42. Regione Abruzzo—Dipartimento infrastrutture, trasporti, mobilità, reti e logistica. Prezzi Informativi Opere Edili nella Regione Abruzzo. 2019. Available online: https://www.regione.abruzzo.it/content/nuovo-prezzario-regionale (accessed on 10 February 2020).

43. Labò, S.; Passoni, C.; Marini, A.; Belleri, A.; Camata, G.; Riva, P.; Spacone, E. Diagrid solutions for a sustainable seismic, energy, and architectural upgrade of European RC buildings. In Proceedings of the XII International Conference on Structural Repair and Rehabilitation, Porto, Portugal, 26–29 October 2016.

44. Labò, S.; Passoni, C.; Marini, A.; Belleri, A.; Riva, P. Diagrid structures as innovative retrofit solutions for existing reinforced concrete buildings. In Proceedings of the 12th Fib International PhD Symposium in Civil Engineering, Prague, Czechia, 29–31 August 2018.

45. MathWorks. *MATLAB®Software,R2019b (Version 9.7)*; Mathworks: Natick, MA, USA, 2019.

46. Inflation.eu. Inflation—Current and Historic Inflation by Country. Available online: https://www.inflation.eu/inflation-rates/italy/historic-inflation/hicp-inflation-italy-2019.aspx (accessed on 30 November 2019).

47. PWC. Low Carbon Economy Index 2018. Available online: https://www.pwc.co.uk/ghost/low-carbon-economy-index-2018.html (accessed on 30 November 2019).

48. Applied Technology Council for the Federal Emergency Management Agency (FEMA). *Performance Assessment Calculation Tool (PACT)*; Applied Technology Council for the Federal Emergency Management Agency (FEMA): Washington, WA, USA, 2012.

49. Welch, D.P.; Sullivan, T.J.; Calvi, G.M. Developing Direct Displacement-Based Procedures for Simplified Loss Assessment in Performance-Based Earthquake Engineering. *J. Earthq. Eng.* **2014**, *18*, 290–322. [CrossRef]

 *buildings* 

Article

Durability and Climate Change—Implications for Service Life Prediction and the Maintainability of Buildings

Michael A. Lacasse *, Abhishek Gaur and Travis V. Moore

National Research Council Canada, Construction Research Centre, 1200 Montreal Road, Building M24, Ottawa, ON K1A 0R6, Canada; abhishek.gaur@nrc.ca (A.G.); Travis.Moore@nrc.ca (T.V.M.)
* Correspondence: Michael.Lacasse@nrc-cnrc.gc.ca; Tel.: +1-613-993-9611

Received: 12 October 2019; Accepted: 27 February 2020; Published: 12 March 2020

Abstract: Sustainable building practices are rooted in the need for reliable information on the long-term performance of building materials; specifically, the expected service-life of building materials, components, and assemblies. This need is ever more evident given the anticipated effects of climate change on the built environment and the many governmental initiatives world-wide focused on ensuring that structures are not only resilient at their inception but also, can maintain their resilience over the long-term. The Government of Canada has funded an initiative now being completed at the National Research Council of Canada's (NRC) Construction Research Centre on "Climate Resilience of Buildings and Core Public infrastructure". The outcomes from this work will help permit integrating climate resilience of buildings into guides and codes for practitioners of building and infrastructure design. In this paper, the impacts of climate change on buildings are discussed and a review of studies on the durability of building envelope materials and elements is provided in consideration of the expected effects of climate change on the longevity and resilience of such products over time. Projected changes in key climate variables affecting the durability of building materials is presented such that specifications for the selection of products given climate change effects can be offered. Implications in regard to the maintainability of buildings when considering the potential effects of climate change on the durability of buildings and its components is also discussed.

Keywords: buildings; building components; building elements; climate change; degradation; durability; maintainability; service life prediction

---

1. Introduction

Sustainable building practices are rooted in the need for reliable information on the long-term performance of building materials; specifically, the expected service-life of building materials, components, and assemblies. This need is ever more evident given the anticipated effects of climate change on the built environment and the many governmental initiatives world-wide focused on ensuring that structures are not only resilient at their inception but also, can maintain their resilience over the long-term.

In this paper, climate resilience pertains to the ability of a building material, component or element to maintain its function if subjected to the effects of climate loads as may occur in the future under different climate change scenarios as compared to those effects arising from loads sustained under current historical climate conditions.

To provide context to the implications for the durability and maintainability of buildings as may be affect by changes in the climate in the future, it is useful to first gain an appreciation of the expected global climate change and thereafter some measure of the climate change of Canada, given that this study focuses on a Canadian perspective. Thereafter, a brief review of the impacts of climate change on buildings is provided as a framework in which the NRC research program on climate resilient buildings is described and thereafter, a review is given of literature pertinent to the degradation of building materials components and elements arising from the effects of climate change.

1.1. Global Climate Change and the Climate Change of Canada

There is evidence around the globe, perhaps also evident in every day weather, that the climate is changing much more drastically than recorded in the past, and the expectation is that climate change will continue into the foreseeable future. Historically observed and projected future global temperature changes based on different emission scenarios are given in Figure 1. The extent of global warming (°C) as provided in this figure is relative to the fifty year period spanning 1850–1900.

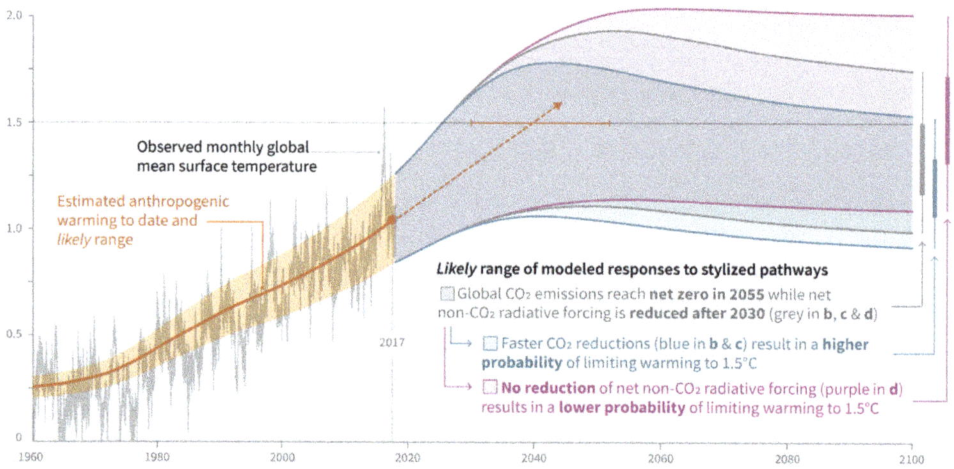

Figure 1. Observed and projected global temperature change based on different emission scenarios; global warming is relative to 1850–1900 (°C) [1].

The overall monthly global mean surface temperature is given, as is the estimated anthropogenic warming to date (in orange tint); the likely temperature range from anthropogenic warming is also provided. The orange dashed line that begins in 2017 projects increases in global warming in the future and indicates that the historically recorded rate of warming global temperatures are projected to increase by 1.5 °C or more in the future. Global warming could be kept at 1.5 °C level provided the CO_2 emissions are gradually reduced to net zero levels. Furthermore, the more rapidly emissions are reduced, the greater the likelihood that by 2100, global temperatures will be maintained at a 1.5 °C increase. The grey portion of the figure that stretches beyond 2017 shows the projected range of warming if CO_2 emissions are mitigated systematically following select "stylized anthropogenic emission and forcing pathways". Whereas, the blue curve shows the resulting effects on temperature change should reductions in CO_2 emissions reach net zero by 2040, the grey curve shows the scenario when the CO_2 emissions reach net zero by 2055. Finally, more rapid CO_2 emission reductions (grey and blue curves) suggest a higher chance of limiting warming to 1.5 °C than the purple curve that represents no decline in CO_2 emissions after 2030 and results in a lower likelihood of limiting warming to 1.5 °C.

Likewise, the climate of Canada is evidently warming, as shown in Figure 2 [2]. The figure shows time series of recorded average annual temperatures across the country that has fluctuated from year to year over the 1948–2018 period. The linear trend indicates that annual temperatures averaged across the nation have warmed continuously during this period adding up to a total warming of 7 °C over the past 71 years. The national average temperature for 2018 (January to December) was 0.5 °C above the mean temperatures over the 1961–1990 period, highlighting significant warming even in the recent decades.

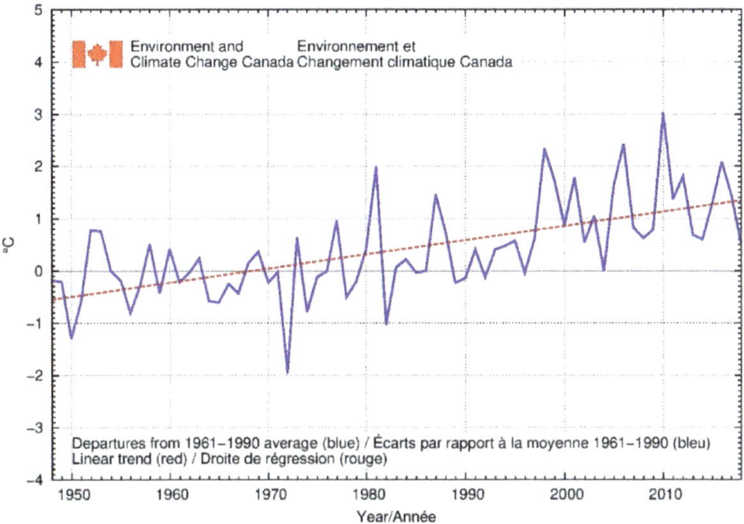

Figure 2. Nationally averaged annual temperature anomalies with reference to recorded temperatures over 1948–2018 [2].

The spatial temperature departures map, provided in Figure 3, shows that the Yukon, most of the Northwest Territories, as well as parts of Nunavut and British Columbia experienced temperatures above the baseline average in 2018, whereas, for some locations of Northern Quebec, temperatures were well below the baseline average. Annual temperatures were generally near the average recorded temperatures over 1961–1990 in the remainder of the country.

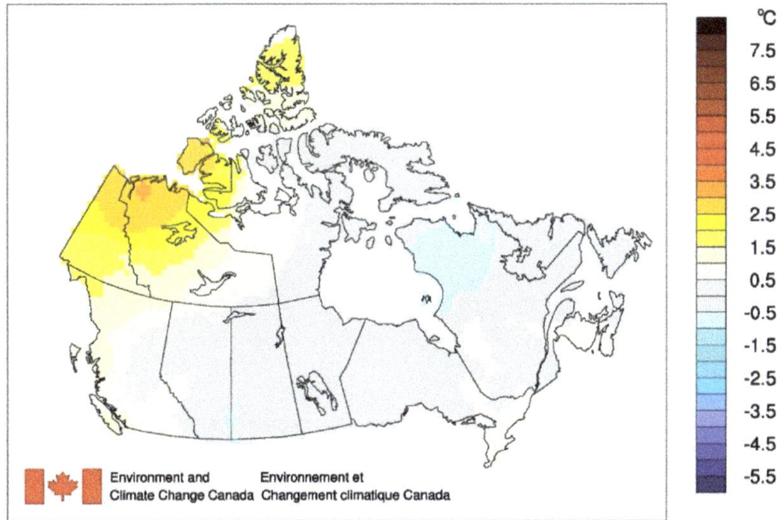

Figure 3. Spatial distribution of temperature departures for 2018 from the 1961–1990 average [2].

1.2. Climate Change and Impacts on Buildings

The primary driver for climate change produced directly by human activities is greenhouse gas (GHG) emissions. Of course, other natural climate determinants such as terrestrial, solar planetary, and orbital effects all affect the global climate. Nonetheless, given the presence of ever increasing amounts of GHG emissions to the atmosphere, climate change is upon us to the extent that there are to be expected changes, depending on the geographical location, in mean values for the primary climate variables that include: temperature, precipitation, humidity, solar radiation and wind speed. Additionally, the variability from mean values of these climate parameters is also expected to increase [3].

In Figure 4, a schematic is given that attempts to depict the complexity of the interrelation amongst climatic factors affecting buildings, land and coastal systems, and the degradation of the environment arising from the potential effects of climate change. This was adapted from that provided by de Wilde and Coley [3]. Buildings provide environmental separation between the outdoor environment, which is subject to the effects of climate change, and the indoor environment to ensure building occupants reside in healthy, comfortable, and functional dwellings. In Figure 4, from left to right, information is provided on: (i) climate change as a driving force; (ii) the environmental effects of climate change that pertain to buildings; and, (iii) the possible impact of those environmental effects on buildings.

As such, it is apparent that changes in mean values for the primary climate variables such as, temperature, precipitation, humidity, solar radiation and wind necessarily arise from natural climate determinants (i.e., solar planetary, terrestrial, orbital), but as well, from man-made greenhouse gas emissions that are driving changes to the future climate. Such changes will also affect the variability of the climate and this is revealed by the anticipated environmental effects in the future where the frequency and severity (duration and intensity) of extreme climate events will increase. Together with these effects, there will be a gradual change in the means of climate variables such as temperature, and an associated rise in sea levels given the warming temperatures.

What are the associated impacts on buildings, as may arise from these various environmental effects? There is anticipated, for certain locations in Canada, increases in the frequency, intensity and duration of precipitation events as well as increase peak wind loads and the frequency of occurrence of extreme winds. Hence, it is clearly expected that extreme wind-driven rain events will be more prevalent in the future. As such, the exterior of the building will be subjected to more intense climate loads of longer duration that will, in turn, increase the risk to premature degradation of building elements, such as roof, wall and fenestration systems, and as well, the risk of water entry of building elements, resulting in moisture-related problems.

Increases in global temperatures will bring about decreases in heating loads, as evident by reductions in heating degree days, and increases in cooling loads, in particular in urban agglomerations where heat island effects may prevail over the summer months. Lack of attention to extreme heat events may bring about overheating in buildings that, in turn, increases health risks to the vulnerable portion of the population such as the elderly, the sick and physically challenged, and the very young. In respect to the operation of buildings, there may be mismatch in the capacity to cool or heat buildings, depending on the season, that could bring about energy-use inefficiencies.

The general climate attributes associated with climate changes as may affect buildings have been known for decades. Hence, the need for climate resilience in buildings is apparent from the number of relevant tools, practices, and guides for increasing the climate resilience of new and retrofit buildings. A few notable examples include: the Public Infrastructure Engineering Vulnerability Committee (PIEVC) assessment protocol [4], the Institute for Catastrophic Loss Reduction (ICLR) Home Builder's Guide to promote the construction of disaster-resilient homes [5], the US Department of Housing and Urban Development (HUD) Climate Change Adaptation Plan [6], and the US Green Building Council's report, Green Building and Climate Resilience [7]. Common to all these resources is providing useful and practical information on how the climate will affect buildings in the future as may arise from a warming climate.

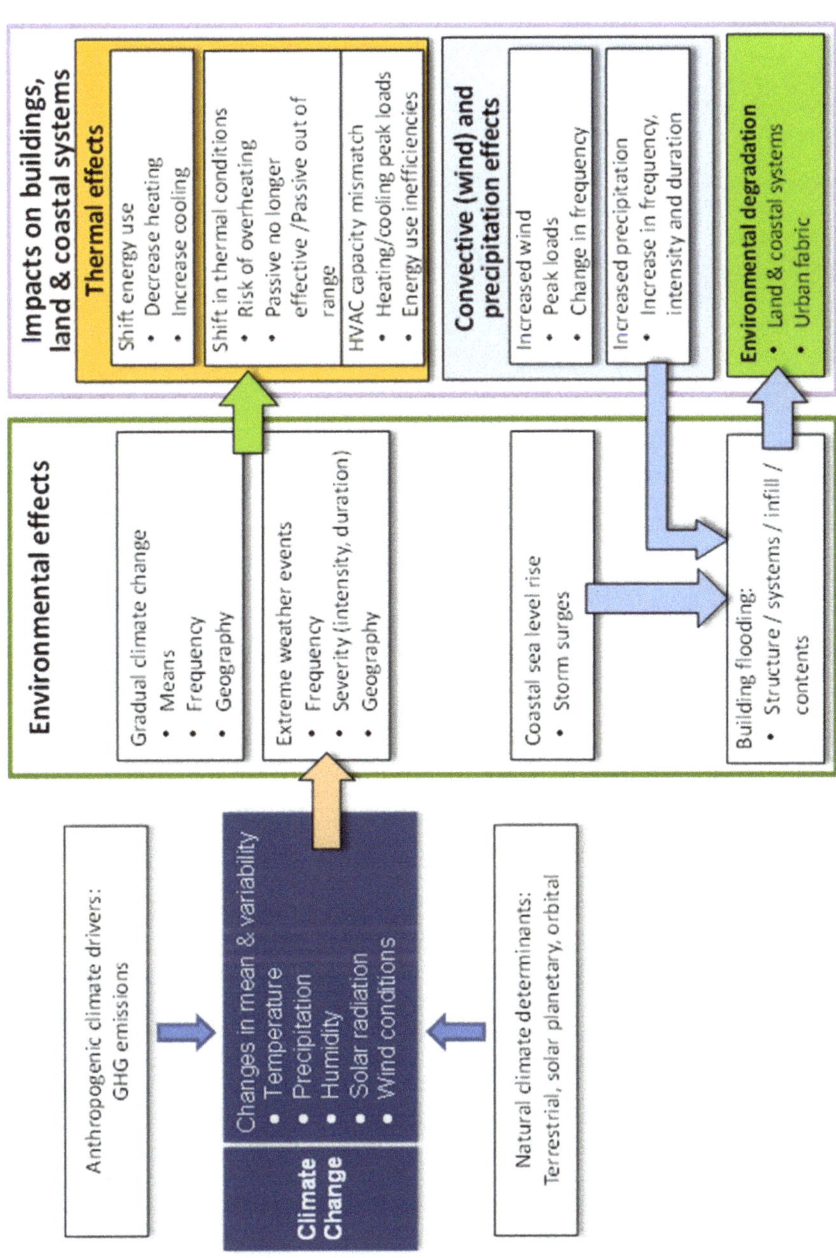

Figure 4. Climate change and impacts on buildings; adapted from [3].

There are also a number of well-developed approaches for the Wildland–Urban Interface (WUI) in North America, including the National Fire Protection Association wildland standards [8], the International Code Council WUI Code [9], as well as standards for buildings on floodplains [10].

Given the past activity globally in respect to mitigating the effects of climate change, what recent Canadian projects have been initiated to permit preparing the built environment for the expected increases in temperature, humidity, precipitation, wind as may arise from climate change? This is discussed in the next sub-section.

1.3. Research Program on Climate Resilient Buildings

The National Research Council of Canada (NRC) has also embarked on a research program related to buildings and climate change, more specifically, a project related to climate-resilient buildings and core public infrastructure [11]. Apart from the many different elements of this project, a key component is the development of climate design data based on different expected climate change scenarios [12]. Access to this climate data will permit determining the resilience of existing building structures, or new buildings, to expected climate change loads when comparing the response of building enclosures to historical loads. This will allow for mitigating any significant changes in response through adoption of suitable building design measures to reduce the risk to premature degradation and from which will evolve a guide to building enclosure design for existing building structures [13]. It will also permit developing hazard maps for different types of building products that will provide practitioners with information on the increased risk to degradation evaluated in relation to local climate conditions.

As part of the NRC research program, there was interest in knowing what previous work had been carried out on the durability of building envelope materials and building elements in respect to expected changes to the climate in the future. A concise summary is provided in Section 2 of previous work as has been completed on climate resilient building materials, components or elements, such as wall and roof assemblies. In regards to the durability of building envelope materials, brief accounts are given of works related to concrete and wood product degradation, the corrosion of metals; and the degradation of plastics due to the effects of solar radiation. Thereafter, studies on the durability of building envelope systems are reviewed for which the frost decay of masonry wall systems and the degradation of wood frame, roof and wall assemblies are briefly described.

In the subsequent section (Section 3), a discussion on the projected changes in key climate variables affecting durability of building materials and on climate issues as may affect durability and service life estimates of building materials and components is given, as this permits suggesting measures for the maintainability of buildings and selection of construction products for climate resilient design (Section 4).

2. Climate Resilient Buildings—Durability of Building Envelope Materials and Elements

The process of weathering of materials at the building exterior leads to their degradation that, in turn, leads to an increase in the rate and severity, of degradation over time [14]. In respect to the effects of climate change, it is unlikely that new mechanisms of degradation will develop. However, the changing climate will affect the built environment through steady changes in weather, increasing variability of and extremes in climate conditions [15]. The significant daily processes of weathering that contribute to premature degradation of the exterior of structures include wind-driven rain, temperature fluctuations, freeze–thaw cycling, frost effects, wetting and drying of porous materials, the action of solar and ultraviolet (UV) radiation, and chemical deposition on metals from the atmosphere.

A number of studies related to this topic specific to the Canadian context have been completed by Auld et al., [16–19]; to briefly summarize: given that buildings are in particular, vulnerable to the effects of weathering that degrade their durability and resilience to extremes conditions over time, it will become increasingly important in the future to ensure that building enclosures are able to resist wind-driven rain and to prevent moisture from penetrating the assembly.

In instances where it is projected that significant increases in degradation rate are to arise, adaptations to the building fabric may be required. In the context of climate change and existing buildings, adaptation is a means to further protect the existing building fabric, to enhance performance and control the rate of degradation. In regards to new buildings in a changing climate, design of buildings must be adapted to consider performance for both current and future climates. As such, building standards and building codes need to be revised to permit consideration of the effects of future climate on building design.

2.1. Selected Studies on Durability of Building Materials and Climate Change

Given that climate change is a global phenomenon, its effects on the durability of materials and building elements has, likewise, been studied broadly and in several countries. Some of the studies that have been carried out in the past decade relate to the effects of climate change on the durability of different materials and building elements, including: wood products, metals directly exposed to the environment (as opposed to imbedded metals), and plastic building products.

2.1.1. Concrete Degradation—Carbonation and Corrosion

(i) Concrete carbonation—increases in CO_2 concentration, and changes in temperature and relative humidity (RH), as my arise from a changing climate over the long-term, will accelerate the degradation processes and consequently, cause a decrease in, serviceability and durability and possibly the safety of reinforced concrete (RC) infrastructure. Peng and Stewart [20], report on an investigation of carbonation-induced degradation of RC under a changing climate for three cities located in China (Kunming, Xiamen and Jinan). A time-dependent analysis was conducted using Monte Carlo simulation, and included the uncertainty of climate projections, deterioration processes, material properties, dimensions and accuracy of the predictive models. Deterioration of RC structures in these cities was represented by the probabilities of initiation and occurrence of damage of reinforcement due to corrosion. It was found that by 2100, the mean depths for carbonation of the RC could increase by up to 45% for RC structures located in these cities due to a changing climate. It was also found that in temperate or cold climate locations in China, climate change can cause an additional 7–20% of carbonation-induced damage of RC buildings by 2100. Such findings permit development of climate adaptation strategies through consideration of improved RC design of structures to ensure their resilience over the long-term.

(ii) Concrete corrosion—Saha and Eckelman [21], report on investigating the effects on RC structures resulting from corrosion through increases in carbonation and chlorination rates. Different climate emission scenarios and (respectively, IPCC A1FI (high) and B1 (low)) were used together with downscaled temperature projections and code-compliant material specifications to model carbonation and chloride-induced corrosion of RC structures in the Northeast United States. Based on these results, it is expected that current RC construction as a result of climate change, will experience depths of penetration that exceed the current code-recommended cover thickness; in respect to the depth of chlorination, this would occur in 2055 and by 2077 for the depth of carbonation. The projected timeline is well within the expected service life of these buildings, indicating the potential for extensive repairs during the building service life.

2.1.2. Degradation of Wood Products

Although few studies have been completed regarding the durability and risk to degradation of wood products used in wood frame building construction, Lisø et al. [22] provided a highly useful overview of the decay potential in wood structures when subject to projected future changes in climatic conditions in Norway. The work consisted of developing a national climate index map that allows for geographically climate-differentiated guidelines for use of protective and preservation measures (e.g., impregnation, surface treatment or design precautions) for wood building elements. Based on this work, it is anticipated that in a changing climate, the vulnerability of wood frame structures in

Norway will increase given that climate change in this country is seen to increase the risk of decay of wood structures. The use of such mapping tools has permitted the development of technical guidelines for wood-based building enclosures, thus allowing a designer to consider both protective design and the preservative treatment of wood in relation to the expected projected climate conditions.

2.1.3. Corrosion of Metals

In Australia, as reported by Trivedi et al. [23], work undertaken by the Commonwealth Scientific Industrial Research Organisation (CSIRO), has shown that new metal structures are expected to last at least 50 years and well past 2064. As such, the effects of climate change must be considered in design and material selection, specifically in regard to changes in climate that increase the rate of corrosion of metal components. From this study, an estimate was provided of the greatest expected change in the rate of corrosion for the year 2070. Changes in corrosion were estimated for 11 coastal and inland locations in Australia. For each station, the climatic data in 2070 was estimated by modifying current data with probable changes based on two global climate models (i) the Meteorological Research Institute model [24]: (MRI-CGCM 3.2.2; most likely median model), and; (ii) the CSIRO model (CSIRO-Mk 3.5 dry climate model). For both models, a high global warming rate was assumed and with a GHG emissions scenario of intensive use of fossil fuel technology (i.e., A1FI scenario). The climatic data was then run through a corrosion "predictor" (a multi-scale process model) to predict corrosion at each location. The predictor revealed a moderate decrease in corrosion at inland locations but a substantial increase in coastal locations. The reduction in corrosion at inland locations was associated with a reduction in RH (i.e., surface wetness), whereas for coastal locations, the increase in corrosion was related to a greater build-up of salt due to fewer rain events.

Tidblad [25] reports on the prediction of atmospheric corrosion of metals in Europe based on climatic parameters derived from climate change scenarios for 2010–2039 and 2070–2099, using chloride deposition data from the project on 'Global Climate Change Impact on Built Heritage and Cultural Landscapes' [26]. For Europe, the future projected atmospheric corrosion of metals show that corrosion is governed by the effects of chloride deposition in coastal and near-coastal areas as is evident from maps portraying the corrosion of carbon steel and zinc. In extreme cases, the change can be as high as an increase in one corrosivity category and the corrosion can be higher than the highest values currently being experienced in Europe for coastal areas of Southern Europe. It is supposed that the reasons for increased coastal corrosion and reduced inland corrosion in Europe are similar to those as were found for Australia.

2.1.4. *Effect of Solar Radiation on Plastics*

Increased solar ultraviolet radiation (UV) reaches the surface of the Earth as a consequence of a depleted stratospheric ozone layer and changes in factors such as cloud cover, land-use patterns and aerosols. Increased levels of UV radiation, especially at high ambient temperatures, are well-known to accelerate the degradation of plastics, rubber and wood materials, thereby reducing their useful lifetimes in outdoor applications. As climate change is expected to result in a 0.9–5.4 °C increase in average temperature by the end of this century, depending on location, this indicates that the degradation of plastics due to exposure to solar radiation will only increase in the future. The useful lifetimes of plastics used routinely outdoors are ensured generally using light-stabilized chemical additives. Although the increased damage to materials due to an increased UV-B (280–315 nm) component in solar radiation reaching the Earth is not well known, such effects can be offset through the use of different approaches; e.g., light-stabilization technologies, surface coatings or, substitution of materials having an enhanced resistance to UV radiation. It is expected, then, that product manufacturers will take measures to ensure the stability of plastics in light of probable increases in UV-B and projected increases in ambient temperature, as may arise in the coming years due to climate change [27].

2.2. Hygrothermal Performance of Building Envelope Systems Affected by Changes in Wind-Driven Rain Loads Arising from Climate Change

The hygrothermal performance of building envelopes to future projected wind-driven rain loads has seldom been investigated in the past, although wind-driven rain and its consequence on buildings has been explored in many studies [28–32]. Only a few studies have been found to comprehensively assess the hygrothermal performance of building enclosures under future projected wind-driven rain loads. These included studies related to the frost decay of masonry cladding materials and the risk to degradation of wood frame, roof and wall assemblies in Sweden; each of these is briefly summarized in the subsequent sections.

2.2.1. Frost Decay of Masonry Materials

Different parts of Europe will necessarily experience different changes in the climate parameters that affect the masonry of heritage buildings. Grossi et al., from the UK [33], indicates that from the point of view of protection of the built heritage, a range of techniques have been developed to interpret climate data in terms of the risk of frost damage and provide meteorological parameters to guide plans for future management of heritage buildings. Emphasis has been placed on the way in which small changes in temperature can be amplified and have large effects on the phase change of moisture within materials where the freeze–thaw effect is a process in which a phase change occurs at an exact temperature. Thus, subtle increases in temperature, even of a few degrees, might positively affect porous building stones. Accordingly, Europe was mapped and divided into areas where the number of freeze–thaw cycles was mapped to which heritage buildings are subjected in a changing climate. From this effort, areas could be identified for likely increases or decreases in the frequency of freeze–thaw events. Thus, Europe is most likely to remain a temperate climate in the future and as such, the effects of temperature and temperature fluctuations on masonry materials are likely to diminish. Consequently, it may be expected that porous stone, as is typically used in monuments, located in future temperate climate zones may be less vulnerable to the degradative effects of freeze–thaw action and temperature change.

It has been shown [34] that the relative risk of degradation due to the effects of frost on porous masonry materials exposed to different climates can be expressed using a simple index incorporating information about the number of freezing events and the total 4-day rainfall occurring prior to freezing for the different months of the year. The index was based on multi-year records of daily air temperatures and rainfall data. The principal advantages of this method are that results are based on readily available series of long-term climate data.

Hygrothermal simulation models of external building envelopes incorporating calcium silicate brick (a frost-sensitive material) were shown to reproduce degradation due to frost action in winter months in the Netherlands [35] as compared to onsite degradation assessments. This model was used to predict frost behavior of such bricks in the future arising from a changing climate. Degradation of masonry and mortar materials caused by frost requires that the average temperature in the material be lower than the freezing point of water, whilst the moisture content of the masonry and mortar should be higher than the capillary saturation point. As such, frost damage may occur if a long-duration rain event is immediately followed by a severe frost given that the length of time over which rain will permeate the masonry will necessarily affect the moisture content. The results from simulations show a reduction in future risk of frost damage by 70%.

2.2.2. Degradation of Wood Frame, Roof and Wall Assemblies

Brischke and Rapp [36] have reviewed the potential impacts of climate change on wood deterioration. Their published work focuses on the effects of global warming and corresponding moistening on the durability of wooden building components. With the use of a mathematical wood decay model an attempt was made to quantify the influence of climatic changes on wood decay rates for various climate change scenarios. The decay model was based on both laboratory and experimental work for which climatic data, wood temperatures, wood moisture contents and decay rates recorded for several years across Europe were correlated. Examples of such predicted changes were provided for specific sites located in Sweden (Uppsala), Germany (Freiburg), the UK (Portsmouth), France (Bordeaux) and Croatia (Zagreb). The results showed that warming and humidification will lead to a significantly reduced service life in wooden building components under climate change. It was further determined that the quantity of climate-induced changes strongly depended on the geographic location and the present climate.

The effects of climate change on the moisture performance on building facades of common wood frame wall constructions in Sweden was assessed in Nik et al. [37]. To do so, the response of a building façade was assessed under historical (1961–1990) and future climates. Two different future time-frames—2021–2050 and 2071–2100—were analyzed. The climate simulated by the RCA3 regional climate model was considered to assess the impacts of climate change. The heat–air–moisture (HAM) simulation model, WUFI, was used to simulate hygrothermal response of building facades in historical and projected future climates. The moisture accumulation in the building façade was assessed for five cases: (a) when three different sets of atmospheric boundary conditions from GCMs were used to simulate regional climate in the RCA3 regional climate model; (b) when three different initial conditions were used to initialize the simulations in the regional climate model; (c) when the regional climate is simulated at two different spatial resolutions—25 and 50 km; (d) when two different methods of calculation of wind-driven rain loads were considered; and e) two different façade materials were considered. The results from the study pointed to an increased risk of moisture-related damage in building facades in Sweden as a consequence of climate change. In addition to this, uncertainties associated with differences in spatial resolutions of the regional climate model were found to be the highest among the five cases analyzed. Another important conclusion of this study was that detailed wind modelling around the buildings was not necessary to accurately model climate change impacts on moisture performance of the buildings in Sweden.

The hygrothermal performance and mould growth potential of a typical and three modified attic constructions in Sweden under potential climate change effects were evaluated in Nik et al. [38]. The historical and future projected climate were obtained from the climate simulations made using the RCA3 regional climate model. Three representative concentration pathways of future greenhouse gas concentrations were considered for analysis. The response of attics to climate was simulated using a whole building heat, air and moisture modelling software, HAM-Tools. The study found future increases in mould growth potential in Sweden as a consequence of future projected climate change. Among the three adaptive designs investigated, the attic with mechanical ventilation was found to be most resilient to the projected climate change effects.

Future projections of climate are associated with uncertainty that can stem from the existence of a wide range of climate models, different downscaling methods, and so on [39]. To account for this uncertainty, typically, a large ensemble of future projections is considered to assess the potential effects of climate change. However, given the computational and time constraints in assessing hygrothermal response under a large ensemble of climate scenarios, it is important to develop methods that can be used to encompass climate-related uncertainty in climate change assessments on buildings. In this regard, Nik [39] quantified the amount of uncertainty encompassed when typical and extreme climate data based on—(a) outdoor dry bulb temperature, (b) equivalent temperature, and (c) rainfall—are selected, and compared it with the uncertainty communicated by the entire ensemble of climate projections. The results of the study indicated that representative typical and extreme projections selected based on the simulated outdoor dry bulb temperature are able to capture the range of hygrothermal responses projected by the entire ensemble of climate projections.

Nik et al. [40] demonstrated that three representative years—typical downscaled year (TDY), extreme cold year (ECY), and extreme warm year (EWY)—selected based on outdoor temperature, are able to capture the range of simulated energy response of the buildings when exposed to long-term (30-year) climate. To demonstrate the approach, several synthesized weather data were created for two European cities: Geneva and Stockholm, considering two RCMs (RCA3 for Stockholm and RCA4 for Geneva), six GCMs (two for Stockholm and four for Geneva) and three emissions pathways (RCP 4.5 and RCP 8.5). Two different building models were exposed to the complete set of historical and future projected climate data, as well as to the climates of selected typical, and extreme warm and cold years. Based on the findings of the study, it was concluded that TDY, ECY, and EWY together are able to capture climatic variations present in the long-term climate, as well as reflect the climate uncertainties from the use of multiple scenarios.

2.3. Summary—Of Building Envelope Materials and Elements

As is evident from that reported in the previous sections, studies have been conducted at various locations around globe that serve their respective countries. In respect to climate change in Northern European countries (i.e., Sweden, Norway), wood degradation is of importance given the predominant use of wood as a construction material. Whereas, the effects of freeze–thaw action on stone and brick masonry materials is more relevant to other European countries where the use of such materials in construction are preferred. In countries such as Australia, where the use of metal roofing is common, corrosion of this roofing component is evidently of importance. To summarize:

- Carbonation of RC in three Chinese cities (Kunming, Xiamen and Jinan) is predicted to increase by 45% by 2100; based on model results of carbonation and chloride-induced corrosion of RC structures located in the Northeast United States, the depths of chloride penetration of these structures will exceed the current code-recommended cover thickness by 2055, and by 2077, the depth of carbonation.
- In Europe, the future projected atmospheric corrosion of metals (carbon steel and zinc) show that corrosion is governed by the effects of chloride deposition in coastal and near-coastal areas; hence, it is foreseen that in the future, corrosion of exposed metals in Europe will increase in coastal zones and decrease inland; similar results were obtained from an Australian metal corrosion predictor that revealed a moderate decrease in corrosion at inland locations but a substantial increase in coastal locations.
- Europe was mapped for the number of freeze–thaw cycles to which heritage buildings will be subjected in a changing climate; Europe is most likely to remain a temperate climate in the future and as such, freeze–thaw effects related to temperature fluctuations and rainfall on masonry materials are likely to diminish; in the Netherlands, results from simulations show a reduction in the future risk of frost damage by 70%
- Warming and humidification will lead to significantly reduced service life in wooden building components under climate change in Europe; examples of predicted changes were provided for specific sites located in Sweden, Germany, the UK, France and Croatia.
- Norway developed a national climate durability index map for risk to degradation of wood products; in a changing climate, the vulnerability of wood frame structures in Norway will increase, as will the risk of decay of wood structures; likewise
- In Sweden, under projected future WDR loads, the amount of water accumulated in the façade of common wood frame wall constructions was projected to increase in the future; the attics of wood frame homes in Sweden are projected for increases in mould growth potential in the future.

3. Discussion on the Durability of Building Materials as Influenced by Climate Change Effects

The discussion first focuses on the magnitude the anticipated effects of climate change for key climate variables such as projected changes to January and July temperatures, total precipitation and average wind speeds under globally averaged warming of 2 °C and 3.5. Thereafter, consideration is given to climate-related effects that will affect the outcome of service life prediction estimates, including an estimation of reliable future WDR loads and their spatial distribution, and, methods as may be used to encompass climate uncertainty. Each of these is discussed in turn.

3.1. Projected Changes in Key Climate Variables Affecting Durability of Building Materials

In Canada, a large fraction of today's infrastructure has been designed using climatic design values calculated from historical climate data without taking into consideration potential impacts of climate change. In other words, an inherent assumption has been made that past extremes will represent future conditions. To design climate-resilient infrastructure, projected future changes in climate will need to be estimated as accurately as possible and reflected in the climatic design values. These design values will also need to be regularly updated to reflect the advances in climate sciences, and climate model development. Additionally, it will be important to determine appropriate return periods for when the data should be recalculated.

Projected changes in temperature-related climate design values at 659 locations identified in Table C-2 of the National Building Code of Canada (NBCC) [41] and S6.1–14 Commentary on CSA S6-14, Canadian Highway Bridge Design Code [42], are presented in Figures 5 and 6. The figures contrast the consequences of two global warming (GW) scenarios, leading to 2° and 3.5° increases in globally averaged temperatures. The changes presented are averaged across a 31-year time-period and show differences between a historical time-period, i.e., 1986–2016, and future time-periods when select levels of globally averaged warming are projected to be reached. According to ECCC [43], a 31-year averaged globally averaged warming of 2° and 3.5° will be reached at future (center) years of 2049 and 2077, respectively.

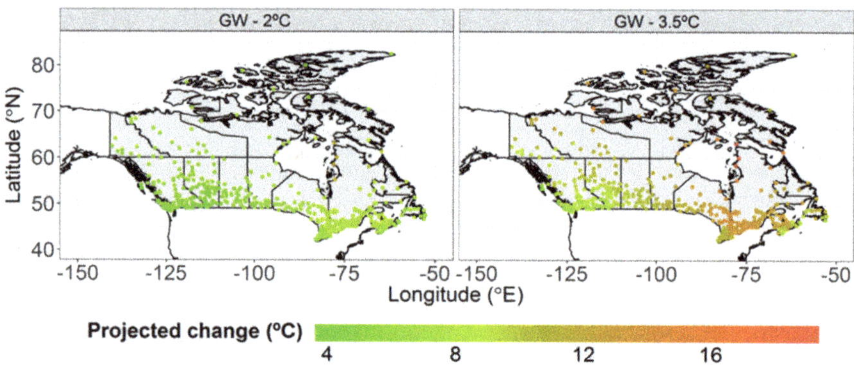

Figure 5. Projected changes in January 1% design temperatures in Canada under globally averaged warming of 2 °C and 3.5°.

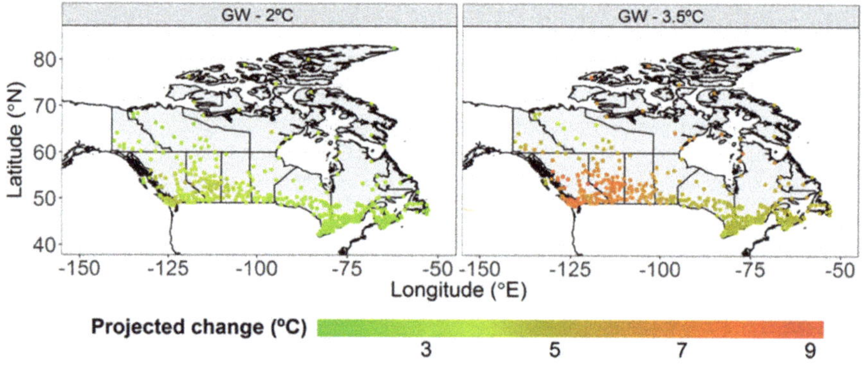

Figure 6. Projected changes in July 2.5% design temperatures (dry) in Canada under globally averaged warming of 2 °C and 3.5°.

The results clearly demonstrate significant increases in January 1% and July 2.5% (dry), temperatures across Canada as a consequence of climate change. The most significant increases are projected for the northernmost regions of Canada. More significant increases in the summertime design temperatures are projected for western regions of Canada than for eastern regions. On the other hand, more significant increases in wintertime design temperatures are projected for eastern regions of Canada than for western regions. As expected, more significant changes are projected under the 3.5 °C global warming scenario than the 2 °C global warming scenario.

In addition to temperatures, climate change is also expected to bring considerable shifts in other key climate variables that affect the durability of building materials. Projected changes in total precipitation and average wind speeds, as simulated by the Canadian Regional Climate Model, CanRCM4 under 2 °C and 3.5 °C global warming scenarios across Canada are presented in Figures 7 and 8, respectively. Higher precipitation totals have been projected for all parts of Canada with the highest increases projected for the northernmost regions. As projected in the case of temperatures, higher precipitation increases have been projected for higher levels of global warming i.e., under 3.5 °C global warming scenario than, for instance, under a 2 °C global warming scenario. However, the sign of change is more uncertain in the case of wind speeds as a larger fraction of the Canadian landmass is projected to have increases in the future as opposed to decreases in wind speeds.

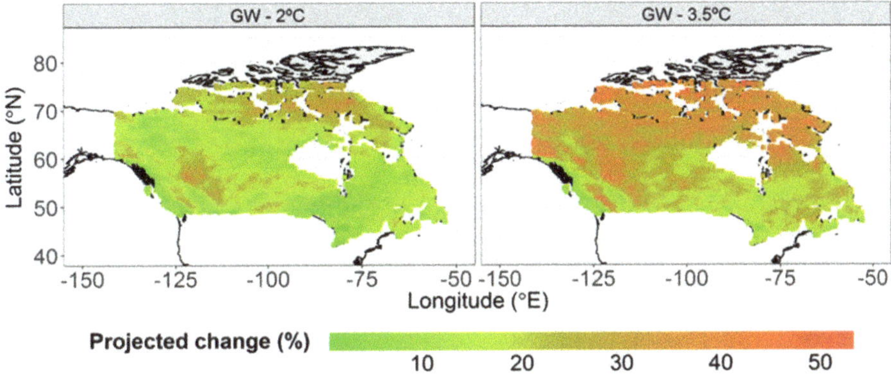

Figure 7. Projected percent changes in total precipitation across Canada under globally averaged warming of 2 °C and 3.5°.

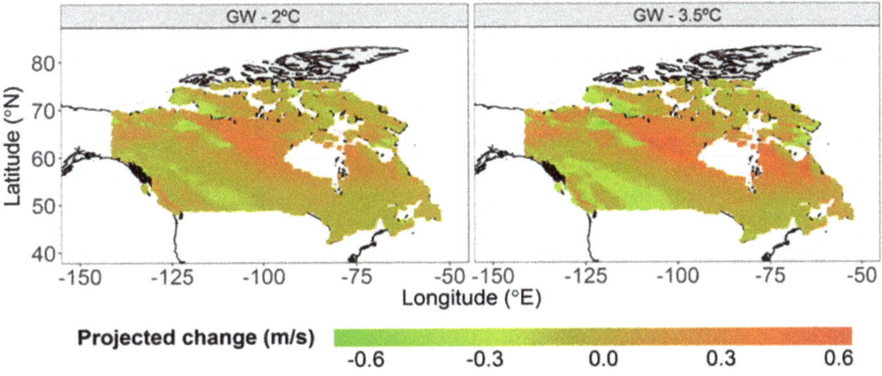

Figure 8. Projected changes in average wind speeds across Canada under globally averaged warming of 2 °C and 3.5.

The above results highlight that buildings across Canada will be exposed to a climate that is expected to be remarkably different from that observed historically during their design life. Therefore, when considering the durability of buildings and their materials and elements, durability evaluations should account for the non-stationarity in climate loads.

There are, indeed, uncertainties associated with the future projections of climate as contributed by the many global climate models developed by climate groups across the globe, different approaches to downscale the climate projections, and range of future emission scenarios in accordance with which the projections are made. Since, however, uncertainty in design data is accepted as a part of construction codes and standards, it should be possible to account for multiple sources of uncertainties associated with future impacts of climate change by reflecting them, for example, in the choice of novel safety factors in building design which take into account the inherent uncertainty in design data. These issues are discussed in the subsequent sub-section.

Although the information presented in these figures for projected values of wind speed, precipitation, winter and summer design temperatures, and heating degree days under different global warming scenarios provides an estimate of the order of magnitude of the expected effects of climate change, it has still to be determined exactly how such information is to be presented in the NBC and how practitioners are to use the additional information to affect design decisions.

3.2. Consideration of Climate Issues as May Affect Durablity and Service Life Estimates

Consideration should be given to two issues as related to climate that will affect the outcome of service life prediction estimates. These include the estimation of reliable future WDR loads and their spatial distribution, and methods as may be used to encompass climate uncertainty

(i)   Reliable estimation of future WDR loads and their spatial distribution: The climate simulations in the GCMs and RCMs are performed at 100–300 km and 25–50 km spatial resolutions, respectively. Climate variables such as wind and rainfall, and their extremes, are not accurately simulated at this spatial resolution as their propagation mechanisms are influenced by local geophysical factors that are not resolved in the GCMs or RCMs. Several studies have used very-high-resolution (sub-4 km) limited-area climate models to simulate these climate variables more accurately [44–47]; however, such simulations are computationally expensive. There is a need to devise a strategy to obtain reliable long-term estimates of wind and rainfall variables which can facilitate more accurate assessment of building response to the effects of climate change.

(ii)   Methods to encompass climate uncertainty: Future climate projections are associated with uncertainty contributed by a wide range of sources such as the choice of GCMs, greenhouse gas emission scenarios, downscaling methods, bias-correction methods [48]. Additionally, climate change impact assessments are performed at time-periods spanning at least 20 years or more. Hygrothermal simulations are computationally expensive and it is impractical to evaluate hygrothermal performance of wall assemblies over time-periods in excess of 20 years, and under a wide array of climate simulations. To reduce the computational costs, it is important to develop methods that can be used to encompass climate-related uncertainty and identify representative climate years/scenarios without compromising on the range of projected changes. Nik [39], for example, obtained an acceptable range of hygrothermal response when only typical and extreme climate projections were used for performing hygrothermal simulations as opposed to the entire set of future projections available. The concept can also be extended to choose the reference years that represent typical and extreme WDR conditions and only consider those years for hygrothermal simulations.

4. Maintainability of Buildings and Selection of Construction Products for Climate Resilient Design

The expectation is that in the coming decades, the general climate of Canada is to become warmer, with some locations experiencing more intense and frequent rain events of longer duration, thus producing heightened wind-driven rain loads. Hence, the atmosphere is not only likely to be warmer but also more humid; however, structural wind loads may increase but only in the far northern latitudes and not likely in current urban areas. What guidance can be offered building practitioners in respect to building maintainability and the selection of construction products to achieve climate resilient performance over the service life of the building?

The maintainability of buildings and related performance indicators have been thoroughly reviewed and conceptually defined by Asmone et al. [49] in a "Green" building context, that is, buildings conforming to the precepts of sustainability whilst accommodating environmental, economic and societal requirements. Maintainability, as opposed to the action of the maintenance of buildings relates to the ability to forecast maintenance costs over the expected life cycle (service life) of the building. Hence, the need for the use of life cycle costing tools, together with service life planning to develop a viable maintenance plan. The "Green" building trend has gained momentum in recent years and it has increasingly been considered as a means of addressing concerns over climate change and the effects of global warming on buildings and their occupants [50]. Such approaches now also include life cycle environmental assessment tools whereby the environmental impact of specific construction design and product choices can be known and considered in the overall service life plan. Nonetheless, in regards to mitigating the effects of climate change, there is little specific information that has been made available relating to the maintainability of buildings over the long-term, irrespective of the availability of rating tools for "Green" building or the development of performance indicators to help achieve highly sustainable and maintainable building projects.

Given that revised climate data that accounts for projections of climate change are not yet available what information can be provided to expert practitioners in respect to the selection of products for new buildings or for the retrofit of existing buildings? Considering that the construction of new buildings and renovation of existing buildings cannot be halted and given the available research on this topic, some recommended specifications for the selection of products arising from climate change effects are provided in Table 1. Two aspects related to the expected climate change effects are considered; those that relate to an increase in: (i) global warming, and (ii) wind-driven rain loads. The notional specifications for selection of products and methods of installation are provided for each of the two aspects in terms of the specific environmental agent acting to cause degradation. For example, should global warming be the anticipated effect arising from climate change for a specific location of interest, then it is expected that higher temperatures and a broader overall range of both annual and diurnal temperature change will occur for that location. Accordingly, one ought to specify dimensionally stable products having a lower coefficient of thermal expansion, thus providing a reduced overall daily and annual dilation; e.g., amongst several possible choices of window frame products, the selection of plastic fenestration components has the lowest coefficient of thermal expansion when directly exposed to solar radiation. Additionally, components should be specified with compatible thermal expansion coefficients to ensure that interaction between components does not also increase the risk to degradation of the assembly.

Global warming will also accelerate the aging process of products directly exposed to solar radiation and the exterior environment due to more prolonged periods of higher temperature and from exposure to higher levels of UV-B radiation. Products such as roofing and cladding components, insulted glass units, plastic fenestration components, polymer-based waterproofing and sheathing membranes, jointing and sealing products, paints and coatings for cladding and comparably exposed components would also be subject to accelerated aging. As such, only products of proven and heightened resistance to heat aging and UV radiation ought to be specified.

Likewise, guidance is provided in Table 1 for specifying products where there is an expected increase in wind-driven rain loads. In this instance, one would expect higher average humidity conditions within windows and door frames, and openings in which the components are installed with an increased incidence of liquid moisture being in prolonged contact with fenestration products. Hence, specification for these types of products would require dimensionally stable products that resist prolonged exposure to both heat and moisture, and more robust design for the installation of such products in wall assemblies. Furthermore, metal products must be resistant to corrosion given that they may be in contract with moisture.

Table 1. Notional specifications for the selection of products given climate change effects.

Climate Change Effects	Environmental Agents	Notional Specifications for Selection of Products, Methods of Installation
Increase in global warming	Higher temperatures and broader overall range of both annual and diurnal temperature change	Dimensionally stable and compatible products having lower coefficient of thermal expansion thus providing a reduced overall dilation (e.g., for: plastic fenestration components directly exposed to solar radiation)
		Products having enhanced elasticity and are resistant to repeated movement cycles (e.g., when considering jointing and sealing products)
	Accelerated aging process due to more prolonged periods of higher temperature and from exposure to higher levels of UV-B radiation	Products of proven and heightened resistance to heat aging and UV radiation (i.e., for products directly exposed to solar radiating and exterior environment, e.g.,: roofing, and cladding products, IG units; plastic fenestration components; polymer-based waterproofing and sheathing membranes; jointing and sealing products, paints and coatings for cladding and similar exposed components)
Increase in wind-driven rain loads	Environmental conditions within window and door frame and in installation openings having higher average humidity conditions together with increased incidence of liquid moisture in more prolonged contact with fenestration products	Select the more robust design approaches that enhance drainage of water from surfaces and minimise the likelihood of retention of water in interstitial spaces (e.g., for: wall assemblies and for window design and installation)
		Dimensionally stable products when wetted and having enhanced resistance to hydrolysis (i.e., degradation from contact with warm liquid water) (e.g., for: insulation products used to ensure continuity of thermal resistance at wall-window and door interfaces; polymer-based waterproofing and sheathing membranes; jointing and sealing products)
		Metal product components having enhanced resistance to corrosion after being wetted (e.g., roof, cladding, window frame, window ties, brick ties products)

5. Summary

A brief overview has been provided related to climate change effects on the durability of building materials and building elements. This overview attempts to provide some perspective on the magnitude of the effects and how such changes in climate variables will affect the durability of building materials and building elements in the coming years. The weathering and degradation of various building materials have been discussed and examples of recent studies are provided that relate to frost decay of masonry materials, the degradation of wood products and concrete elements, the corrosion of metals, and the effect of solar radiation on plastics.

In the final section, the development of climate change design data is reviewed whereby examples are given of expected changes in climate design data (e.g., wind speed; precipitation; design temperatures) for two climate change scenarios. The information presented in these figures for different global warming scenarios provides an order of magnitude of the expected effects of climate change. It has yet to be determined exactly how such information is to be presented in the NBC and how practitioners are to use the additional information to affect design decisions. Irrespective of how such information is to be implemented in the NBC, when considering the durability of buildings and their materials and elements, durability evaluations should account for the non-stationarity of the future climate.

Author Contributions: M.A.L. conceived the paper ourline; A.G. and M.A.L. wrote the first draft; T.V.M. revised the first draft; all authors worked on addressing the reviewer and editor comments. All authors have read and agreed to the published version of the manuscript.

Funding: This research described in this paper has been sponsored by the National Research Council Canada and in part by Infrastructure Canada.

Acknowledgments: The authors acknowledge constructive feedback from two anonymous reviewers and the editor that greatly helped in improving the quality of the manuscript.

Conflicts of Interest: The authors declare no conflict of interest.

References

1. Masson-Delmotte, V.P.; Zhai, H.-O.; Pörtner, D.; Roberts, J.; Skea, P.R.; Shukla, A.; Pirani, W.; Moufouma-Okia, C.; Péan, R.; Pidcock, S.; et al. Summary for Policymakers. In *Global Warming of 1.5 °C*; World Meteorological Organization: Geneva, Switzerland, 2018.

2. ECCC. *Climate Trends and Variations Bulletin: Annual 2018*; Environment and Climate Change Canada: Toronto, Canada, 2018. Available online: https://www.canada.ca/en/environment-climate-change/services/climate-change/science-research-data/climate-trends-variability/trends-variations/annual-2018-bulletin.html (accessed on 9 January 2020).

3. de Wilde, P.; Coley, D. Editorial, Implications of a changing climate for buildings. *Build. Environ.* **2012**, *55*, 1–7. [CrossRef]

4. Engineers Canada. *Public Infrastructure Engineering Vulnerability Committee (PIEVC) Assessment Protocol*; Engineers Canada: Ottawa, ON, Canada, 2011.

5. Institute for Catastrophic Loss Reduction (ICLR). *Home Builder's Guide to promote the construction of disaster-resilient homes. Institute for Catastrophic Loss Reduction*; Institute for Catastrophic Loss Reduction: Toronto, Canada, 2010.

6. US Department of Housing and Urban Development. *Climate Change Adaptation Plan*; United States Department of Housing and Urban Development: Washington, DC, USA, 2014.

7. Larsen, L.; Rajkovich, N.; Leighton, C.; McCoy, K.; Calhoun, K.; Mallen, E.; Bush, K.; Enriquez, J. *Green Building and Climate Resilience*; US Green Building Council: Washington, DC, USA, 2011.

8. NFPA. *NFPA 1141: Standard for Fire Protection Infrastructure for Land Development in Wildland, Rural, and Suburban Areas*; National Fire Protection Association: Quincy, MA, USA, 2017.

9. International Code Council. *International Wildland-Urban. Interface Code*; ICC: New York, NY, USA, 2012.

10. American Society of Civil Engineers. *Flood Resistant Design and Construction*; American Society of Civil Engineers: Reston, VA, USA, 2014.

11. Infrastructure Canada. *Climate-Resilient Buildings and Core Public Infrastructure Initiative*; Government of Canada: Ottawa, ON, Canada, 2019.

12. National Research Council Canada. *The National Research Council Canada and Infrastructure Canada Take the Lead in Preparing Canada's Buildings and Infrastructure for Climate Resiliency*; Government of Canada: Ottawa, ON, Canada, 2018.

13. National Research Council Canada. *Updating Climatic Data and Loads for Codes, Standards and Guides*; Government of Canada: Ottawa, ON, Canada, 2019.

14. Chu-Tsen, L. Study on Exterior Wall Tile Degradation Conditions of High-rise Buildings in Taoyuan City. *J. Asian Archit. Build. Eng.* **2018**, *17*, 549–556. [CrossRef]

15. Phillipson, M.C.; Emmanuel, R.; Baker, P.H. The durability of building materials under a changing climate. *WIREs Clim. Chang.* **2016**, *7*, 590–599. [CrossRef]

16. Auld, H.; Klaassen, J.; Comer, N. *Weathering of Building Infrastructure & the Changing Climate: Adaptation Options*; Adaptation & Impacts Research Division; Environment Canada: Toronto, ON, Canada, 2007.

17. Auld, H.; MacIver, D. *Changing Weather Patterns, Uncertainty and Infrastructure Risks: Emerging Adaptation Requirements*; Adaptation & Impacts Research Division; Environment Canada: Toronto, ON, Canada, 2007.

18. Auld, H. Adaptation by design: The impact of changing climate on infrastructure. *J. Public Work. Infrastruct.* **2008**, *1*, 276–288.
19. Auld, H.; Waller, J.; Eng, S.; Klaassen, J.; Morris, R.; Fernandez, S.; Cheng, V.; MacIver, D. The changing climate and national building codes and standards. In Proceedings of the 9th Symposium on the Urban Environment, American Meteorological Society, Keystone, CO, USA, 2–6 August 2010.
20. Peng, L.; Stewart, M.G. Climate change and corrosion damage risks for reinforced concrete infrastructure in China. *Struct. Infrastruct. Eng.* **2016**, *12*, 499–516. [CrossRef]
21. Saha, M.; Eckelman, M.J. Urban scale mapping of concrete degradation from projected climate change. *Urban. Clim.* **2014**, *9*, 101–114. [CrossRef]
22. Lisø, K.R.; Hygen, H.O.; Kvande, T.; Thue, J.V. Decay potential in wood structures using climate data. *Build. Res. Inf.* **2006**, *34*, 546–551. [CrossRef]
23. Trivedi, N.S.; Venkatraman, M.S.; Chu, C.; Cole, I.S. Effect of climate change on corrosion rates of structures in Australia. *Clim. Chang.* **2014**, *124*, 133–146. [CrossRef]
24. Yukimoto, S.; Adachi, Y.; Hosaka, M.; Sakami, T.; Yoshimura, H.; Hirabara, M.; Tanaka, T.Y.; Shindo, E.; Tsujino, H.; Deushi, M.; et al. A New Global Climate Model/Meteorological Research Institute: MRI-CGCM3. *J. Meteorol. Soc. Jpn.* **2012**, *90A*, 23–64. [CrossRef]
25. Tidblad, J. Atmospheric corrosion of metals in 2010–2039 and 2070–2099. *Atmos. Environ.* **2012**, *55*, 1–6. [CrossRef]
26. Sabbioni, C.; Bonazza, A. How mapping climate change for cultural heritage? The Noah's Ark project. In *Climate Change and Cultural Heritage, Proceedings of the Ravello International Workshop*; Lefèvre, R.-A., Sabbioni, C., Eds.; Centro Universitario Europeo per i Beni Culturali: London, UK, 2009.
27. Andrady, A.L.; Hamid, H.; Torikaic, A. Effects of solar UV and climate change on materials. *Photochem. Photobiol. Sci.* **2011**, *10*, 292. [CrossRef] [PubMed]
28. Baheru, T.; Chowdhury, A.G.; Pinelli, J.-P.; Bitsuamlak, G. Distribution of wind-driven rain deposition on low-rise buildings: Direct impinging raindrops versus surface runoff. *J. Wind Eng. Ind. Aerodyn.* **2014**, *133*, 27–38. [CrossRef]
29. Foroushani, S.S.M.; Ge, H.; Naylor, D. Effects of roof overhangs on wind-driven rain wetting of a low-rise cubic building: A numerical study. *J. Wind Eng. Ind. Aerodyn.* **2014**, *125*, 38–51. [CrossRef]
30. Perez-Bella, J.M.; Domínguez-Hernandez, J.; Rodríguez-Soria, B.; del Coz-Díaz, J.J.; Cano-Sunen, E. Combined use of wind-driven rain and wind pressure to define water penetration risk into building façades: The Spanish case. *Build. Environ.* **2013**, *64*, 46–56. [CrossRef]
31. Tang, W.; Davidson, C.I. Erosion of limestone building surfaces caused by wind-driven rain: 2. Numerical modeling. *Atmos. Environ.* **2004**, *38*, 5601–5609. [CrossRef]
32. Tariku, F.; Simpson, Y.; Iffa, E. Experimental investigation of the wetting and drying potentials of wood frame walls subjected to vapor diffusion and wind-driven rain loads. *Build. Environ.* **2015**, *92*, 368–379. [CrossRef]
33. Grossi, C.M.; Brimblecombe, P.; Harris, I. Predicting long term freeze-thaw risks on Europe built heritage and archaeological sites in a changing climate. *Sci. Total Environ.* **2007**, *377*, 273–281. [CrossRef]
34. Lisø, K.R.; Kvande, T.; Hygen, H.O.; Thue, J.V.; Harstveit, K. A frost decay exposure index for porous, mineral building materials. *Build. Environ.* **2007**, *42*, 3547–3555. [CrossRef]
35. Aarle van, M.; Schellen, H.; Schijndel van, J. Hygro Thermal Simulation to Predict Risk of Frost Damage in Masonry; Effects of Climate Change. *Energy Procedia* **2015**, *78*, 2536–2541. [CrossRef]
36. Brischke, C.; Rapp, A.O. Potential impacts of climate change on wood deterioration. *Int. Wood Prod. J.* **2010**, *1*, 85–92. [CrossRef]
37. Nik, V.M.; Mundt-Petersen, S.; Kalagasidis, A.S.; Wilde, P. Future moisture loads for building facades in Sweden: Climate change and wind-driven rain. *Build. Environ.* **2015**, *93*, 362–375. [CrossRef]
38. Nik, V.M. Application of typical and extreme weather data sets in the hygrothermal simulation of building components for future climate—A case study for a wooden frame wall. *Energy Build.* **2017**, *154*, 30–45. [CrossRef]
39. Nik, V.M.; Kalagasidis, A.S.; Kjellström, E. Assessment of hygrothermal performance and mould growth risk in ventilated attics in respect to possible climate changes in Sweden. *Build. Environ.* **2012**, *55*, 96–109. [CrossRef]
40. Nik, V.M. Making energy simulation easier for future climate-Synthesizing typical and extreme weather data sets out of regional climate models (RCMs). *Appl. Energy* **2016**, *177*, 204–226. [CrossRef]

41. NBCC. *National Building Code of Canada*; National Research Council Canada: Ottawa, ON, Canada, 2015.

42. CAN/CSA Standard. S6.1-14-Commentary on S6-14. In *Canadian Highway Bridge Design Code*; CSA Group: Toronto, ON, Canada, 2014.

43. Environment and Climate Change Canada. *Memorandum of Understanding between National Research Council and Environment and Climate Change Canada on Climate Resilient Buildings and Core Public Infrastructure: Future Projections of Climate Design Values. Tier 1 Climate Projections*; Environment and Climate Change Canada: Gatineau, QC, Canada, 2018.

44. Prein, A.F.; Langhans, W.; Fosser, G.; Ferrone, A.; Ban, N.; Goergen, K.; Keller, M.; Tölle, M.; Gutjahr, O.; Feser, F.; et al. A review on regional convection-permitting climate modeling: Demonstrations, prospects, and challenges. *Rev. Geophys.* **2015**, *53*, 323–336. [CrossRef]

45. Brisson, E.; Van Weverberg, K.; Demuzere, M.; Devis, A.; Saeed, S.; Stengel, M.; van Lipzig, N.P. How well can a convection-permitting climate model reproduce decadal statistics of precipitation, temperature and cloud characteristics? *Clim. Dyn.* **2016**, *47*, 3043–3061. [CrossRef]

46. Prein, A.F.; Gobiet, A.; Suklitsch, M.; Truhetz, H.; Awan, N.K.; Keuler, K.; Georgievski, G. Added value of convection permitting seasonal simulations. *Clim. Dyn.* **2013**, *41*, 2655–2677. [CrossRef]

47. Saeed, S.; Brisson, E.; Demuzere, M.; Tabari, H.; Willems, P.; van Lipzig, N.P.M. Multi-decadal convection permitting climate simulations over Belgium: Sensitivity of future precipitation extremes. *Atmos. Sci. Lett.* **2017**, *18*, 29–36. [CrossRef]

48. Gaur, A.; Simonovic, S.P. Towards Reducing Climate Change Impact Assessment Process Uncertainty. *Environ. Process.* **2015**, *2*, 275–290. [CrossRef]

49. Asmone, A.S.; Conejos, S.; Chew, M.Y.L. Green maintainability performance indicators for highly sustainable and maintainable buildings. *Build. Environ.* **2019**, *163*, 12. [CrossRef]

50. Khan, J.S.; Zakaria, R.; Aminuddin, R.; Abidin, N.I.; Sahamir, S.R.; Ahmad, R.; Abas, D.N. Web-based automation of green building rating index and life cycle cost analysis, IOP Conf. *Series: Earth Environ. Sci.* **2018**, *143*, 012062. [CrossRef]

Article

Profitability of Various Energy Supply Systems in Light of Their Different Energy Prices and Climate Conditions

Elaheh Jalilzadehazhari [1],*, Georgios Pardalis [2] and Amir Vadiee [3]

[1]   Forestry and Wood Technology Department, Linnaeus University, 351 95 Växjö, Sweden
[2]   Department of Built Environment and Energy Technology, Linnaeus University, 351 95 Växjö, Sweden; georgios.pardalis@lnu.se
[3]   School of Business Society and Engineering, Division of Civil Engineering and Energy Systems, Mälardalen University, 721 23 Västerås, Sweden; amir.vadiee@mdh.se
*   Correspondence: elaheh.jalilzadehazhari@lnu.se

Received: 27 April 2020; Accepted: 23 May 2020; Published: 28 May 2020

Abstract: The majority of the single-family houses in Sweden are affected by deteriorations in building envelopes as well as heating, ventilation and air conditioning systems. These dwellings are, therefore, in need of extensive renovation, which provides an excellent opportunity to install renewable energy supply systems to reduce the total energy consumption. The high investment costs of the renewable energy supply systems were previously distinguished as the main barrier in the installation of these systems in Sweden. House-owners should, therefore, compare the profitability of the energy supply systems and select the one, which will allow them to reduce their operational costs. This study analyses the profitability of a ground source heat pump, photovoltaic solar panels and an integrated ground source heat pump with a photovoltaic system, as three energy supply systems for a single-family house in Sweden. The profitability of the supply systems was analysed by calculating the payback period (PBP) and internal rate of return (IRR) for these systems. Three different energy prices, three different interest rates, and two different lifespans were considered when calculating the IRR and PBP. In addition, the profitability of the supply systems was analysed for four Swedish climate zones. The analyses of results show that the ground source heat pump system was the most profitable energy supply system since it provided a short PBP and high IRR in all climate zones when compared with the other energy supply systems. Additionally, results show that increasing the energy price improved the profitability of the supply systems in all climate zones.

Keywords: single-family house; energy supply system; payback period; internal rate of return; energy price; Swedish climate zones

---

1. Introduction

The energy renovation of single-family houses remains a challenging task throughout Europe due to threats posed by climate change on different regions' environments and economies. The current range of energy renovations among member of the European union is between 0.5% and 2.5% per year, with an average of 1% per year [1]. To achieve the European parliament's target for 40% energy efficiency by 2030, the renovation rate should increase to about 2.5% to 3% of the housing stock per year [1]. Although Sweden set a more ambitious target of 50% more efficient energy use by 2030, compared to the levels from the reference year 1995 [2], the renovation rate in this country was about 0.8% per year in 2016 [1]. At this point, the Swedish residential sector could be a major contributor to achieving the national target for reduction in energy use, since it is responsible for 22% of the country's total energy consumption, from which 12% comes from single-family houses [3]. Single-family houses account for almost 50% of the total building stock in Sweden [4]. The total heated area (applying the Swedish Atemp definition) of the single-family houses in Sweden was about 293 million square meter in 2016, about 54% larger than the total area of multi-family houses [5]. According to the Swedish Statistics Central Bureau (SCB), 86% of single-family houses are about 30 years old, while 50% of them use electricity to support their heating demands [6]. Moreover, technical installations of single-family houses and insulation layers within building envelopes are likely to be near the end of their expected lifetime [7]. In addition, the existing single-family houses in Sweden primarily exhibit poor energy performance due to technical deterioration in heating, ventilation and air conditioning systems [7].

The energy renovation of single-family houses should be based on two main considerations. First, the energy use of single-family houses after renovation should meet the national energy target. Installing renewable energy supply systems was mainly considered as a solution to reduce the total energy use of single-family houses in Sweden. From 2010 until 2015, the market share of installed heat pumps and photovoltaic solar panels in Sweden increased to about 30% [4] and 19% [8], respectively. The second consideration concentrates on the cost of such an energy supply system for the homeowner. This is because installing renewable energy supply systems requires mainly high-cost investments, setting a great challenge for homeowners [9]. Financing such a supply system often involves taking out a loan, which will only be granted if the renovation increases the value of the property or reduces operational costs to offset the interest costs of the loan. The value of property in Sweden is highly dependent on its location. Accordingly, there is a risk as to the value of the property after installing energy supply systems will not cover the loan. Homeowners need to prioritize the adoption of energy supply systems that will allow them to save a high amount of energy and minimize their operational costs. However, the profitability of renewable energy supply systems is mainly determined by several different factors: Their technical characteristics [10], energy prices [11], interest rates [12], buildings' energy use patterns [10], climate conditions [13], lifespan [14] and performance gap between actual and predicted energy savings [15], which is considered when analysing the profitability of the supply systems. There are several reasons for the inconstancies between actual and predicted energy savings; but "rebound effect" was frequently discussed by former studies [16,17]. The rebound effect refers to occupants' tendency to consume more energy due to economic gains provided by energy-efficient renovations.

Although previous studies have analysed the implications of the abovementioned factors on the profitability of renewable energy supply systems in Sweden [12,18–20], they provided no information on how simultaneous changes in energy prices, interest rates, climate conditions and lifespan affect their profitability. Towards this direction, this study takes a novel approach and provides with an analysis of the profitability of a ground source heat pump (GSHP), photovoltaic solar panels (PVs) and an integrated ground source heat pump with a PV system (GSHP-PV) for a single-family house in Sweden. The profitability of the energy supply systems was analysed by calculating the payback period (PBP) and internal rate of return (IRR) for these systems. The IRR represents the interest rate, which, for a given time, can be obtained from the investment [21]. A high IRR signifies the profitability of an investment. On the other hand, PBP corresponds to the time in which, for a given discount rate, the investment cost will be repaid [22]. As homeowners prefer investments with low-risk exposure [23], a short PBP is more preferable. The profitability of the GSHP, PV and GSHP-PV systems was analysed for four Swedish climate zones, all with different climate conditions. In addition, three different energy prices, three different interest rates and two different lifespans were considered when calculating the IRR and PBP. Any such approach allows one to analyse how simultaneous changes in energy prices, interest rates, climate conditions and lifespan affect the cost-effectiveness of energy supply systems. The evaluation of the PBP and IRR assists in projecting energy efficiency policies while informing house owners of the outcomes of the investments they decide to make.

2. Methodology

The single-family house had a total heated area of 140 m^2, with a ventilated volume of about 350 m^3. The heated area was divided over 2 floors above the ground (Figure 1). In addition, the single-family house included an unheated attic with a wooden pitched roof on the 3rd floor. The attic area was equipped with a mechanical ventilation system.

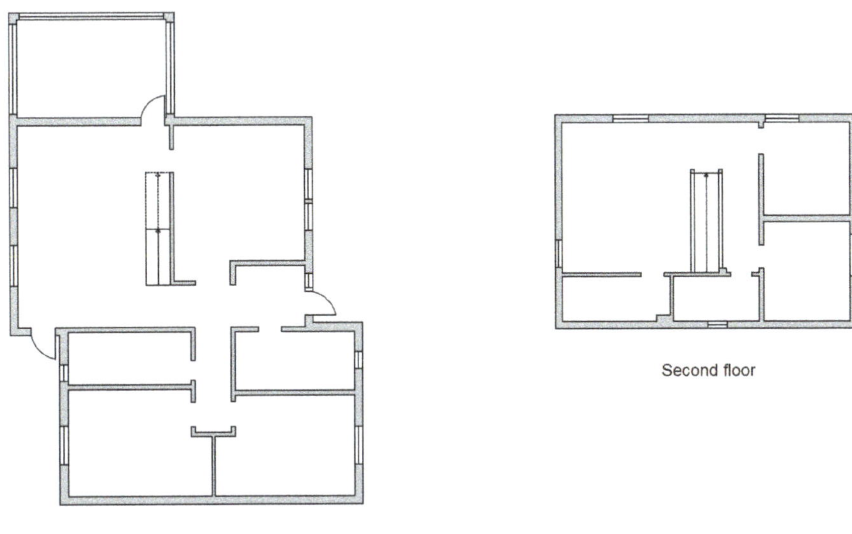

Figure 1. The first and second floors of the single-family house.

The EnergyPlus simulation tool (8.5.0) was used to evaluate the energy performance of the house. The thermal specifications of the single-family house were in accordance with the national codes at the time of construction in 1979 (Table 1). The heat transfer between the ground and external floor slab followed ISO 13370 [24].

Table 1. The thermal specifications of the single-family house.

Building Envelopes	U-Value
U-value of external walls	0.25 (W/m^2 K)
U-value of attic roof	0.08 (W/m^2 K)
U-value of external floor	0.27 (W/m^2 K)
U-value of external windows	1 (W/m^2 K)

The heating system was an electrical boiler, which was connected to water-based underfloor and radiator distribution systems. The supply and return temperatures of the underfloor and radiator systems were set to 60/45 °C. The domestic hot water consumption was assumed to be 14 m^3/person per year, which corresponded to 800 kWh/person per year [25]. The airtightness was assumed to be 1.6 L/s m^2 at the pressure of ±50 (Pa), which complied with national codes at the time of construction. In addition, the single-family house was equipped with a mechanical ventilation system, which was comprised of a supply fan and a heat exchanger with an efficiency of 80% and 85%, respectively. The ventilation system was operated 24 h a day. The airflow rate was set to ±0.35 L/m^2 to comply with national codes [26]. To ensure comfortable thermal conditions, the upper and lower temperature limits were set to 22 °C and 18 °C, respectively [27]. The occupancy density was about 0.02 persons per 1 m^2, while the occupancy internal heat gain was assumed to be 80 W/person, following the recommendations provided by Levin [25] for Swedish housing. The occupancy schedule follows Figure 2 below.

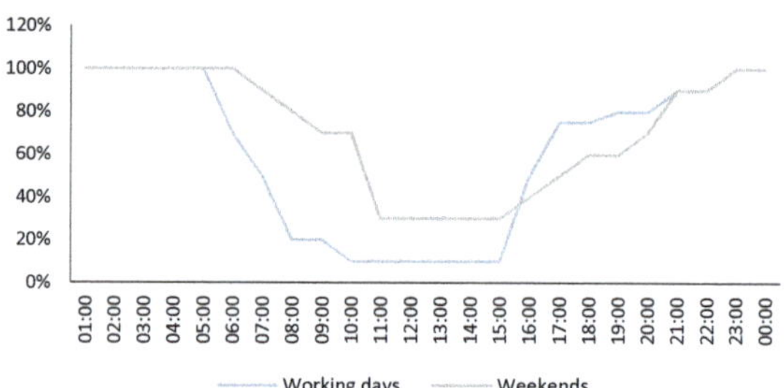

Figure 2. The occupancy schedule applied when running simulations.

The clothing thermal insulation followed ISO7730-Standard [28], as it was set to be 1 (clo) in winter and 0.5 (clo) in summer. The heat gain from artificial lighting and electric appliances was assumed to be 3.6 W/m^2 [12]. The visible reflection of the interior walls, ceilings, and floors was set to be 60%, 80% and 20%, respectively, which complied with the Swedish standard [29].

2.1. Weather and Climate Zones in Sweden

Following the Köppen-Geiger climate map, Sweden is located in the cold climate zone (D) [30]. In addition to the abovementioned categorization, the National board of housing building and planning [31] further classified Sweden into 4 climate zones, which allowed for the performing of detailed energy demand calculations based on the different climate conditions. The climate zones vary mainly in their outside design temperatures, annual average temperatures, wind velocity, relative humidity and solar radiation. The Swedish climate zone I represent the coldest and northmost zone, while climate zone IV represents the warmest and furthest south zone [31].

The single-family house was modelled in 4 different cities: Kiruna in climate zone I, Sundsvall in climate zone II, Stockholm in climate zone III, and Gotheborg in climate zone IV (Figure 3). This decision was made to analyse the contribution of the climate zones in terms of the cost-effectiveness of the 3 different energy supply systems in Sweden. Weather data files, required for performing simulations in the respective cities, were collected from Climate OneBuilding [32].

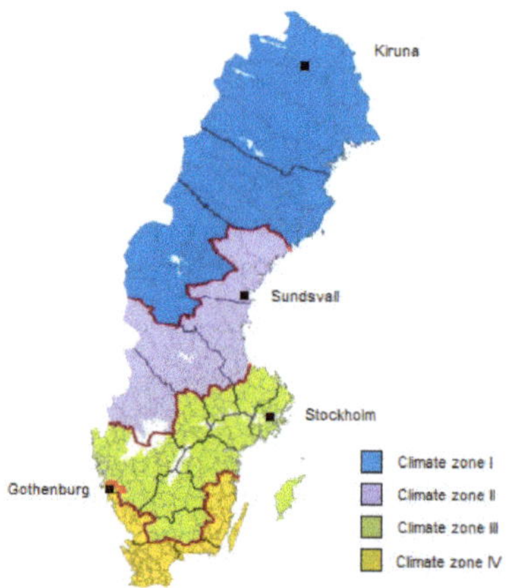

Figure 3. Four Swedish climate zones and the position of four regions analysed.

2.2. Energy Supply Systems

Three distinct energy supply systems were considered to improve the energy efficiency of the single-family house. The first considered supply system was a ground source heat pump (GSHP) with a coefficient of performance (COP) of 3 and a total power of 50 kW. The lifespan of the GSHP was 15 years. The 2nd supply system was a solar photovoltaic (PV) system with a lifespan of 20 years. Each PV panel had a power output of 285 W [33]. In total, 31 PV panels with a total area of 43.7 m^2 were installed on a south-sloping roof with a 45° tilt toward the south. The third supply system was comprised of an integrated GSHP and PV system (GSHP-PV). The COP, total power and lifespan of the GSHP used in the integrated supply system was similar to those of the first energy supply system. However, the total number of PV panels to be installed in each climate zone varied in order to avoid overproduction issues. Although the overproduced electricity of the system can be sold back to the grid [34], PV panels had significantly higher embodied CO2 when compared with other supply systems [35]. Louwen, Van Sark [36] discussed about great downward trends of the environmental impact of PV production. Although previous studies included the evaluation of embodied CO_2 impact of PV systems in their analyses, they used different data sources, technologies, system boundaries, locations and lifespans. The inconsistency in results, presented by previous studies, limits one to derive a robust conclusion about PV systems' performance in reducing carbon footprints. Table 2 presents the specifications of the PV systems installed in each region.

Table 2. Photovoltaic solar panels (PV) systems, used when installing ground source heat pump-PV (GSHP-PV) system in four climate zones.

	Total Number of Panels	Total Area (m^2)	Max. Power (kW)	Tilt Toward South
Climate zone I	31	51.2	8.8	45°
Climate zone II	25	41.3	7.1	45°
Climate zone III	23	38	6.6	45°
Climate zone IV	21	33	5.7	45°

The schedule for operating the abovementioned supply systems was set to 24/7 since they were designed to support both heating and domestic hot water demands.

2.3. Cost-Effectiveness Evaluations

The IRR and PBP were quantified using Equations (1) and (2). The IRR is the interest rate of "i", which, for a given lifespan of "t", the net present value (NPV) is zero. Meanwhile, PBP is the lifespan of "t", which, for a given interest rate of "i", makes NPV zero.

$$NPV = \sum_{t=0}^{n} (D'_t) * \frac{1}{(1+r)^t} - (I+U) \tag{1}$$

$$D'_t = (E_0 - E_t) * \alpha (1+\beta)^t \tag{2}$$

where:

NPV is the net present value during the lifespan of n year;

D'_t is the annual energy saving cost;

E_0 is the initial total energy use before installing the supply systems;

E_t is the total energy use after installing the supply systems;

r is the interest rate;

t is the lifespan of n years;

α is the energy price per kwh/m^2;

β is the inflation in energy price (%);

I_0 is the investment cost;

U is the maintenance and installation costs.

The IRR was calculated for the lifespans of 30 and 50 years. In quantifying PBP, the interest rates of 1%, 3% and 5% were considered. This decision was made to analyse the implication of lifespan and interest rate on the IRR and PBP, respectively. In addition, 3 different energy prices were considered when calculating the IRR and PBP (Table 3). Furthermore, the sensitivity of the IRR and PBP to rises in energy prices and changes in lifespan was analysed.

Table 3. Three different energy prices considered.

Energy Price	Energy Price for Electricity
Lowest energy price among European countries in 2019 * [37]	0.48 (SEK/kWh)
Highest energy price among European countries in 2019 * [37]	3.36 (SEK/kWh)
Energy price in Sweden in 2019 * [37]	1.45 (SEK/kWh)

* Including tax and levies.

Table 4 presents the investment, maintenance, installation and labour cost of the three renovation packages. In calculating the IRR and PBP, the energy supply systems were replaced when they reached the end of their lifespan.

Table 4. Investment, maintenance and installation costs of energy supply systems.

Supply Systems	Investment Cost	Maintenance Cost (SEK/kW.Y)	Installation Cost
GSHP [38]	6000 (SEK/kW)	150	24,000 (SEK/kW)
PV system [33]	19,000 (SEK/kW)	342	3800 (SEK/kW)
GSHP-PV system [33,38]	25,000 (SEK/kW)	492	27,800 (SEK/kW)

3. Results and Discussions

Table 5 shows the variation in the total energy consumption of the single-family house before and after installing the energy supply systems in the four regions. The GSHP and GSHP-PV systems reduced the total energy consumption to below 69 (kWh/m^2), thereby satisfying the national energy target in reducing total energy use by 50% when compared with 1995. However, the PV system fulfilled the national energy code only in climate zone IV. In addition, the analyses of the results showed that the GSHP-PV system had the best performance in the four regions, as it reduced the total energy use by about 91% in climate zone I, 93% in climate zone II, and 98% in climate zone III and climate zone IV.

Table 5. The total energy consumption of the single-family house.

	Initial (kWh/m^2)	After Installing GSHP (kWh/m^2)	After Installing PV (kWh/m^2)	After Installing GSHP + PV (kWh/m^2)
Climate zone I	197	53	164	21
Climate zone II	124	41	86	9
Climate zone III	119	32	79	3
Climate zone IV	113	29	67	2

3.1. Analysing Variations in the IRR

Figure 4 shows the variations in IRR when installing energy supply systems with three different energy prices and lifespans of 30 years. Considering the GSHP system, when the energy price was 0.48 (SEK/kWh), the IRR was positive only in climate zone I. This was because the aggregate levels of monetary savings due to the installation of GSHP were greater in climate zone I than in the other climate zones. Accordingly, with a low energy price of 0.48 (SEK/kWh), the GSHP was profitable only in climate zone I, while it lost its value at its respective IRR in all climate zones. A higher energy price had a significant effect on the convenience of GSHP installation in all four climate zones, as it augmented the IRR across them all. However, IRR was augmented at a different degree of increase. When changing the energy price from 0.48 (SEK/kWh) to 1.45 (SEK/kWh), the IRR in climate zone I rose by a 10-fold increase; however, it was augmented by an 8-fold increase in climate zone II and by a 7-fold increase in climate zones III and IV. Similarly, when the energy price was changed from 0.48 (SEK/kWh) to 3.36 (SEK/kWh), the IRR rose 25-fold, 19-fold, 17-fold, and 16-fold in climates I, II, III and IV, respectively.

Considering the PV system, when the energy price was 0.48 (SEK/kWh) or 1.45 (SEK/kWh), all IRR values were negative. A negative IRR implied that the PV system was not profitable as it lost value at its respective IRR in all climate zones. This was because, when the energy prices were low, the investment cost of the PV system outweighed the cost of the saved energy. With an energy price of 3.36 (SEK/kWh), the lowest IRR was obtained in climate zone I. This was because, at higher latitudes, like Kiruna in climate zone I, the amount of electricity produced by the PV system was less than the electricity production in other climate zones, which increased the use of electricity from the grid. Accordingly, the aggregate levels of monetary savings due to the installation of the PV system were smaller in climate zone I. The highest IRR of 3% was obtained in climate zones III and IV, signifying that the PV system was most profitable in these climate zones because the PV system received high solar irradiation at lower latitudes, such as Gothenburg and Stockholm; accordingly, it exhibited better performance in producing electricity.

In terms of the GSHP-PV system, when the energy price was 0.48 (SEK/kWh), all IRR values were negative. Although the GSHP-PV system had the best performance in reducing the total energy use of the single-family house, its high investment costs exceeded the cost of the saved energy. With an energy price of 1.45 (SEK/kWh) or 3.36 (SEK/kWh), all IRR values were positive, but the highest IRR was obtained in climate zone I, while the lowest IRR was observed in climate zone IV. This occurred since the amount of saved energy due to the installation of the GSHP-PV system was higher in climate zone I, which added to the aggregate levels of monetary savings. Accordingly, the GSHP-PV system was most profitable in climate zone I when compared with the other climate zones. When changing the energy price from 1.45 (SEK/kWh) to 3.36 (SEK/kWh), the IRR in climate zone I was augmented by an 8-fold increase, while it rose 6-fold in climate zone II and 5-fold in climate zones III and IV.

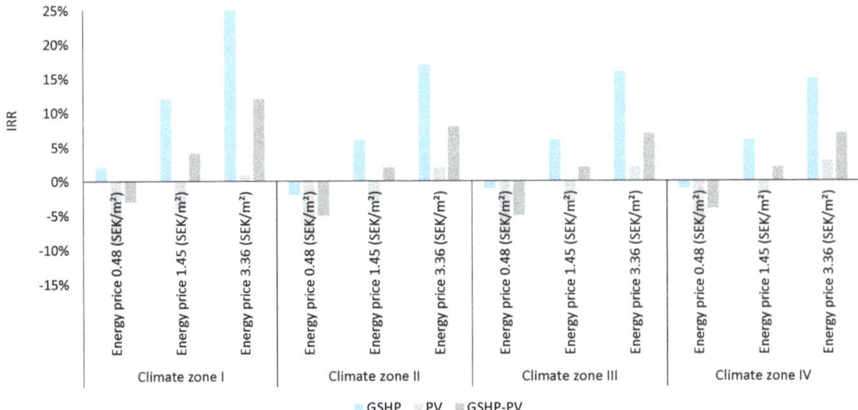

Figure 4. Internal rate of return (IRR) values in four climate zones when installing energy supply systems with three different energy prices and a lifespan of 30 years.

Figure 5 presents the variations in IRR when installing energy supply systems with 3 different energy prices and lifespans of 50 years. The GSHP system was profitable in all climate zones only when the energy price was 1.45 (SEK/kWh) and 3.36 (SEK/kWh). When changing the energy price from 1.48 (SEK/kWh) to 3.36 (SEK/kWh), the IRR in climate zone I rose 9-fold, while it was augmented 8-fold, 7-fold, and 6-fold in climate zones II, III and IV, respectively.

With an energy price of 0.48 (SEK/kWh) or 1.45 (SEK/kWh), the PV system lost value at its respective IRR in all climate zones. Increasing the energy price to 3.36 (SEK/kWh) led to positive growth in the IRR in all climate zones; however, the IRR turned positive only in climate zone IV. Accordingly, the PV system was profitable in climate zone IV only when the energy price was high.

With a low energy price of 0.48 (SEK/kWh), the investment costs for the GSHP-PV system devalued at its respective IRR in all climate zones. The devaluation rate was lower in climate zone I, while it further deteriorated in other climate zones. This was because of the amount of costs of the saved energy due to the installation of the GSHP-PV system being greater in climate zone I than in the other climate zones. Increasing the energy price to 1.45 (SEK/kWh) had a positive effect on the IRR in all climate zones; however, the IRR became positive only in climate zone I. With a high energy price of 3.36 (SEK/kWh), all IRR values were positive. The highest IRR was obtained in climate zone I, while the lowest was observed in climate zones III and IV. Increasing the energy price from 1.48 (SEK/kWh) to 3.36 (SEK/kWh) augmented the IRR by a 6.7-fold increase in climate zone I, while it rose by 6-fold in climate zone II and 5-fold in climate zones III and IV.

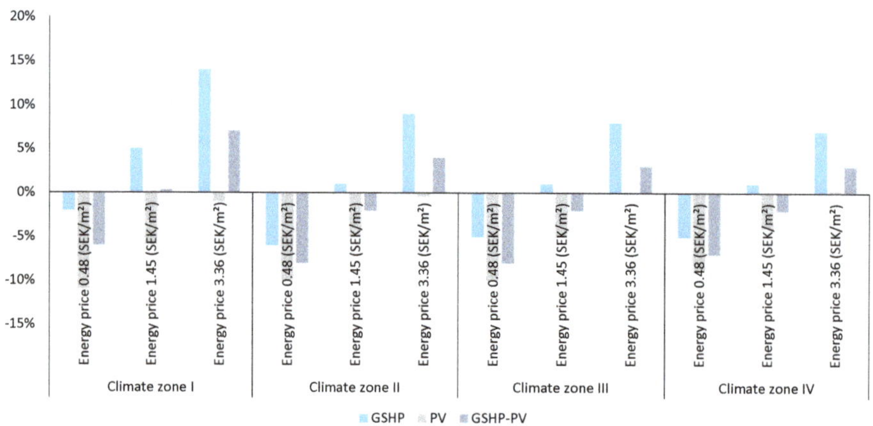

Figure 5. IRR values in four climate zones when installing energy supply systems with three different energy prices and a lifespan of 50 years.

Sensitivity Analyses

Figure 6a–d show the variations in the IRR in the four climate zones when installing the GSHP in terms of changes in energy price and lifespan. Regardless of whether energy price was low or high, changing the lifespan from 30 years to 50 years had a negative impact on the IRR in all climate zones. Accordingly, funding investments in the GSHP system during a lifespan of 30 years was more convenient way to invest in all climate zones. The analysis of the results showed that the IRR was more sensitive to increases in energy prices when the lifespan was 30 years. This was because, for a lifespan of 30 years, the difference between the costs of saved energy and investment costs was larger than for a lifespan of 50 years. In addition, the IRR was more sensitive to changes in energy prices in climate zone I, followed by climate zones II, III and IV. This was because the GSHP system had the best performance in climate zone I, and, accordingly, an increase in energy price augmented the IRR more significantly in this climate zone.

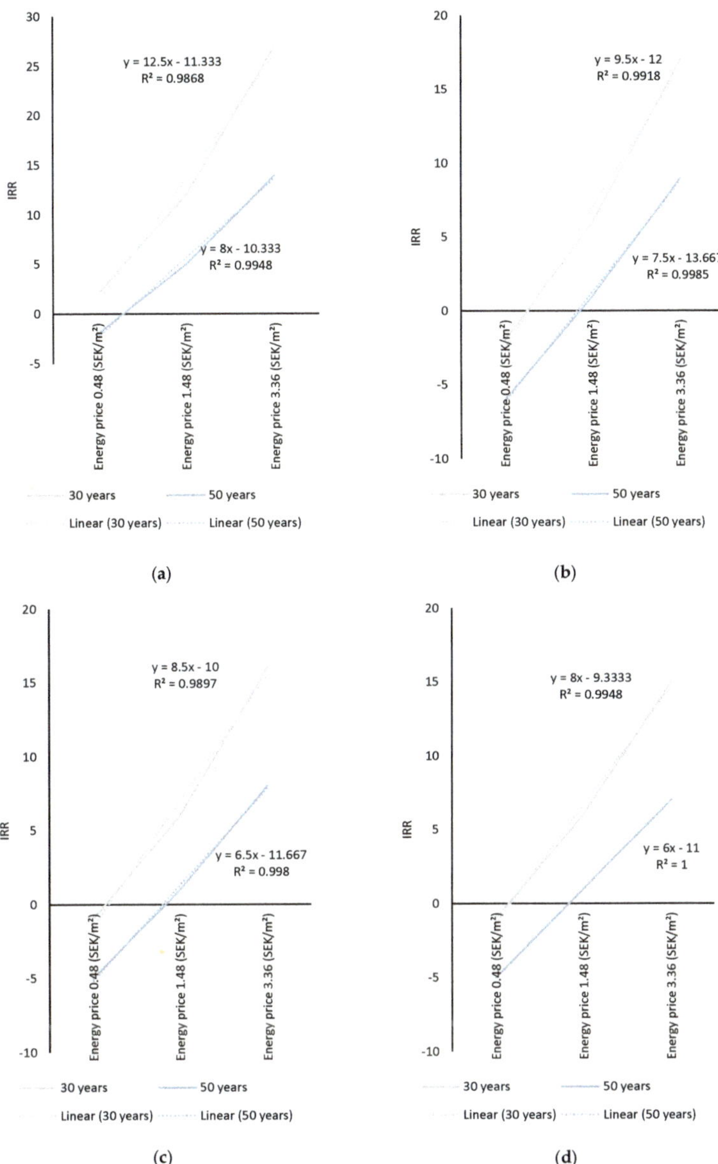

Figure 6. (**a**) Variation in IRR with respect to changes in lifespan and energy price in climate zone I. (**b**) Variation in IRR with respect to changes in lifespan and energy price in climate zone II. (**c**) Variation in IRR with respect to changes in lifespan and energy price in climate zone III. (**d**) Variation in IRR with respect to changes in lifespan and energy price in climate zone IV.

Figure 7a–d show the variation in the IRR in the four climate zones with respect to energy price and lifespan when installing the PV system. The analysis of results showed that, when the lifespan was 30 years, the IRR was equally sensitive to changes in energy prices in all climate zones. In addition, with a lifespan of 30 years, the IRR was more sensitive to changes in energy price, than for a lifespan of 50 years.

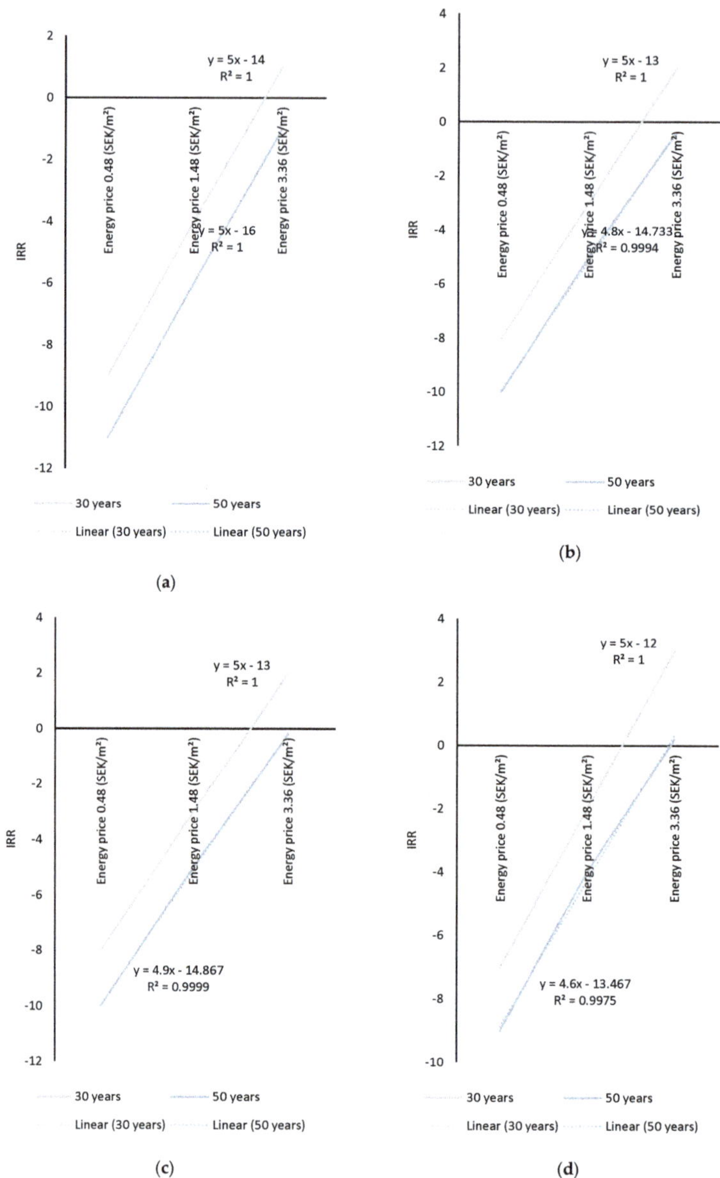

Figure 7. (**a**) Variation in IRR with respect to changes in lifespan and energy price in climate zone I. (**b**) Variation in IRR with respect to changes in lifespan and energy price in climate zone II. (**c**) Variation in IRR with respect to changes in lifespan and energy price in climate zone III. (**d**) Variation in IRR with respect to changes in lifespan and energy price in climate zone IV.

Figure 8a–d show the variation in the IRR in four climate zones when installing the GSHP-PV system in terms of changes in energy price and lifespan. Similarly, the IRR was more sensitive to changes in energy prices when the lifespan was 30 years in all climate zones. In addition, the IRR was more vulnerable to changes in energy prices in climate zone I, followed by climate zones II, III and IV. This was because the amount of monetary savings due to the installation of the GSHP-PV system in climate zone I was greater than those of the other climate zones; hence, an increase in energy price raised the IRR to a greater degree in this climate zone.

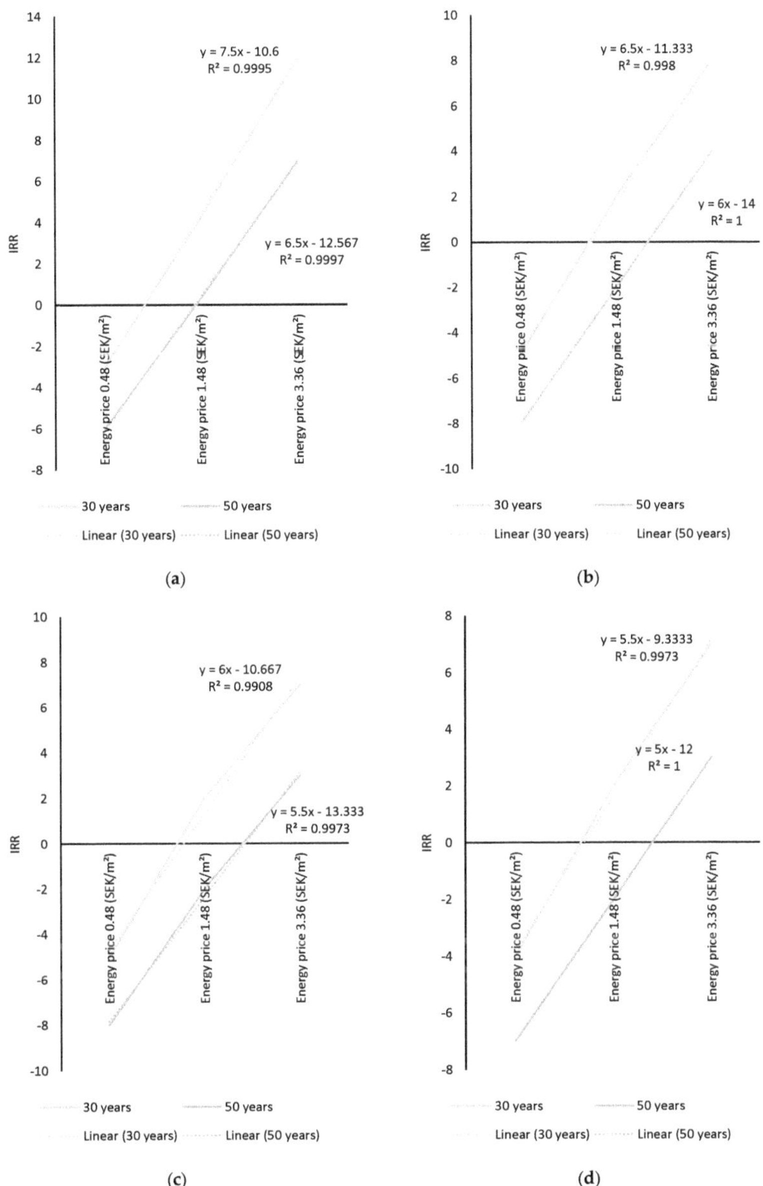

Figure 8. (**a**) Variation in IRR with respect to changes in lifespan and energy price in climate zone I. (**b**) Variation in IRR with respect to changes in lifespan and energy price in climate zone II. (**c**) Variation in IRR with respect to changes in lifespan and energy price in climate zone III. (**d**) Variation in IRR with respect to changes in lifespan and energy price in climate zone IV.

3.2. Analysing Variations in PBP

Figure 9 shows the variations in the PBP when installing energy supply systems with three different energy prices and three different interest rates. With interest rates of 1% and an energy price of 0.48 (SEK/kWh), the GSHP was the only profitable system in climate zone I, because it provided the highest energy saving rate of 73% in climate zone I, which accumulated in the monetary savings toward investment and maintenance costs. When the interest rate was increased to 3% and 5%, the investment costs for installing the GSHP system were never returned. However, the GSHP system was profitable in all climate zones when the energy prices increased to 1.45 (SEK/kWh) and 3.36 (SEK/kWh). Although increasing the interest rate usually augmented the PBP values, the increase in energy price mitigated the negative effect of a high interest rate on the PBP in all climate zones. Regardless of fluctuations in interest rate, when energy prices were 1.45 (SEK/kWh) or 3.36 (SEK/kWh), the lowest PBP was observed in climate zone I, while the highest was obtained in climate IV.

The profitability of the PV system was strongly dependent on a low interest rate of 1%, along with a high energy price of 3.36 (SEK/kWh). This was because a low interest rate, together with a high energy price, elevated the aggregate levels of monetary savings in all climate zones.

The profitability of the GSHP-PV system was determined by both the interest rate and energy price. Regardless of fluctuations in interest rate, when the energy price was 0.48 (SEK/kWh), funding investments in the GSHP-PV system was inconvenient, since the investment costs for installing this supply system was never repaid. Increasing the energy price mitigated the negative effect of the interest rates on PBP. With an interest rate of 1% and an energy price of 1.45 (SEK/kWh), the GSHP-PV was profitable only in climate zones I and II. Meanwhile, increasing the interest rate to 3% decreased the monetary savings, as the investment costs of the GSHP-PV system were only repaid in climate zone I. With an interest rate of 5%, the GSHP-PV system yielded no financial gain in any climate zones. However, with energy prices rising to 3.36 (SEK/kWh), the difference between the costs of the saved energy due to the installation of the GSHP-PV system and the investment costs turned positive in all climate zones, which made the GSHP-PV system a profitable solution for reducing total energy use. Regardless of fluctuations in interest rate, when the energy price was 3.36 (SEK/kWh), the lowest and highest PBPs were obtained in climate zone I and IV, respectively.

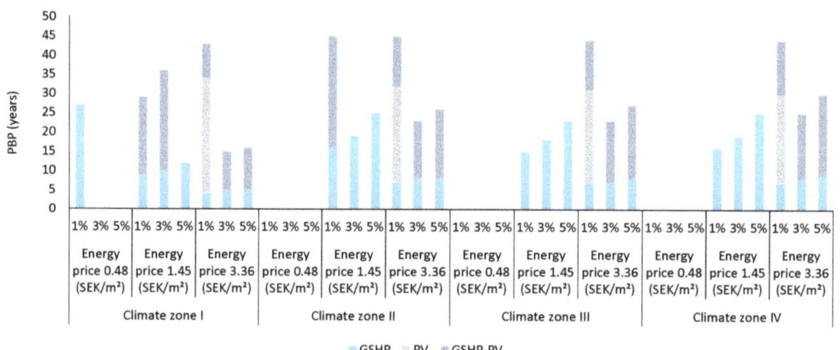

Figure 9. Payback period (PBP) values in four climate zones when installing energy supply systems with three different energy prices and three interest rates.

Sensitivity Analyses

Figure 10a–d show the variations in PBP when installing the GSHP in terms of changes in energy price and interest rate. The analysis of results showed that PBP was more sensitive to changes in energy prices when the interest rate was high. In other words, rising energy prices reduced the PBP in all climate zones but more significantly when the interest rate was high, which signified the effectiveness of a higher energy price in terms of the profitability of energy supply systems when interest rates were high. In addition, the sensitivity of PBP to changes in energy prices varied among the four climate zones. Regardless of the fluctuations in interest rate, the PBP was less vulnerable to changes in energy price in climate zone I, while it was highly sensitive to changes in energy price in climate zone IV.

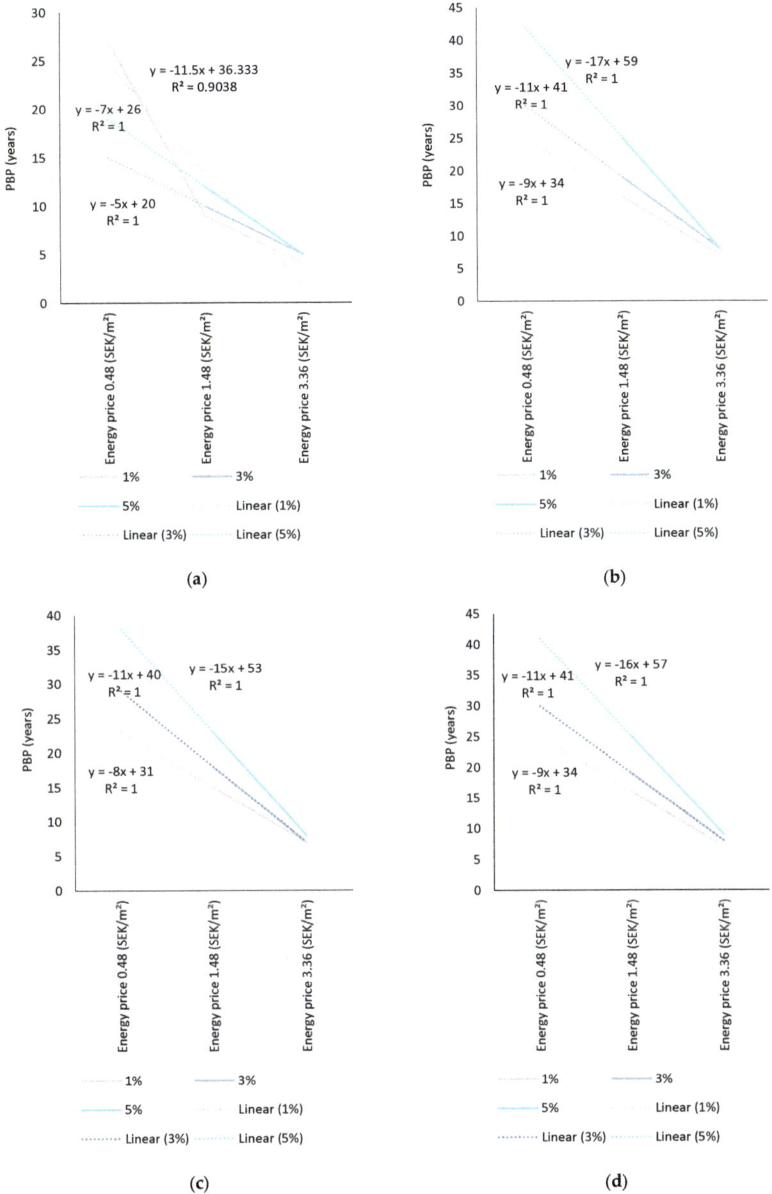

Figure 10. (**a**) Variation in PBP with respect to changes in the interest rate and energy price in climate zone I. (**b**) Variation in PBP with respect to changes in the interest rate and energy price in climate zone II. (**c**) Variation in PBP with respect to changes in the interest rate and energy price in climate zone III. (**d**) Variation in PBP with respect to changes in the interest rate and energy price in climate zone IV.

No analyses were performed to study the sensitivity of PBP when installing the PV system since this system was only profitable with a high energy price of 3.36 (SEK/kWh) and a low interest rate of 1%.

Figure 10a–d show the variations in PBP when installing the GSHP-PV in terms of changes in energy price and interest rate. With an interest rate of 1%, PBP was less sensitive to changes in energy prices in climate zone I than in other climate zones. This was because the amount of monetary savings due to the installation of the GSHP-PV system was greater in climate zone I; accordingly, increases in energy price had a smaller effect on PBP in this climate zone. In addition, PBP was more sensitive to change in energy prices in climate zone I when the interest rate was increased to 3%. With an interest rate of 5%, the GSHP-PV system was profitable only when the energy price was 3.36 (SEK/kWh); accordingly, no analyses were performed to study whether or not the PBP was sensitive to change in energy price.

4. Final Remarks

The following remarks were concluded from the analyses performed.

The GSHP-PV system exhibited the best performance in reducing the total energy use of the single-family house.

In general, the GSHP system was the most profitable energy supply system, since it provided a short PBP and high IRR in all climate zones when compared with the other energy supply systems.

Increasing the energy price improved the profitability of the supply systems in all climate zones.

With an energy price of 1.45 (SEK/kWh) or 3.36 (SEK/kWh), funding investments in the GSHP and GSHP-PV systems during a lifespan of 30 years was more convenient in all climate zones.

The IRR was more sensitive to changes in energy price when the lifespan was 30 years in all climate zones.

5. Conclusions

The European parliament's target for the energy efficiency of buildings mandates an increase in the rate of house renovations. The single-family houses in Sweden can be seen as highly relevant in fulfilling this target, as they account for almost half the residential building stock in this country. In addition, the single-family houses in Sweden are responsible for a large amount of total energy consumption within the residential sector. Installing renewable energy supply systems has thus been seen as a solution to meet the European parliament's target. However, the rate of energy renovations performed on single-family houses in Sweden has been low, reaching only 0.8% in 2016. One of the reasons explaining this low renovation rate is the high investment costs required for installing renewable energy supply systems that pose a great challenge to house owners. Therefore, there is a need for evaluating the cost-effectiveness of different supply systems that could be applied to single-family houses. This study took a novel approach to analyse how simultaneous changes in energy prices, interest rates, climate conditions and lifespan affect the cost-effectiveness of energy supply systems. For this purpose, the internal rate of return (IRR) and the payback period (PBP) of three potential energy supply systems that could be possibly applied in the case of a single-family house in Sweden. The first energy supply system included the installation of a ground source heat pump (GSHP), while the second and third packages comprised the installation of photovoltaic panels (PV system) and the mounting of an integrated GSHP and PV system (GSHP-PV), respectively. The IRR was calculated for lifespans of 30 and 50 years, while the PBP was obtained for interest rates of 1%, 3% and 5%. Furthermore, the implications of the three different energy prices and four climate zones on IRR and PBP were analysed.

The analyses of the results showed that the GSHP provides a higher IRR; accordingly, it yields the highest profit during the lifespan of a house when compared to other energy supply systems. This was because of the high performance of the GSHP in reducing total energy use and its relatively low investment cost. Furthermore, the results showed that raising the energy cost can increase the IRR of the supply systems because it adds to costs related to saved energy and thus offset the investment costs. Comparatively, the GSHP provides the lowest PBP when compared with the PV and GSHP-PV systems. In addition, the results showed that increasing the interest rate adds to PBP of the supply systems since it depreciates the cost of saved energy. In contrast, increasing the energy price can ease the investment cost of the renovation package more effectively, thereby reducing the PBP. With a lifespan of 30 years, the IRR was more sensitive to changes in energy price, than for a lifespan of 50 years.

Although the renewable energy supply systems hold promise for providing energy savings, this promise can be misleading mainly due to the rebound effect. The rebound effect comprises two main categories: Temporal rebound and spatial rebound [15]. The temporal rebound is associated with phenomena such as temperature take-back, in which occupants keep a higher indoor temperature, practically over longer periods, thereby counteracting the energy efficiency provided by renewable energy supply systems. The spatial rebound refers to occupants' tendency to expand the heated space.

Results, considering the effectiveness of the energy supply systems in reducing the total energy consumption, can be generalised to countries with similar climate conditions. However, the cost-effectiveness of the renewable energy resources depends on how investment costs, energy prices and energy policies evolve. Effective energy policy formulations can facilitate renewable energy transitions. Any such policy may include financial incentives or taxation regimes applying to non-renewable energy resources [39]. However, the scope of the energy policies may vary in different countries, since they are strongly influenced by social, political and economic factors [40].

The results can be expanded to provide an aid when developing energy efficiency policies with a goal of increasing the rate of implementation of energy renovations. In addition, the analyses performed could be expanded for the further quantification of the IRR and PBP in a more comprehensive manner, including the replacement of windows and insulation layers in building envelopes.

Author Contributions: Conceptualization, E.J. and G.P.; methodology, E.J.; software, E.J.; validation, E.J.; formal analysis, E.J.; investigation, E.J. and A.V.; resources, E.J.; data curation, E.J.; writing—original draft preparation, E.J. and G.P.; visualization, E.J.; supervision, A.V.; project administration, E.J.; funding acquisition, E.J. All authors have read and agreed to the published version of the manuscript.

Funding: This research was funded by Interreg Europe, grant number PGI05809.

Acknowledgments: This study was accomplished as part of LC-Districts project, financed by Interreg Europe. Authors appreciate greatly for their contributions.

Conflicts of Interest: The authors declare no conflict of interest.

References

1. Artola, I.; Rademaekers, K.; Williams, R.; Yearwood, J. Boosting building renovation: What potential and value for Europe. In *Study for the iTRE Committee, Commissioned by DG for Internal Policies Policy Department A*; European Parliament: Brussels, Belgium, 2016; p. 72.

2. Goverment Offices of Sweden. Available online: https://www.government.se/ (accessed on 1 March 2019).

3. Swedish Energy Agancy. *Energy Situation [Title in Swedish: Energiläget]*; Swedish Energy Agancy: Bromma, Sweden, 2017; pp. 1–86.

4. Swedish Energy Agancy. *Summary of Energy Statistics for Dwellings and Nonresidential Premises [Title in Swedish: Energistatistik för Småhus, Flerbostadshus och Lokaler 2016]*; Swedish Energy Agancy: Bromma, Sweden, 2016; pp. 1–30.

5. Swedish Energy Agency. *Energy in Sweden—2018*; Swedish Energy Agancy: Bromma, Sweden, 2019. Available online: http://www.energimyndigheten.se/en/news/2018/energy-in-sweden---facts-and-figures-2018-available-now/ (accessed on 1 March 2019).

6. Sköldberg, H.; Ryden, B. The heating market in Sweden-an overall view. In *Sverige Värmemarknad*; 2014. Available online: http://www.varmemarknad.se/pdf/The_heating_market_in_Sweden_141030.pdf (accessed on 1 March 2019).

7. National Board of Housing Building and Planning. *The Technical Status of the Existing Building Stock*; National Board of Housing Building and Planning: Karlskrona, Sweden, 2010.

8. Lindahl, J. *National Survey Report of PV Power Applications in Sweden*; Uppsala University and International Energy Agency: Uppsala, Sweden, 2014.
9. Bjørneboe, M.G.; Svendsen, S.; Heller, A. Initiatives for the energy renovation of single-family houses in Denmark evaluated on the basis of barriers and motivators. *Energy Build.* **2018**, *167*, 347–358. [CrossRef]
10. Shen, P.; Lukes, J.R. Impact of global warming on performance of ground source heat pumps in US climate zones. *Energy Convers. Manag.* **2015**, *101*, 632–643. [CrossRef]
11. Shah, I.H.; Hiles, C.; Morley, B. How do oil prices, macroeconomic factors and policies affect the market for renewable energy? *Appl. Energy* **2018**, *215*, 87–97. [CrossRef]
12. Gustafsson, M.; Gustafsson, M.S.; Myhren, J.A.; Bales, C.; Holmberg, S. Techno-economic analysis of energy renovation measures for a district heated multi-family house. *Appl. Energy* **2016**, *177*, 108–116. [CrossRef]
13. Bonakdar, F.; Sasic Kalagasidis, A.; Mahapatra, K. The Implications of Climate Zones on the Cost-Optimal Level and Cost-Effectiveness of Building Envelope Energy Renovation and Space Heat Demand Reduction. *Buildings* **2017**, *7*, 39. [CrossRef]
14. Jalilzadehazhari, E.; Mahapatra, K. The Most Cost-Effective Energy Solution in Renovating a Multi-family House. In Proceedings of the Cold Climate HVAC Conference, Kiruna, Sweden, 12–15 March 2018; Springer: Berlin/Heidelberg, Germany, 2018; pp. 203–216.
15. Tweed, C.; Humes, N.; Zapata-Lancaster, G. The changing landscape of thermal experience and warmth in older people's dwellings. *Energy Policy* **2015**, *84*, 223–232. [CrossRef]
16. Sorrell, S.; Dimitropoulos, J.; Sommerville, M. Empirical estimates of the direct rebound effect: A review. *Energy Policy* **2009**, *37*, 1356–1371. [CrossRef]
17. Winther, T.; Wilhite, H. An analysis of the household energy rebound effect from a practice perspective: Spatial and temporal dimensions. *Energ. Effic.* **2015**, *8*, 595–607. [CrossRef]
18. Bonakdar, F.; Dodoo, A.; Gustavsson, L. Cost-optimum analysis of building fabric renovation in a Swedish multi-story residential building. *Energy Build.* **2014**, *84*, 662–673. [CrossRef]
19. Ekström, T.; Bernardo, R.; Blomsterberg, Å. Cost-effective passive house renovation packages for Swedish single-family houses from the 1960s and 1970s. *Energy Build.* **2018**, *161*, 89–102. [CrossRef]
20. Ekström, T.; Blomsterberg, Å. Renovation of Swedish Single-family Houses to Passive House Standard–Analyses of Energy Savings Potential. *Energy Procedia* **2016**, *96*, 134–145. [CrossRef]
21. Fowlie, M.; Greenstone, M.; Wolfram, C. Do energy efficiency investments deliver? Evidence from the weatherization assistance program. *Q. J. Econ.* **2018**, *133*, 1597–1644. [CrossRef]
22. Amstalden, R.W.; Kost, M.; Nathani, C.; Imboden, D.M. Economic potential of energy-efficient retrofitting in the Swiss residential building sector: The effects of policy instruments and energy price expectations. *Energy Policy* **2007**, *35*, 1819–1829. [CrossRef]
23. Friedman, C.; Becker, N.; Erell, E. Retrofitting residential building envelopes for energy efficiency: Motivations of individual homeowners in Israel. *J. Environ. Plan. Manag.* **2018**, *61*, 1805–1827. [CrossRef]
24. ISO 13370, E. *Thermal Performance of Buildings-Heat Transfer via the Ground-Calculation Methods*; ISO 13370: 2007; CEN: Brussels, Belgium, 2007.
25. Levin, P. *User Data for Housing Version 1.0 [In Swedish: Brukarindata Bostäder Version 1.0]*; Sveby: Stockholm, Sweden, 2012.
26. National Board of Housing Building and Planning. *Building Regulation, Regulations and General Advices [In Swedish: Boverkets Byggregler, Föreskrifter och Allmänna råd, BBR. BFS 2011:6 med Ändringar till och med BFS 2018:4 BBR 26]*; National Board of Housing, Building and Planning: Stockholm, Sweden, 2018.
27. ASHRAE Standard 55-2010. *Standard 55-2010: "Thermal Environmental Conditions for Human Occupancy"*; ASHRAE: Atlanta, GA, USA, 2010.
28. ISO7730-Standard. *7730. Ergonomics of the Thermal Environment—Analytical Determination and Interpretation of Thermal Comfort Using Calculation of the PMV and PPD Indices and Local Thermal Comfort Criteria*; International Organization for Standardization: Geneva, Switzerland, 2005.
29. Månsson, L. *Light & Room, Guide for Planning of Indoor Lighting*; Ljuskultur: Stockholm, Sweden, 2003.
30. Peel, M.C.; Finlayson, B.L.; McMahon, T.A. Updated world map of the Köppen-Geiger climate classification. *Hydrol. Earth Syst. Sci. Discuss.* **2007**, *4*, 439–473. [CrossRef]
31. National Board of Housing Building and Planning. *Building Regulation, [In Swedish: Boverkets Byggregler, BFS 2011: 6 med ändringar till och med 2015: 3]*; National Board of Housing, Building and Planning: Stockholm, Sweden, 2015; (In Swedish: Boverket).

32. Climate OneBuilding. 2020. Available online: http://climate.onebuilding.org/WMO_Region_6_Europe/SWE_Sweden/index.html (accessed on 1 March 2020).

33. Solhybrid. 2018. Available online: http://www.solhybrid.se/ (accessed on 1 August 2018).

34. Galimshina, A.; Engström, J. *Development of a Solar Strategy for Helsingborgshem*; EEBD: Lund, Sweden, 2017.

35. Finnegan, S.; Jones, C.; Sharples, S. The embodied CO2e of sustainable energy technologies used in buildings: A review article. *Energy Build.* **2018**, *181*, 50–61. [CrossRef]

36. Louwen, A.; Van Sark, W.G.; Faaij, A.P.; Schropp, R.E. Re-assessment of net energy production and greenhouse gas emissions avoidance after 40 years of photovoltaics development. *Nat. Commun.* **2016**, *7*, 1–9. [CrossRef] [PubMed]

37. Eurostat. 2020. Available online: http://appsso.eurostat.ec.europa.eu/nui/submitViewTableAction.do, (accessed on 1 January 2020).

38. Fedrizzi, R.; Dipasquale, C.; Bellini, A. D6.5—Position Paper on Systemic Energy Renovation. Available online: http://inspirefp7.eu/wp-content/uploads/2017/01/WP6_D6.5_20161201_P1_Position-Paper-on-Systemic-Energy-Renovation.pdf (accessed on 1 August 2018).

39. National Research Council. *Electricity from Renewable Resources: Status, Prospects, and Impediments. Chapter 4: Economics of Renewable Electricity*; National Academies Press: Washington, DC, USA, 2010.

40. Burke, M.J.; Stephens, J.C. Political power and renewable energy futures: A critical review. *Energy Res. Social Sci.* **2018**, *35*, 78–93. [CrossRef]

Article

Influence of Reinforcing Steel Corrosion on Life Cycle Reliability Assessment of Existing R.C. Buildings

Pietro Croce *, Paolo Formichi * and Filippo Landi *

Department of Civil and Industrial Engineering-Structural Division, University of Pisa, 56122 Pisa, Italy
* Correspondence: p.croce@ing.unipi.it (P.C.); p.formichi@ing.unipi.it (P.F.); filippo.landi@ing.unipi.it (F.L.)

Received: 13 April 2020; Accepted: 26 May 2020; Published: 28 May 2020

Abstract: Time-dependent reliability assessment is a crucial aspect of the decision process for rehabilitation of existing reinforced concrete structures. Since the assessment strongly depends on degradation of materials with time, the paper focuses on the influence of corrosion in reinforcing steel on time-reliability curves of relevant reinforced concrete (r.c.) structures, built in Italy in the 1960s, belonging to different building categories. To realistically represent the probability distribution functions (*pdf*s) of the relevant properties of reinforcing steel and concrete commonly adopted in the 1960s, stochastic models for steel yielding and concrete compressive strength have been derived, by means of a suitable cluster analysis, from secondary databases of test results gathered at that time in Italy on concrete and steel rebar specimens. This cluster analysis, based on Gaussian mixture models, provides a powerful tool to "objectively" extract material classes and associated probability density functions from databases of experimental test results. In the study, different degradation conditions and several reinforcing steel and concrete classes are considered, also aiming to scrutinize their influence on the time-dependent reliability curves. Finally, to stress the significance of the study, the time-dependent reliability curves so obtained are critically examined and discussed also in comparison with the target reliability levels currently adopted in the Eurocodes.

Keywords: existing structures; reinforced concrete; time-dependent reliability; life cycle; Gaussian mixture models; strength degradation; steel corrosion; secondary databases

---

1. Introduction

Life cycle management is a topical research subject of modern structural engineering [1–3]. At present, many research studies are focused on the development of practical methods for the time-dependent reliability assessment of existing structures and infrastructures [4].

Despite the fact that reinforced concrete structures are generally associated with high durability [5], during their service life they are subjected to several deterioration processes, which are responsible for the decay of structural performance and safety over time. The most relevant causes of ageing effects in r.c. structures are, inter alia, steel corrosion and concrete carbonatation.

In the reliability assessment of existing structures, a key issue is the appropriate estimate of the mechanical properties of materials and the evaluation of their relevant statistical parameters [6]. In principle, that issue can be tackled by carrying out a suitable experimental campaign devoted to assessing the statistics of these mechanical properties, but the number of the tests is often limited by practical reasons, associated on the one hand with the difficult accessibility of some relevant points or elements to be investigated and on the other hand with the need to preserve the structural or architectural integrity of the construction. For these reasons, the most common and effective approach to the problem is based on an a priori characterization of materials, drawn from historical data obtained by documents, literature and in-situ or laboratory tests referring to coeval constructions, which is subsequently consolidated and updated by limited in-situ non-destructive or semi-destructive tests, if any.

In reinforced concrete of existing structures, the most relevant mechanical parameters are the compressive strength of the concrete and the yielding stress of the steel reinforcement.

Concrete compressive strength is generally estimated by means of non-destructive tests, such as sclerometric in-situ tests, ultrasonic tests, Sonreb combined tests or semi-destructive tests, such as extraction of cores [7] or pull-out tests. Test procedures are standardized in international standards (e.g., EN13791 [8], EN12504 series [9–12], ACI214-4.R10 [13]), which are often further specified and supplemented at national level through ad-hoc guidelines, such as the recent "Italian guidelines for the evaluation of in-situ concrete properties" [14].

As already remarked in general terms, since the safeguard of the static performances and integrity of the construction as well as its finishes and installed plants is a primary need, also in existing reinforced concrete structures the number of in-situ destructive or semi-destructive tests is limited. Although in the most favorable cases test results allow the mean values of the relevant mechanical parameters and the material's degree of homogeneity throughout the structure to be assessed, they are usually not sufficient to derive accurate statistics of the mechanical parameters needed for the reliability assessments themselves.

Difficulties in performing in-situ investigations are even more evident for steel rebars. Despite the fact that experimental evaluation of the yield and ultimate strength of reinforcing bars requires a sufficient number of samples, as suggested, e.g., in [15], the in-situ extraction of rebars from existing reinforced concrete structures is frequently unwise, since the repair is often not easy, for example, due to poor weldability of steel grades used in the past for the reinforcement. For these reasons, the main sources of information are commonly the original design documentation, if available, or the structural codes in force at the time of construction.

To overcome the lack of information about the mechanical properties of buildings' structural materials in terms of *pdf*s and relevant statistical parameters, suitable methods for the analysis of secondary material properties databases are needed. Concerning this aspect, a very sound and efficient methodology to identify homogenous material classes in databases of raw test results commonly stored in testing laboratory's archives has been previously proposed in [6,16]. Starting from standard acceptance test results obtained at the time of the construction, this methodology, based on a Gaussian mixture model, briefly described in Section 3, allows identification, in an "objective" way, of the relevant clusters of data, as well as their statistical properties.

By introducing appropriate degradation models, the outcomes of the above mentioned analyses can be further implemented to assess the time-dependent reliability of existing structures.

The present work focuses on time-dependent reliability of reinforced concrete structures built in Italy during the 1960s, also aiming to check how the concrete and reinforcing steel classifications influence the results. The time interval considered in the study is particularly significant since it is associated with the most intensive construction activity in Italy in the framework of post-war reconstruction; in fact, a relevant part of the built Italian environment dates back to it.

To illustrate the practical application of the method, some relevant time-dependent reliability curves are discussed, referring to three relevant case studies consisting of r.c. reference buildings belonging to three different categories: residential, commercial and storage buildings. The study aims to assess how corrosion effects in reinforcing steel influence the structural reliability over time, depending on the degradation conditions as well as on the reinforcing steel and concrete classes.

To realistically represent the actual pdfs of the relevant material properties, the stochastic models for steel and concrete classes are those derived from the cluster analyses of the historical test data of the Laboratory of the Department of Civil and Industrial Engineering of the University of Pisa illustrated in the already mentioned references [6,16].

Finally, to further underline the significance of the study, time-reliability curves are compared with the target reliability levels currently given over time by the Eurocodes [17].

2. Time-Dependent Reliability Analysis

A methodology to estimate time-dependent reliability of aging structures, taking into account the non-stationarity of loads and the degradation of structural resistance, has been proposed and previously applied in [18] and [19]. Truly, that methodology can also be applied to assess the influence on structural reliability of non-stationarity of climatic actions due to climate change effects; in fact, it only requires the knowledge of suitable factors of change, expressing the variations over time of the relevant parameters of the extreme values distribution of the investigated action [20–22].

According to [23], the degradation of structural resistance over time can be represented by a suitable deterministic degradation function, $D(t)$. Saying that $R(t)$ is the resistance at time t and R_0 is the initial resistance of the structure, $R_0 = R(0)$, it is:

$$R(t) = R_0 \ D(t) \tag{1}$$

Let F_S be the cumulative distribution function (CDF) of the load intensity; the time dependent reliability $L(t)$ can be thus expressed by

$$L(t) = \exp\left\{-\int_0^t \lambda(\bar{t})\left[1 - F_S\left(R_0 \ D(\bar{t})\right)\right]d\bar{t}\right\} \tag{2}$$

where $\lambda(\bar{t})$ is the mean occurrence rate of the action, and $\left[1 - F_S\left(R_0 \ D(\bar{t})\right)\right]$ is the probability that the load intensity exceeds the resistance.

Assuming an extreme value type I (EVI) distribution for the leading action, the time-dependent structural reliability trivially results

$$L(t) = \exp\left\{-\int_0^t \lambda(\bar{t})\left[1 - \exp\left\{-\exp\left[-Z(\bar{t})\right]\right\}\right]d\bar{t}\right\}, \tag{3}$$

where

$$Z(\bar{t}) = \frac{R_0 \ D(\bar{t}) - \mu(\bar{t})}{\sigma(\bar{t})} \tag{4}$$

and $\mu(\bar{t})$ and $\sigma(\bar{t})$ are the location and the scale parameters of the distribution.

The variation of the EVI parameters as a function of the time is described in [18,19] by means of factors of change derived from the analysis of climate projections and weather generated series, also considering the uncertainty in the prediction [21]. That method is much more general and can be applied, for example, for imposed loads as well.

Recalling that the probability of failure $P_f(t)$ is the complement to one reliability,

$$P_f(t) = 1 - L(t) \tag{5}$$

the reliability index $\beta(t) = -\Phi^{-1}\left(P_f\right)$ can be easily derived, as function of the time t, from Equation (5).

The reliability index so obtained can be compared with the target reliability level given in structural codes for a specific reference or observation period; for example, in EN1990 [17] it is usually assumed $\beta_t = 3.8$ for an observation period of 50 years, that means $P_f \approx 1/\left(1.4{\cdot}10^4\right)$, or equivalently, for structural designs mainly governed by variable actions, $P_f \approx 1/\left(7{\cdot}10^5\right)$ in one year, roughly corresponding, for the same structure, to $\beta_t = 4.7$ for an observation period of one year.

With the above mentioned methods, the variation over time of the structural reliability can be assessed not only for new structures, designed according to modern codes, such as Eurocodes [17,24] and ACI specifications [25], but also for the assessment of existing structures, provided that adequate information about material properties and degradation effects is available.

In the following section, the general methodology for the evaluation of statistical parameters of material properties from secondary material tests databases, which was adopted in [6] for concrete classes and in [16] for reinforcing steel classes, is briefly recalled.

3. Use of Gaussian Mixture Models in Cluster Analysis of Test Results on Building Materials

As already said, the reliability assessment of existing structures requires suitable information about statistical distribution of relevant properties of building materials, also taking into account their degradation over time.

In reliability-based assessment of existing structures, the statistical distributions of relevant mechanical properties are often selected based on literature data [26], introducing further uncertainties in the evaluations. To reduce these uncertainties, an "a priori" assessment of statistical information about the material's properties could be very helpful, also in view of possible application of Bayesian updating techniques. This "a priori" estimate should rely on historical investigations, taking into account codes and standards, scientific and historical documentations, or even architectural canons, in force at the construction time.

Anyway, the most suitable approaches are irrefutably based on experimental test results.

As anticipated, starting from experimental results, the assessment of mean values of relevant mechanical properties is not particularly demanding; on the contrary, a sound evaluation of their standard deviation, or more generally of their inherent uncertainty, requires the availability of many in-situ test results. However, owing to the fact that the number of samples is generally too limited to derive that information, it is often necessary to rely on databases of standard laboratory or in-situ test results obtained for similar coeval structures.

The elaboration of raw data stored in testing laboratory archives is not trivial. The major difficulty frequently originates from the uncertain association of the tested sample to the corresponding material class; in fact, the declared material classes are often untrustworthy or wrong, also for the presence of downgraded material. For that reason, the database results in a mixture of different individual populations, each one typifying a material class, which should be duly identified.

A general and user independent methodology allowing the identification of material classes and the evaluation of the statistical parameters of their relevant mechanical properties is discussed in [6]. The method, which is based on a Gaussian mixture model (GMM) [27], allows the identification of different clusters, representing homogenous statistical populations, in secondary databases of experimental tests data derived, for example, from standard acceptance tests.

3.1. Mixture Models

Mixture models (MMs) approach the statistical modeling of heterogeneity in a cluster analysis on a mathematical basis.

MMs are able to analyze quite complex distributions, composed of several homogenous populations belonging to the same distribution family, allowing different individual components to be identified. In fact, such kinds of complex distributions cannot be described by single probability density functions, which are unable to provide a satisfactory model for local variations in the observed data. On the contrary, assuming the distribution composed by one single homogenous population could lead to erroneous statistical information.

Due to their flexibility, mixture models are applied for the statistical modeling of a wide variety of random phenomena.

When all the distributions of the mixture belong to the normal family, the mixture model is said to be a Gaussian mixture model (GMM) [27]. As discussed in [6] and [16], they have been successfully adopted to identify material resistance classes in a whole database of test results, even if the origin of individual data is unknown.

The above mentioned procedure, briefly described in the following, has been applied to identify the relevant statistical properties of concrete and reinforcing steel classes used in Italy during the 1960s, as illustrated in Section 4.

3.2. Cluster Analysis

To identify individual clusters and their relevant statistical properties, the following method, based on the already cited Gaussian mixture model, can be adopted:

- a database of tests results, carried out on standardized specimens adequately representing the examined building materials, should be collected. Data can be gathered from pertinent laboratory or in-situ test results available in the archives of laboratories for testing building materials. To be adequately significant, tested specimens should be consistent with the investigated structural material in terms of raw materials, workmanship and age; therefore, they should refer to coeval structures as the considered structure and be located within the same geographical region;
- a cluster analysis is then carried out based on Gaussian mixture models, with the objective to identify different individual homogenous populations included in the whole database, as well as their probability density functions (*pdf*s). As said before, the mixture model (MM) is a probabilistic model suitable to identify a subpopulation within an overall population by means of its pdf. In a MM with k components, the density $f(y)$ of the random variable Y can be written in the form [28].

$$f(y) = \sum_{i=1}^{k} w_i f_i (y) \tag{6}$$

where $f_i (y)$ is the *pdf* of the i-th component of the mixture, and $0 < w_i < 1$ its mixing proportion, or weight:

$$\sum_{i=1}^{k} w_i = 1. \tag{7}$$

The mixture model (MM), which could be interpreted as a method of classification by unsupervised learning [28], provides results that can be directly used for the reliability assessment of existing buildings. In performing the analysis, it must be considered that preliminary manipulations of the collected data, like a priori assignment of some test results to a given class, on the basis of recorded information available in the test report or on the basis of engineering judgments, are not necessary and should be avoided, since this practice can lead to misleading results.

4. Concrete and Reinforcing Steel Properties of Typical R.C. Structures Built in Italy in the 1960s

In order to identify building material classes and the associated pdfs to be used in the time dependent reliability analysis discussed in Section 5, the above mentioned methodology has been applied to databases of material test results concerning concrete and reinforcing steel used in Italy during the 1960s. As said before, that historical period is extremely relevant, since a large part of the Italian built environment was erected at that time, in the framework of the so-called "economic boom" following the post-Second World War reconstruction.

4.1. Database of Material Acceptance Tests from the 1960s

The mechanical properties of the historical building materials were derived from the materials' acceptance tests carried out in the 1960s in the Laboratory of the Department of Civil and Industrial Engineering of the University of Pisa. The laboratory has been active, as an official laboratory, since 1939 [29], being one of the oldest Italian facilities operating in this field. During the 1960s, samples of reinforcing bars came from all over Italy, from construction sites concerning not only buildings, but also bridges and other civil and industrial engineering works.

The collected databases, derived from official test reports, consist of about 6000 tensile tests on steel rebars used in the years 1961, 1963, 1965 and 1967 [16] and on 18,222 compressive tests of cubic specimens for the years 1961, 1963, 1965, 1967 and 1969 [6].

The cluster analysis of the aforementioned databases allowed homogenous classes and associated probability distribution functions for the fundamental mechanical parameters to be identified. In the following sections, relevant results concerning the concrete compressive strength and the yielding strength of rebars are summarized.

4.2. Concrete Material Classes

The material acceptance tests performed on concrete cubic specimens have been already analyzed in [6], therefore in the following, only some key results relevant for this work are recalled.

Considering typical production of concrete in Italy during the 1960s [30], as well as code requirements for material acceptance in force at that time [29], preliminary information was collected about concrete strength variation [31]. The study highlighted that no standardized concrete classes existed at that time and that frequently concrete was produced directly at the building site, adopting mix designs mainly based on the producer's past experience.

The cluster analysis performed on the collected database leaded to a Gaussian mixture model made of eight components. Indeed, different number of components, k, were considered ($k = 7$, 8, 9 and 10). The value $k = 8$ was chosen being associated to lower values of the Bayesian information criterion (BIC). It must be underlined that, except the extreme classes, which are not particularly significant, relevant information on most common classes are quite independent on the number of components.

The identified concrete classes are shown in Figure 1, and their relevant statistical parameters are given in Table 1, together with the probability associated to each component. Obviously, the significance of class 1 is very limited for practical applications, due to the very low concrete resistance values and the associated high *COV*.

4.3. Reinforcing Steel Classes in Italy during 1960s

During 1960s, different types of steel rebars were commonly used in Italy. Indeed, further to the common plain rebars, the use of ribbed rebars, to improve steel-to-concrete bond performance, increased over the years and different shapes were proposed before a complete standardization.

An illustration of the main different typologies of ribbed bars used at that time is summarized in Figure 2.

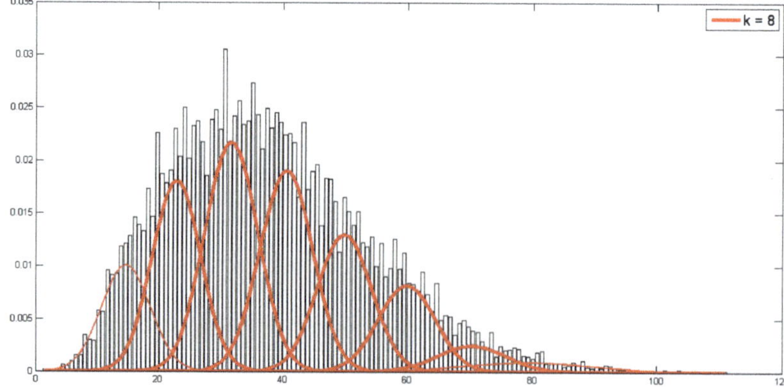

Figure 1. Results of the cluster analysis for concrete compressive strength f_c, identification of eight different sub-classes.

Table 1. Concrete compressive strength—statistical parameters μ and COV of the Gaussian Mixture Model (GMM) and probability of each component.

Material Class	μ (N/mm^2)	COV (%)	Component Probability (%)
1	14.7	27	10
2	22.9	17	18
3	31.8	13	23
4	40.6	10	20
5	49.9	9	14
6	59.9	8	10
7	70.2	8	3
8	79.5	13	2

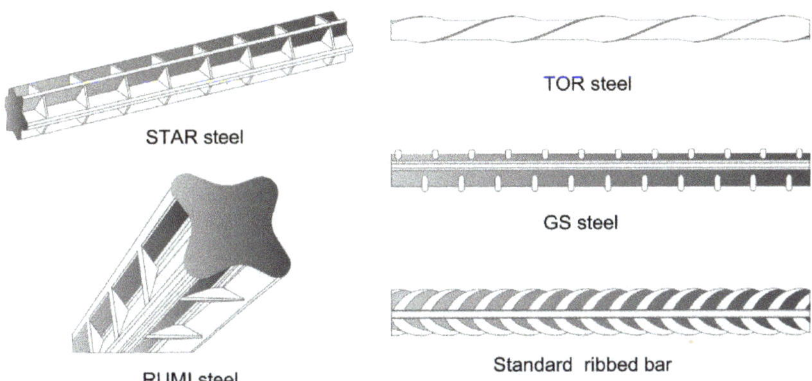

Figure 2. Steel ribbed bar typologies commonly used in Italy during the 1960s.

In the collected database, 84% of the samples are plain bars, 13% are ribbed bars, while only 3% are not labelled as plain or ribbed. The analysis of the evolving percentages of samples during the years confirms that the time period under examination represents a transition phase; in fact, while in the years 1961, 1963, 1965 the percentage of plain rebars was almost 88% of the tested rebars, from 1967 on it decreased to 66%, due to the increasing use of ribbed bars.

In the collected database, the main categories of rebars are *Aq*-steel plain rebars (Aq42, Aq50 and Aq60), high elastic limit steel (known in Italy as ALE), GS steel, RUMI, TOR and STAR rebars. However, a relevant percentage of unclassified steel samples is also registered.

In Figure 3, histograms of yielding strength for the main steel categories, Aq, ALE, RUMI, GS, STAR and "unclassified common steel", are shown.

In each of these macro-categories, subcategories have been recorded, in turn: Aq 42, Aq45, Aq 42/50, Aq 50, Aq 50/60, hard 60/70, ALE, ALE 4400, ARES, RUMI, RUMI LU, RUMI 400, RUMI LU3 or RUMI LU3 4000, RUMI4400, RUMI LU3/4400, RUMI LU3/5000, GS, GS3600, GS4400, GS 4400 ALE, GS4500, GS5000, STAR, STAR4400, STAR4500, STAR5000, TNT 60, RUMI TNT, FERRO BOX 4400, TOR, Thor Aq52/60, special steel.

With the aim of a preliminary classification, three main steel categories have been considered:

- *Aq-category*, including plain rebars only, incorporating all the *Aq* subcategories;
- *Ribbed bars*, including all the steel classes for ribbed rebars (RUMI, GS, STAR and "ribbed not classified" steel).
- *Not classified steel*, including all plain rebars not specifically labelled.

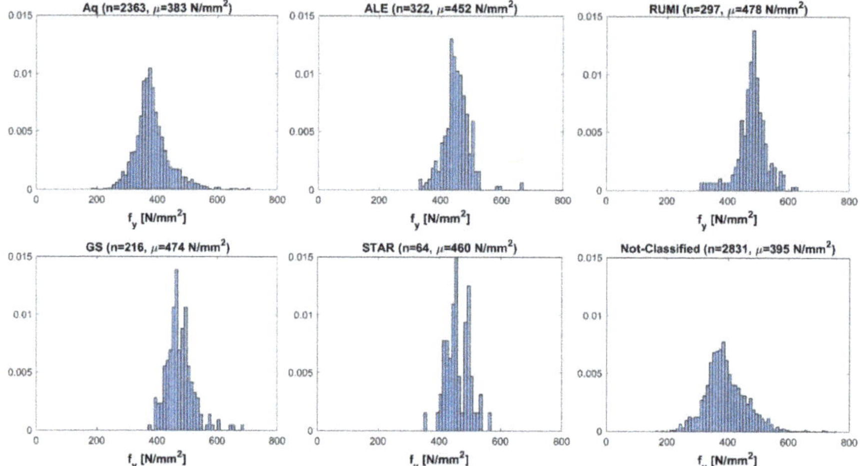

Figure 3. Histogram of yielding strength of main steel rebars used in Italy in 1960s: *Aq*, high elastic limit steel (known in Italy as ALE), RUMI, GS, STAR and not classified.

In Figure 4a the bar charts of yielding strength for the three main steel categories are shown. The *Aq-category* and *Not classified steel* show similar distributions with the mean value of the yield strength $\mu = 382$ N/mm^2 and $COV = 15\%$ and $\mu = 393$ N/mm^2 and $COV = 18\%$, respectively. It is worth noting that the Italian regulation in force in those years (Ministerial Memorandum 1957 [32]) recommended the use of the macro-category "*Aq*" plain rebars, with subclasses (Aq42, Aq50, Aq60) corresponding to the steel typologies identified by the previous regulation (R.D. 1939 [29]), i.e., mild, medium, and high strength steel. Ribbed bars were introduced in Italy by the same Memorandum [32]; therefore they represent only a marginal part of the database, characterized by an average value of the yield strength $\mu = 472$ N/mm^2 and by $COV = 10\%$.

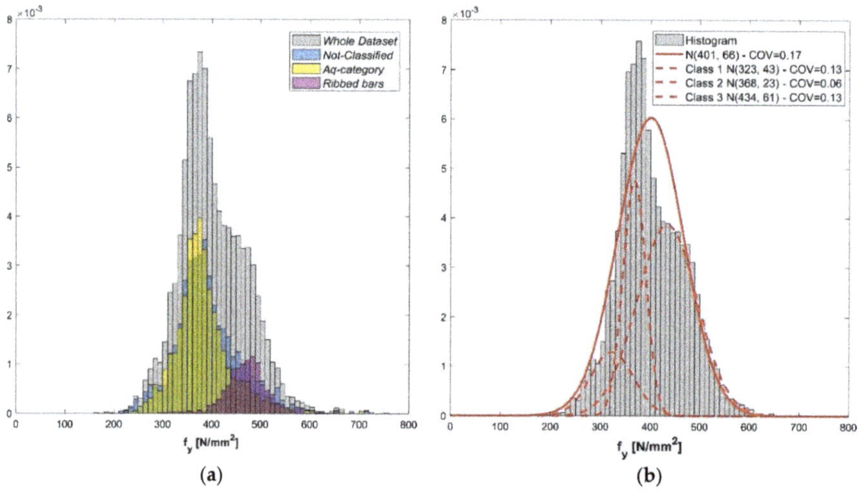

Figure 4. (a) Bar charts of yielding strength of steel rebars for the whole dataset and the sub-divisions of "*Not classified steel*", "*Aq-category*" and "*Ribbed bars*"; (b) results of the cluster analysis for yielding strength f_y and identification of three different sub-classes.

As it is evident in Figure 4, the two main sub-populations *"Not classified steel"* (in blue) and *"Aq-category"* (in yellow) are overlapped, complicating the identification of sub-classes. Moreover, although the steel category for ribbed rebars (in magenta) is clearly identified as a subpopulation in the descending part of the frequency bar chart, corresponding to the higher strength values, it partly overlaps the higher values registered for plain bars. As a consequence, a robust methodology is needed to identify homogenous sub-populations, like the one discussed above for concrete.

Analyzing once again the whole dataset by means of the cluster analysis based on Gaussian mixture models (GMM), it is thus possible to identify three sub-classes of steel rebars. It is important to stress that the advantage of the proposed method is its "blindness", since it does not require any preliminary assumptions about the classes.

The results of the analysis for yielding strength are illustrated in Figure 4b, where the pdfs associated to the three classes (dashed red lines) are compared with the pdf related to the whole dataset (solid red line). The relevant statistical parameters, i.e., the mean value of yield strength, μ, and coefficient of variation (COV), are given in Table 2 together with the probability associated to each individual component.

Table 2. Steel yielding strength f_y—statistical parameters μ and COV of the GMM and probability of each identified component.

Material Class	μ (N/mm^2)	COV (%)	Component Probability (%)
1	323	13	14
2	368	6	27
3	434	13	59

As already noted for concrete, it is crystal clear that cluster analysis leads to a better evaluation of the statistical parameters for steel rebars rather than the analysis of the whole dataset. Indeed, the proper identification of sub-classes allows the estimate of the coefficient of variation associated with each class to be significantly improved: the COV, which is 6% for the intermediate class and 13% for the upper and lower class, is sensibly smaller than the one resulting from the analysis of the whole dataset ($COV = 17\%$). However, the coefficients of variation of the lower and upper classes are sensibly higher than the values usually suggested in the literature [26] for reliability analyses of existing structures, generally varying between 6% and 8%.

For the higher steel class, to which the ribbed bars belong, the increased scattering of data could be ascribed to the different production processes of various ribbed bar typologies adopted at that time. On the contrary, in the lower steel class, a positive bias can be appreciated with respect to the expected nominal resistances as codified in the structural standards in force at the time. This could be justified by the inclusion in the lower resistance steel classes of downgraded high strength steel rebars, not complying with requirements for upper steel classes.

5. Reliability Assessment of Existing R.C. Buildings under Different Degradation Conditions

Starting from the stochastic models for concrete and steel rebars derived in the previous section, this study focuses on the reliability of existing reinforced concrete buildings built in Italy during the 1960s and designed according the already mentioned R.D. 1939 [29].

At that time, the "allowable" or "permissible stress method" was the reference method for structural design [33]. As known, the method is based on a global safety factor affecting material strengths. The assessment consists of checking that the maximum calculated stresses, induced by the applied loads, f_{G+Q}, are lower than the allowable stress, f_{adm}, obtained by dividing the material's resistance (e.g., yield stress in the rebars or compressive strength in concrete) by the pertinent global safety factor:

$$f_{adm} \geq f_{G+Q}. \tag{8}$$

In [29], a permissible stress, $f_{adm,s}$, equal to 140 N/mm^2 was recommended for mild steel and equal to 200 N/mm^2 for medium and high strength steel rebars. For concrete, the allowable stress was generally derived by dividing the cubic resistance after 28 days, $\sigma_{r,28}$, by a safety factor equal to three, except for the "high strength" concrete, as defined at that time, characterized by $\sigma_{r,28} \geq 22.5$ N/mm^2, for which a higher stress value was allowed, according to the following equation:

$$f_{adm.,c} = 7.5 + \frac{\sigma_{r,28} - 22.5}{9} \text{ N/mm}^2. \tag{9}$$

It must be underlined that the use of medium and high strength steel rebars was allowed only in association with the above defined "high strength" concrete.

Assuming the use of a "high strength" concrete ($\sigma_{r,28} = 25$ N/mm^2, $f_{adm,c} = 7.8$ N/mm^2) and the two categories "mild steel" ($f_{adm,s} = 140$ N/mm^2) and "medium" or "high strength" steel rebars ($f_{adm,s} = 200$ N/mm^2), the design of a simply supported r.c. slab subjected to both permanent and variable load is first carried out and then the time-dependent reliability is evaluated.

In the spirit of [19], in the design exercises an allowable stress $f_{adm,s} = 140$ N/mm^2, like for mild steel, is adopted when reinforcing steel class 1 is considered for reliability analysis, while an allowable stress $f_{adm,s} = 200$ N/mm^2, like for medium and high strength steel, is adopted when reinforcing steel classes 2 and 3 are taken into account.

The permanent loads (self-weight of the slab and remaining permanent loads, such as finishes etc.) acting on the simply supported slab with a span $L = 6$ m is assumed to be equal to 7 kN/m^2, while for variable loads, depending on the building's category, values in line with the common design assumptions adopted at the time are considered. These values are, obviously, not significantly different from those adopted in the modern standards for imposed loads on structures.

Making reference to previous engineering experience, the total height of the slab is set to $h = 0.25$ m for an effective height $d = 0.22$ m [34].

From the statistical point of view, the permanent load G and the variable load Q are modelled as in Table 3, in line with the models recommended by the Joint Committee on Structural Safety (JCSS) [35]. The pdfs derived in the previous section are adopted for concrete and steel classes for initial structural resistance.

Table 3. Statistical description of structural resistance and loads.

	Variable	Mean μ	COV	*pdf*
Initial Resistance		Material Classes defined in Section 4		Normal
Permanent Load	($G_k = 7$ kN/m^2)	G_k	0.10	Normal
Variable Load	Residential ($Q_k = 2$ kN/m^2)	0.21 Q_k	1.42	EVI
	Shopping ($Q_k = 5$ kN/m^2)	0.52 Q_k	0.35	EVI
	Storage ($Q_k = 7.5$ kN/m^2)	0.72 Q_k	0.15	EVI

For the reliability calculation the following limit state function, as given in [34], is adopted.

$$Z(x) = A_s f_s \left(d - \frac{A_s f_y}{2 b f_c} \right) - \frac{(G + Q)L^2}{8}. \tag{10}$$

The degraded resistance $R(t)$ is obtained by multiplying the initial resistance for a suitable degradation function $D(t)$, according to Equation (1).

For time-dependent reliability assessment of aging structures, structural deterioration is often modeled using equations of the form [23,36–38]

$$D(t) = 1 - a(t - T_1)^b + \varepsilon(t), \qquad t > T_1 \tag{11}$$

where a and b are the parameters governing the deterioration process, which can be determined from an analysis of experimental data, $\varepsilon(t)$ is a zero-mean stochastic process, which takes into account the randomness of the observed data, and T_1 denotes the random time required to initiate the deterioration.

In Equation (11), the exponent b can be assumed $b = 0.5$, if the damage process is strictly diffusion controlled, or $b = 1$, if the process is strictly rate controlled, such as in case of the corrosion of the reinforcement.

Obviously, relations such as Equation (11) are essentially empirical and depend on environmental conditions, and it is very difficult to generalize them to cover situations where no experimental data are available.

In the following reliability analysis, a sensitivity study is carried out taking into account four degradation conditions, associated with zero, low, medium and high rate of the corrosion, respectively [36].

In the present study, it has been assumed that degradation affects mainly the steel reinforcement. The adopted parameters for the steel degradation conditions, which are based on the parametric studies of reinforced concrete beams subjected to corrosion attack carried out in [39], are summarized in Table 4.

Table 4. Typical parameters governing different steel degradation rates (Equation (10)).

Degradation Condition	Degradation Rate	T_1 (years)	a	b
0	Non-degrading	—	—	—
I	Low	10	0.0005	1
II	Medium	5	0.005	1
III	High	2.5	0.0065	1

The time-degradation curves, representing the four investigated degradation conditions given in Table 4, are illustrated in Figure 5. It must be underlined that in 70 years the three degrading conditions, I, II and III, lead to a reduction of the steel section of about 3%, 33% and 44%, respectively.

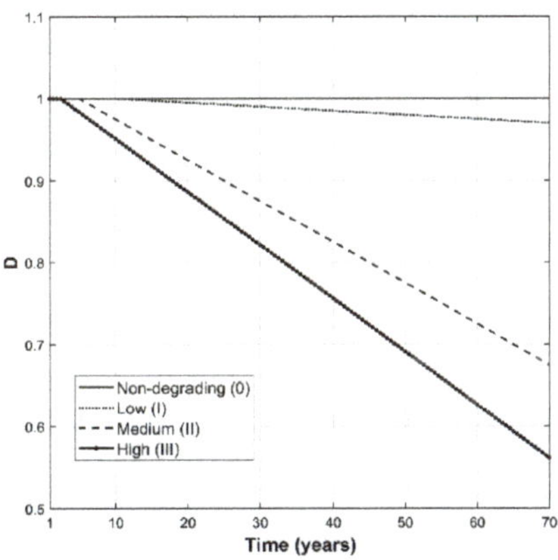

Figure 5. Investigated degradation conditions for steel reinforcement.

The results of the time-dependent reliability analysis are illustrated, for each of the investigated deterioration conditions, in Figures 6–8 for typical residential buildings. More precisely, Figure 6 refers to steel class 1, Figure 7 to steel class 2 and in Figure 8 to steel class 3.

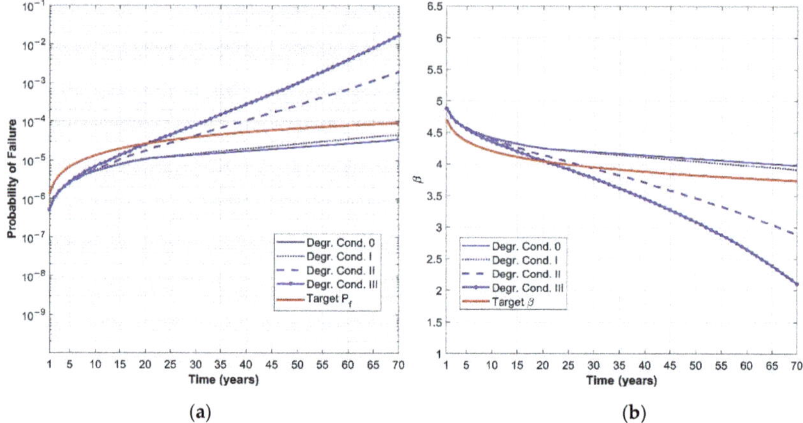

Figure 6. Time-dependent probability of failure (**a**) and reliability index β (**b**) for residential buildings—reinforcing steel class 1 and concrete class 3 for different steel degradation conditions.

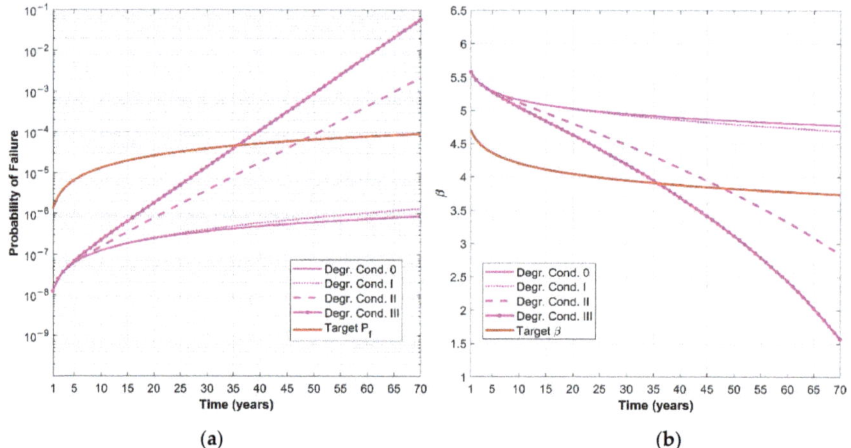

Figure 7. Time-dependent probability of failure (**a**) and reliability index β (**b**) for residential buildings—reinforcing steel class 2 and concrete class 3 for different steel degradation conditions.

In the assessment the three different steel classes previously derived are taken into account, while for concrete, strength class 3 (μ = 31.8 N/mm^2, COV = 13%) is adopted. In particular, the variations of the probability of failure, P_f, and of the reliability index, β, over time are shown.

For the sake of a direct comparison with target reliability values intended to represent modern design approaches for new structures, the curves are compared with the variation of target reliability levels over time as defined in the Eurocodes, represented in the diagrams by red curves [17], but it is obvious that for existing structures some limited reduction in the reliability index is often accepted. As previously remarked, the plotted time variation of the reference probability of failure (as well as the reference target reliability index) refers to structures whose design is governed by the effects of the time-variant components in the limit state equation.

Inspecting the diagrams in Figures 6–8, it clearly appears that when the reinforcing steel is subjected to degradation condition I, the rate of resistance loss is so small that the failure probability is comparable to that obtained for non-degrading steel.

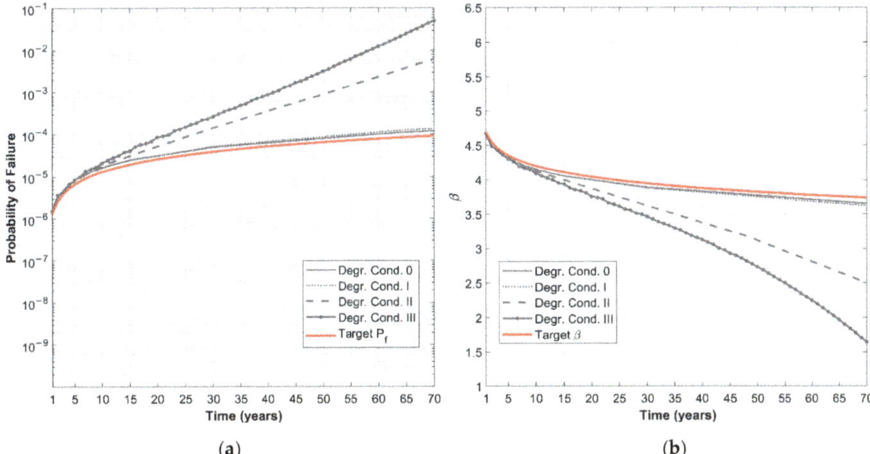

Figure 8. Time-dependent probability of failure (**a**) and reliability index β (**b**) for residential buildings—reinforcing steel class 3 and concrete class 3 for different steel degradation conditions.

From the examination, it can be seen that:

- Under degradation conditions 0 and I:

  - if the rebars belong to reinforcing steel class 1, the structural reliability satisfies the minimum modern requirements;
  - if the rebars belong to reinforcing steel class 3, the structural reliability is in accordance with the required minimum target levels;
  - while if the rebars are reinforcing steel class 2, even higher reliability than that in the Eurocodes for new structures can be achieved.

- Under degradation conditions II and III, confirming similar observations for some r.c. bridge girders [36], the failure probability increases significantly, reaching unacceptable values, in particular approaching the end of the service life, thus stressing the importance of maintenance interventions.

Another interesting observation regards the performance of reinforcing steel class 3 compared to reinforcing steel class 2. Notwithstanding the higher mean strength value for class 3, the reliability analysis indicates that a lower reliability index (i.e., higher failure probability) is expected to be found with respect to structures where reinforcing steel class 2 is used, and this is motivated by the higher coefficient of variation (13% instead of 6%). Moreover, as the relevant literature generally reports $COV \approx 7\%$ for steel rebars, the reliability indexes thus evaluated are unsafe sided also for steel class 1, characterized by $COV = 13\%$.

These results confirm the relevance of the proposed methodology: the reliability assessment of existing structures, based on limited in-situ test results, may result in the adoption of high mean strength values, which can lead, when associated with small values of COV deduced from the literature, to an unjustified overestimation of the reliability index of the investigated structure.

The remarks are particularly relevant for the correct implementation of modern codes and standards for the assessment of existing structures. Regarding the assessment, these codes (e.g., EN1998-3 [40]) generally deserve great attention and importance should be given to the mean values of the materials' relevant properties, measured by means of in-situ tests on the structure being investigated. This approach is clearly justified in the design phase by the need to assess the actual properties of structural materials and not to pre-judge them only on the basis of available documentation or the practice at the time of the construction, also considering that, on the other hand, as discussed previously, the possibility of carrying out an extensive testing campaign to get an adequate representation of the statistical parameters of the properties of the materials is frequently precluded.

It must be stressed again that the use of these estimations of the mean value of the relevant properties is mainly devoted to back-calculating, applying suitable reduction factors, the design values to be adopted in the assessments. To clarify the remark, an additional case has been analyzed for the residential building considered here, hypothesizing that steel class 1 is used.

Assuming that the mean value previously determined for the reinforcing steel class 1 represents the actual properties of the reinforcing steel of the structure, $\mu = 382$ N/mm^2 is the mean value of yielding strength derived from five or six tests on samples directly extracted from the structure.

According to the already mentioned EN1998-3 [40], these results lead to an allowable stress $f_{adm,s} \approx 200$ N/mm^2, much higher than the allowable stress permitted at the time of construction, which was 140 N/mm^2. This assumption, generally acceptable when the *COV* of the yield stress complies with reference values (i.e., *COV* $\approx$ 7%), can produce unsafe results, when higher *COV* characterizes the actual distribution. Indeed, if adopting $f_{adm,s} \approx 200$ N/mm^2, the verification would have led to an unjustified overestimation of the element's reliability. Consequently, for that element it would be possible evidently unsafe-sided increases of permanent and variable loads for refurbishment purposes or change in use. The increase of the total load $(G + Q)$ resulting from the increase from 140 N/mm^2 to 200 N/mm^2 of the allowable stress of the reinforcement is about 40%. Considering that increased load, the time-dependent reliability curves for a r.c. element with reinforcing steel class 1, previously illustrated in Figure 6, become those reported in Figure 9.

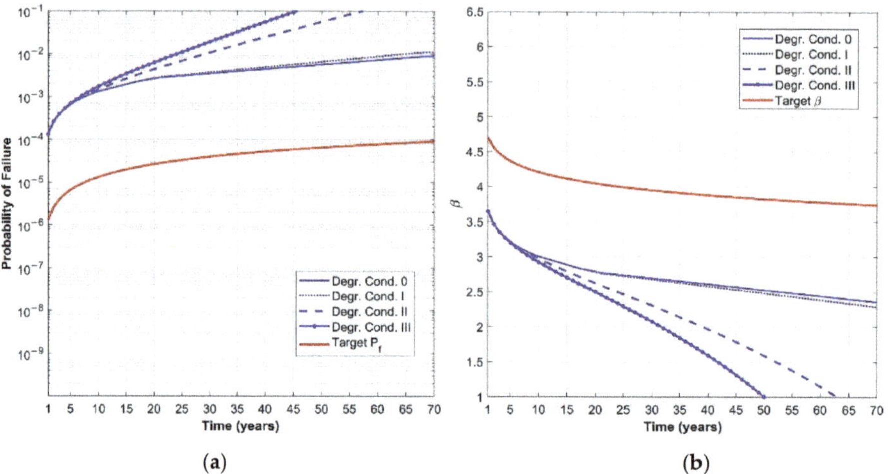

(a) (b)

Figure 9. Time-dependent probability of failure (**a**) and reliability index β (**b**) for residential buildings—reinforcing steel class 1 and concrete class 3, designed considering increased design loads.

Inspecting Figure 9 it clearly appears that the element's reliability is much lower than expected following the possible indications of modern assessment standards.

Time-dependent reliability curves, summarized for residential buildings in Figures 6–8, have been derived for shopping and storage buildings too as illustrated in Figures 10 and 11 respectively.

In the figures the four different degradation conditions are considered for each one of the three reinforcing steel classes derived from the cluster analyses. The diagrams confirm that also for these building's categories, the influence of degradation conditions on different reinforcing steel classes produces similar effects as for the residential buildings.

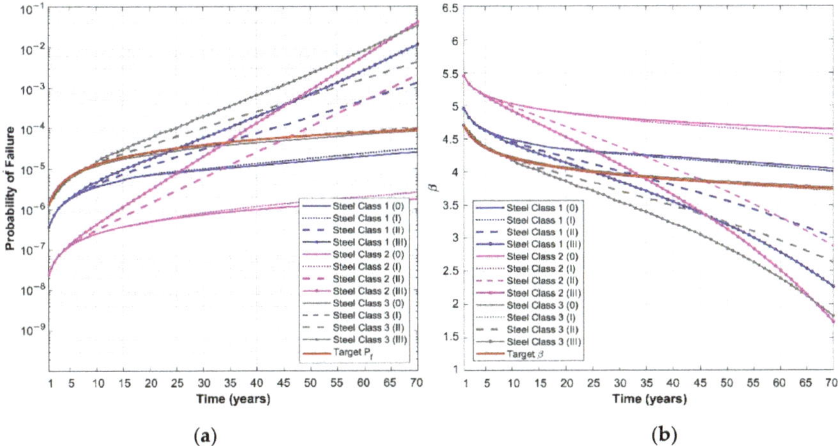

Figure 10. Time-dependent probability of failure (**a**) and reliability index β (**b**) for shopping buildings (concrete class 3).

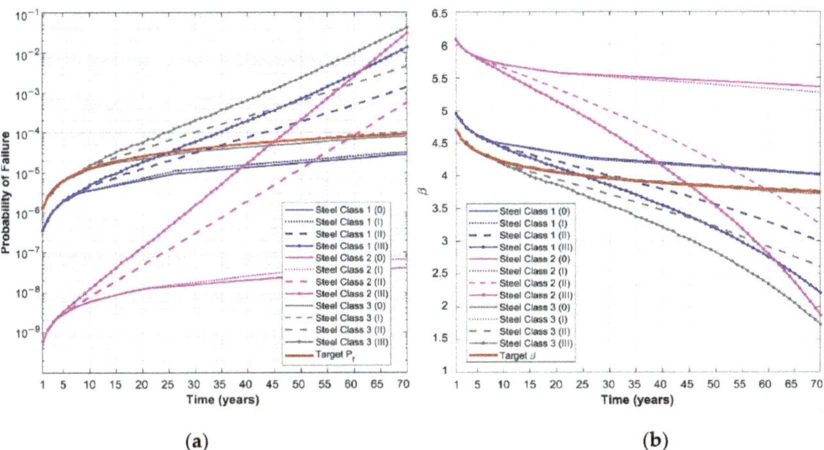

Figure 11. Time-dependent probability of failure (**a**) and reliability index β (**b**) for storage building (concrete class 3).

In order to evaluate the influence of different concrete classes on the time-dependent reliability, the analyses over time can be easily extended also to r.c. structures characterized by higher or lower quality concrete class than that assumed above (concrete class 3).

The study is concentrated on shopping buildings, for which concrete class 2 ($\sigma_{r,28} \cong 20 \text{ N/mm}^2$, $f_{adm,c} \cong 6.7 \text{ N/mm}^2$, $COV = 17\%$) and concrete class 4 ($\sigma_{r,28} \cong 35 \text{ N/mm}^2$, $f_{adm,c} \cong 8.9 \text{ N/mm}^2$, $COV = 10\%$) are also considered in the analysis.

The time-dependent reliability curves obtained for the various steel degradation conditions and reinforcing steel classes are reported in Figure 12 for concrete class 2 and in Figure 13 for concrete class 4, to be compared with those illustrated in Figure 10 for concrete class 3.

The comparison shows that, as expected, the reliability increases as the COV decreases, i.e., moving from concrete class 2 to concrete class 4. On the contrary, the differences in reliability depend only on the degradation condition, being nearly constant over time for a given degradation condition.

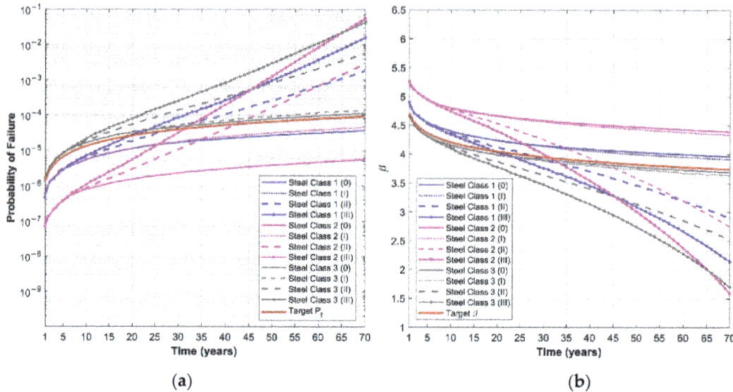

Figure 12. Time-dependent probability of failure (**a**) and reliability index β (**b**) for shopping buildings (concrete class 2).

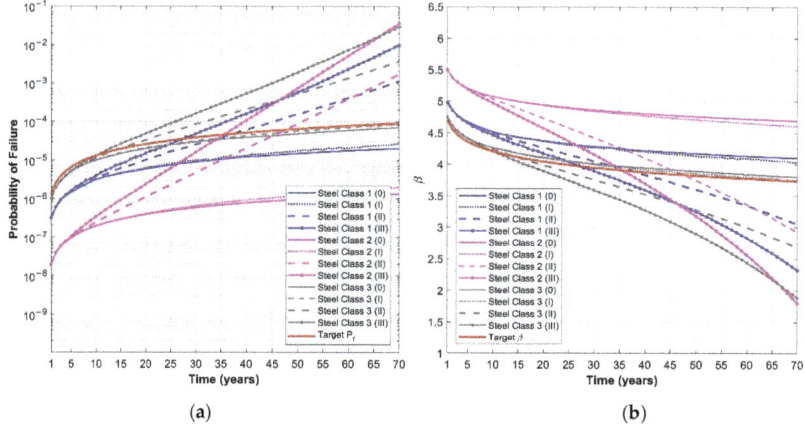

Figure 13. Time-dependent probability of failure (**a**) and reliability index β (**b**) for shopping buildings (concrete class 4).

6. Conclusions

Life cycle assessment is a fundamental tool to support decisions on retrofitting interventions on existing structures and for their prioritization.

Since reliability assessment is highly influenced by the statistical parameters of material properties, a specific preliminary study is needed, covering both historical data and in-situ test results. In real cases, the possibility to carry out enough tests is generally very limited and they are unlikely to obtain an adequate statistical description of the mechanical parameters.

In this study, a methodology for the time-dependent reliability analyses of existing structures is presented, also referring to relevant cases studies.

The implementation of the method relies also on a special procedure to evaluate material properties and their statistical parameters, based on cluster analysis. The cluster analysis, based on a Gaussian mixture model, is generally applicable, and it allows individual components belonging to a mixed population to be identified. This methodology is particularly adequate when databases of raw test results need to be analyzed. In fact, it allows homogenous material classes and the probability density functions associated to their mechanical properties to be identified and does not require any a priori assignment of relevant clusters, resulting in it being independent from engineering judgment or any other influence related to the user's skill.

The method has been applied to thousands of historical test results, dating back to the 1960s, concerning concrete compressive strength and yield stress of steel rebars, allowing resistance classes for both materials to be identified and a sound estimation of the related statistical parameters to be obtained. Thanks to the great number of analyzed test results, the outcomes of the proposed methodology are particularly relevant for the estimation of the COV of mechanical properties in the reliability assessment of r.c. structures coeval with the samples in the database. Results presented in this work are particularly relevant taking into account that a large part of the built environment in Italy and, more generally, in Europe dates back to the decade 1960–1970. In fact, according to the Italian National Institute of Statistics [41], around 30% of Italian r.c. buildings were built in the period 1950–1970 (19% in the decade 1960–1970).

To illustrate the application of the methodology, time-dependent reliability analyses for significant case studies, consisting of r.c. elements part of residential, shopping and storage buildings, have been carried out, focusing on the effects of corrosion in steel rebars, under different environmental conditions, resulting in no degradation to high degradation effects.

Time-dependent reliability trends have been compared with the reliability targets in the Eurocodes, showing quite satisfactory results under no or low-degradation conditions, with an exception made for steel class 3, which despite being associated to the highest mean yielding stress among the examined classes shows a higher COV; these results confirm the importance of a proper evaluation of statistical parameters for the variables in the limit state function. Results for medium and high corrosion rates show that the reliability becomes quite soon too low, even accepting a reduction of the target reliability values for existing structures.

Concerning the influence of the concrete class, the study confirms that, as expected, the reliability increases as the COV decreases, i.e., moving from concrete class 2 to concrete class 4, while differences in reliability depend only on the degradation condition, being nearly constant over time for a given condition.

Further studies are being developed, based on the promising achievements in this work.

Author Contributions: Conceptualization, P.C., P.F. and F.L.; methodology, P.C., P.F. and F.L.; software, P.C., P.F. and F.L.; validation, P.C., P.F. and F.L.; writing—original draft preparation, P.C., P.F. and F.L.; writing—review and editing, P.C., P.F. and F.L.; resources, P.C. All authors have read and agreed to the published version of the manuscript.

Funding: This research received no external funding.

Conflicts of Interest: The authors declare no conflicts of interest.

References

1. Strauss, A.; Hoffmann, S.; Wendner, R.; Bergmeister, K. Structural assessment and reliability analysis for existing engineering structures, applications for real structures. *Struct. Infrastruct. Eng.* **2009**, *5*, 277–286. [CrossRef]
2. Bastidas-Arteaga, E. Reliability of Reinforced Concrete Structures Subjected to Corrosion-Fatigue and Climate Change. *Int. J. Concr. Struct. Mater.* **2018**, *12*, 10. [CrossRef]
3. Frangopol, D.M.; Soliman, M. Life-cycle of structural systems: Recent achievements and future directions. *Struct. Infrastruct. Eng.* **2015**, *12*, 1–20. [CrossRef]
4. Frangopol, D.M. Life-cycle performance, management, and optimization of structural safety under uncertainty: Accomplishments and challenges. *Struct. Infrastruct. Eng.* **2011**, *7*, 389–413. [CrossRef]
5. Yanweerasak, T.; Pansuk, W.; Akiyama, M.; Frangopol, D.M. Life-cycle reliability assessment of reinforced concrete bridges under multiple hazards. *Struct. Infrastruct. Eng.* **2018**, *14*, 1011–1024. [CrossRef]
6. Croce, P.; Marsili, F.; Klawonn, F.; Formichi, P.; Landi, F. Evaluation of statistical parameters of concrete strength from secondary experimental test data. *Constr. Build. Mater.* **2018**, *163*, 343–359. [CrossRef]
7. Beconcini, M.L. *Laboratorio Delta: Indagini sui Materiali Sulle Strutture e Sul Sottosuolo*; Pacini Fazi: Lucca, Italy, 2018. (In Italian)
8. *EN13791 Assessment of In-Situ Compressive Strength in Structures and Precast Concrete Components*; CEN: Brussels, Belgium, 2019.

9.   *EN12504-1, Testing Concrete in Structures. Cored Specimens. Taking, Examining and Testing in Compression*; CEN: Brussels, Belgium, 2019.

10. *EN12504-2, Testing Concrete in Structures. Non-Destructive Testing. Determination of Rebound Number*; CEN: Brussels, Belgium, 2012.

11. *EN12502-3, Testing Concrete in Structures. Determination of Pull-Out Force*; CEN: Brussels, Belgium, 2005.

12. *EN12502-3, Testing Concrete. Determination of Ultrasonic Pulse Velocity*; CEN: Brussels, Belgium, 2005.

13. *ACI214.4 R10: Guide for Obtaining Cores and Interpreting Compressive Strength Results*; American Concrete Institute: Farmington Hills, MI, USA, 2016.

14. Italian Public Works Council. *Guidelines for the Evaluation of In-Situ Concrete Properties*; Ministry of Infrastructures: Rome, Italy, 2017. (In Italian)

15. Italian Public Works Council. *Memorandum for Application of Italian Building Code, N. 7*; Italian Official Journal, Istituto Poligrafico e Zecca dello Stato: Rome, Italy, 2019. (In Italian)

16. Croce, P.; Formichi, P.; Landi, F.; Marsili, F.; Puccini, B.; Zotti, V. Statistical Parameters of Steel Rebars of Reinforced Concrete Existing Structures. In Proceedings of the ESREL2020-PSAM15, Venice, Italy, 21–26 June 2020. (accepted for publication).

17. *EN 1990—Eurocode—Basis of Structural Design*; CEN: Brussels, Belgium, 2002.

18. Croce, P.; Formichi, P.; Landi, F. Climate Change: Impacts on Climatic Actions and Structural Reliability. *Appl. Sci.* **2019**, *9*, 5416. [CrossRef]

19. Croce, P.; Formichi, P.; Landi, F. Probabilistic methodology for the assessment of the impact of climate change on structural safety. In Proceedings of the ESREL2020-PSAM15, Venice, Italy, 21–26 June 2020. (accepted for publication).

20. Croce, P.; Formichi, P.; Landi, F. Structural safety and design under climate change. In *IABSE Congress New York: The Evolving Metropolis*; IABSE: Zurich, Switzerland, 2019; ISBN 978-385748165-9.

21. Croce, P.; Formichi, P.; Landi, F.; Marsili, F. Evaluating the effect of climate change on thermal actions on structures. In *Life-Cycle Analysis and Assessment in Civil Engineering: Towards an Integrated*; Vision, R., Caspeele, L., Taerwe, D., Frangopol, M., Eds.; Taylor & Francis Group: Oxfordshire, UK, 2019; pp. 1751–1758. ISBN 978-1-138-62633-1.

22. Croce, P.; Formichi, P.; Landi, F.; Mercogliano, P.; Bucchignani, E.; Dosio, A.; Dimova, S. The snow load in Europe and the climate change. *Clim. Risk Manag.* **2018**, *20*, 138–154. [CrossRef]

23. Li, Q.; Wang, C.; Ellingwood, B.R. Time-dependent reliability of aging structures in the presence of non-stationary loads and degradation. *Struct. Saf.* **2015**, *52*, 132–141. [CrossRef]

24. *EN1993-1-1—Eurocode 3: Design of Steel Structures—Part 1-1: General Rules and Rules for Buildings*; CEN: Brussels, Belgium, 2005.

25. *ACI Standard 318-89 Building Code Requirements for Reinforced Concrete*; ACI: Detroit, MI, USA, 1989.

26. JCSS Probabilistic Model Code, Joint Committee on Structural Safety. 2002. Available online: https://www.jcss-lc.org/jcss-probabilistic-model-code/ (accessed on 26 May 2020).

27. MacLachlan, G.; Peel, D. *Finite Mixture Models*; John Wiley and Sons: New York, NY, USA, 2000.

28. Press, W.H.; Tevkolsky, S.A.; Vetterling, W.T.; Flannery, B.P. *Numerical Recipes, the Art of Scientific Computing*, 3rd ed.; Cambridge University Press: New York, NY, USA, 2007.

29. *Royal Decree 16/11/1939 nr. 2229, Code for the Execution of Concrete and Reinforced Concrete Structures*; Italian Official Journal, Istituto Poligrafico e Zecca dello Stato: Rome, Italy, 1939. (In Italian)

30. *Santarella, Reinforced Concrete, Design and Behavior*; Hoepli: Milano, Italy, 1969. (In Italian)

31. Neville, L. Book Reviews Ethics in clinical practice: An interprofessional approach Edited by Georgina Hawley. Pearson Education Limited Price: £19.99 Pp 410 ISBN: 0132018276. *Br. J. Heal. Assist.* **2008**, *2*, 289. [CrossRef]

32. Ministry of Public Works. *Memorandum 23/05/1957, nr. 1472, Steel Rebars for Reinforced Concrete Structures*; Italian Official Journal, Istituto Poligrafico e Zecca dello Stato: Rome, Italy, 1957. (In Italian)

33. Ellingwood, B.R. Probability-based codified design: Past accomplishments and future challenges. *Struct. Saf.* **1994**, *13*, 159–176. [CrossRef]

34. Leonardo Da Vinci Pilot Project CZ/02/B/F/PP-134007. Implementation of Eurocodes: Handbook 2—Reliability Background. Guide to the Basis of Structural Reliability and Risk Engineering Related to Eurocodes, Supplemented by Practical Examples. 2005. Available online: https://eurocodes.jrc.ec.europa.eu/showpublication.php?id=63 (accessed on 26 May 2020).

35. Vrouwenvelder, T. The JCSS probabilistic model code. *Struct. Saf.* **1997**, *19*, 245–251. [CrossRef]
36. Enright, M.P.; Frangopol, D.M. Service-Life Prediction of Deteriorating Concrete Bridges. *J. Struct. Eng.* **1998**, *124*, 309–317. [CrossRef]
37. Melchers, R.E. Probabilistic Models for Corrosion in Structural Reliability Assessment—Part 2: Models Based on Mechanics. *J. Offshore Mech. Arct. Eng.* **2003**, *125*, 272–280. [CrossRef]
38. Ellingwood, B.R. Risk-informed condition assessment of civil infrastructure: State of practice and research issues. *Struct. Infrastruct. Eng.* **2005**, *1*, 7–18. [CrossRef]
39. Enright, M.P.; Frangopol, D.M. Probabilistic analysis of resistance degradation of reinforced concrete bridge beams under corrosion. *Eng. Struct.* **1998**, *20*, 960–971. [CrossRef]
40. *EN1998-3—Eurocode 8: Design of Structures for Earthquake Resistance—Part 3: Assessment and Retrofitting of Buildings*; CEN: Brussels, Belgium, 2005.
41. Italian National Institute of Statistics (ISTAT). Population Housing Census. 2011. Available online: http://dati-censimentopopolazione.istat.it/Index.aspx?DataSetCode=DICA_EDIFICIRES# (accessed on 26 May 2020).

 *buildings*

Article

Climate Adaptation in Maintenance Operation and Management of Buildings

Steinar Grynning [1,*], Klodian Gradeci [1], Jørn Emil Gaarder [1], Berit Time [1], Jardar Lohne [2] and Tore Kvande [2]

1   SINTEF Community, 7034 Trondheim, Norway; Klodian.Gradeci@sintef.no (K.G.); jorn.gaarder@sintef.no (J.E.G.); berit.time@sintef.no (B.T.)
2   Department of Civil and Environmental Engineering, Norwegian University of Science and Technology, 7491 Trondheim, Norway; jardar.lohne@ntnu.no (J.L.); tore.kvande@ntnu.no (T.K.)
*   Correspondence: steinar.grynning@sintef.no

Received: 30 March 2020; Accepted: 22 May 2020; Published: 4 June 2020

Abstract: The aim of this paper is to analyze the basic criteria, trends, applications, and developments related to climate adaptation in building maintenance and operation management (MOM) practices in Norway. Investigations conducted as part of the study include an analysis of current literature addressing climate adaptation in relation to MOM practices, supplemented by a review of existing research projects and initiatives in this field. Three case studies involving different Norwegian building owner organizations were examined in order to investigate the current status of the application and extent of climate adaptation practices in relation to MOM. The study has revealed a significant gap between theory and practice when it comes to integrating MOM in relation to climate adaptation. The concept of climate adaptation is only addressed as a high-level strategic issue. The case studies thus emphasize the need for a structured process that can enable the incorporation of climate adaptation in current MOM practices. This proposes a generic and structured climate-adaptive MOM framework that will enable the incorporation of climate adaptation in into corporate MOM practices at different scales and organizational levels. Implementation of this flexible and transferable framework is expected to provide a basis for accruing further knowledge on climate adaptation. Further work with the framework should include the introduction of more tangible and tailored tools and processes, including checklists or scoring systems accompanied by relevant climate adaptation factors and plans.

Keywords: climate change; climate adaptation; buildings; maintenance; operation; management

1. Introduction

1.1. Background

It is clearly demonstrated that climate change is increasing the amount and intensity of precipitation in Norway, and that this will have a major impact on future built environments. As a result, we will probably have to make changes to the ways in which we construct, maintain, operate, and manage our buildings. According to the Intergovernmental Panel on Climate Change (IPCC) [1], an average increase of 20% in precipitation volumes has occurred over the last 100 years, and an additional 20% increase is expected before the year 2100. Moreover, projections for climate change in Norway [2] indicate that a warmer climate can be anticipated, accompanied by an increase in the frequency of extreme weather events and more intense precipitation in certain parts of the country. Changes in temperature will result in an increase in the number of freeze-thaw cycles, and a greater proportion of winter precipitation will fall as rain. These changes will lead to greater demands to protect buildings from the effects of climate change.

Since Norway is already situated in a region characterized by relatively high levels of precipitation, moisture-related damage has always posed significant challenges to the Norwegian built environment. Current projections of climate change impacts indicate that these challenges will increase in the years to come. Moisture, damp building structures, and precipitation all combine to stress the building envelope (the facades and roofs) and can cause significant damage. Adapted and improved technical systems are required, and must be accompanied by comprehensive supervisory and maintenance regimes. Increased precipitation will also present challenges in terms of the floodwater threat, and factors such as drainage, water retention, and building-adjacent terrain considerations will become increasingly significant during the design phase of future buildings.

Approximately 80 percent of Norway's current buildings will be standing in the year 2050 [3], but not all of these are designed to meet the challenges resulting from progressive climate change [4,5]. This fact underlines the importance of achieving a balance between mitigation and adaptation during the design of new, and the retrofitting of old, buildings. It also emphasizes that appropriate maintenance, operations, and management (MOM) strategies will be vital tools in ensuring adequate climate adaptation of Norway's existing building stock.

The adaptation of the built environment to climate change has received significant attention in Norway during the last two decades [6–8]. A prior study carried out within the framework of the Norwegian research program Klima 2050 [9] has recognized the need for research-based knowledge related to building maintenance [10]. The study identified a significant knowledge gap in the field of building maintenance and renovation, especially in relation to technical solutions and related components.

In 2014, Flyen et al. [11] recorded a significant time lag in the maintenance of Norwegian public buildings, arguing that this lag increases the vulnerability of the built environment to the stresses imposed by climate change. Furthermore, a national condition status report published by the Norwegian Consulting Engineers' Association [3] indicated that the entire built environment in Norway is worth approximately NOK 5800 billion, and that the cost of renovating this building stock to an adequate standard is estimated to be NOK 2800 billion. Even though this is the grand total for renovation due to all degradation aspects, climate related renovation needs can be argued to be substantial. To put this in perspective, the Norwegian gross national product (GNP) in 2016 was NOK 3100 billion [12]. Previous research has estimated that the total annual costs related to building repair in Norway are approximately EUR 1.65 billion [13].

1.2. Definitions

1.2.1. Climate Change

Climate change in the context of this article is defined as *"a change in the state of the climate that can be identified (e.g., by using statistical tests) by changes in the mean and/or the variability of its properties and that persists for an extended period, typically decades or longer"* [14]. Depending on which of the various IPCC scenarios we select, we can expect a global temperature increase of between 1.5 and more than 6 °C before 2100. In a Norwegian context, a temperature increase of between 2.3 and 4.6 °C is anticipated during the same period [2]. Future climate change will thus have a significant impact on the built environment. Estimates carried out on behalf of the Norwegian government show that the anticipated impact on the building and infrastructure sectors will be high both in terms of monetary and societal costs [15]. A conservative estimate indicates immediate costs of 0.1–0.2% of the GNP for the European countries. The estimate is uncertain and excludes, i.e., accumulated costs over time. If these accumulation effects are accounted for, the figure will be considerably higher [15].

1.2.2. Climate Adaptation

Two fundamental response approaches to climate change are recognized in the literature: climate mitigation and climate adaptation. Climate mitigation is defined by the IPCC [14] as *"the notion of limiting or controlling emissions of greenhouse gases so that the total accumulation is limited"*. The terminology used in this paper is based on definitions also provided by the IPCC, which define **adaptation as** *"the notion of making changes in the way we do things to respond to changes in climate"*. **Adaptation** in the context of climate change is defined by the United Nations as *"adjustment in natural or human systems in response to actual or expected climatic stimuli or their effects, which moderates harm or exploits beneficial opportunities"* [16]. The term *"***climate adaptation costs***"* is not within the scope of this paper.

1.2.3. Maintenance and Operation Management (MOM)

This paper presents work related to Maintenance and Operation Management (MOM). Terminology related to this term is based on definitions set out in the Norwegian industry standards NS 3424:2012 [17] and NS 3456:2010 [18]. Here, **maintenance** is defined as a *"combination of technical, administrative and managerial actions with the intention of maintaining or re-allocating the condition to a level that fulfils the functional requirements throughout the life-cycle of an item"* [17]. **Corrective maintenance** *"is maintenance carried out after fault detection and intended to put an item into a state in which it can perform a required function"* (EN 13306 [19]) **Preventive maintenance** is *"maintenance carried out at predetermined intervals or according to prescribed criteria and intended to reduce the probability of failure or the degradation of the functioning of an item"* (EN 13306). **Predictive maintenance** is condition-based *"maintenance carried out following a forecast derived from repeated analysis or known characteristics and evaluation of the significant parameters of the degradation of the item"* (EN 13306). **Operation** is a *"combination of all technical, administrative and managerial actions, other than maintenance actions, that result in the item being used"* [17]. **Upgrade** is related to *"work that is to be carried out on a building or its technical installations in such a manner that the building fulfils new, stricter demands and/or that the buildings' area or capacity of installations are increased"* [18].

The development of MOM strategies and the technical systems considered suitable for the implementation of said strategies are crucial to achieving appropriate climate adaptation of existing buildings to ensure that they meet future functional requirements. In this context, the term **functional** refers to *the expression of how the physical building works according to its purpose and fits the need of the user organization* [20].

1.3. Aims and Scope

The aims of this paper are firstly, to analyze the trends, applications, and development of climate adaptation as it applies to MOM practices; and secondly, to propose a climate-adaptive MOM framework for public and large building owners. It has the following objectives:

- to examine relevant and current literature addressing climate adaptation in MOM practices (Section 3),
- to review existing research projects in an attempt to understand the research landscape in this field in a Norwegian context (Section 3)
- to examine the application and extent of climate adaptation from a day-to-day MOM perspective (Section 4)
- to propose a generic framework that facilitates a structured process for developing climate-adaptive MOM practices for professional and public building owners that can reduce climate change risk and increase the resilience of the Norwegian built environment (Section 5).

2. Methods

2.1. General Overview

A multimethod-based research approach is adopted here to ensure coordination and interdependency between Norwegian research efforts and everyday practice (see Table 1 and Figure 1).

Table 1. Overview of methods, objectives, and research questions.

Research Method	Scoping Review	Review of Norwegian Projects/Initiatives	Case Studies	Collaborative Workshop Series
Objective	1. To examine relevant and current literature addressing climate adaptation from a MOM perspective	2. To review existing research projects and understand the research landscape in this field	3. To examine the application and extent of climate adaptation from a day-to-day MOM perspective.	4. To propose a generic framework that facilitates a structured process for developing climate-adaptive MOM practices that can reduce climate change risk and increase the resilience of the Norwegian building environment
Research questions	1.1 What trends are emerging in current MOM-related literature?	2.1 What research-based initiatives or projects are associated with MOM and upgrade?	3.1 What are the characteristics of current MOM-systems for climate adaptation at different scales in Norwegian public sector institutions?	4.1 What should be added to or modified in terms of MOM practice in order to meet the challenges presented by climate change?
	1.2 What are the implications of climate change, and where is the need for climate adaptation?	2.2 What is the main purpose of the projects/initiatives?	3.2 What challenges do these systems face, and how can they be improved?	4.2 How should we structure a process for identifying climate-adaptive measures in MOM practices?

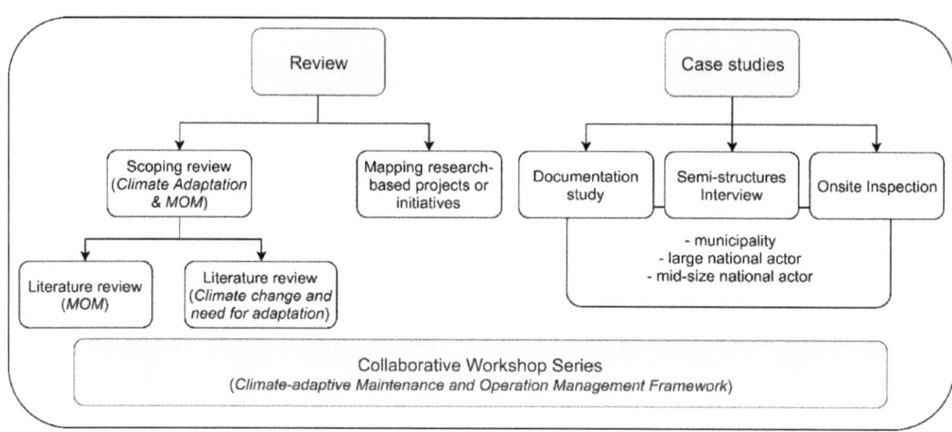

Figure 1. General overview and structure of the methodology.

2.2. Literature Review

An initial scoping review was carried out in order to examine the nature, extent and range of research activities that have addressed the incorporation of climate adaptation in MOM practice. The review was based in an established research methodology [21] by which the queries used to search the databases were taken from experts working on major research projects in the field, and based on the CIMO (context intervention mechanisms outcomes) framework [22]. The search string, keywords, and Boolean operators are shown in Table 2. Two electronic databases containing peer-reviewed literature were used; SCOPUS and Web of Science, which revealed 22 and 15 potentially relevant documents, respectively. However, based on their titles and abstracts alone, most of this literature was of little relevance to the scope of this paper. Nevertheless, we did identify some articles that discussed climate adaptation and MOM in contexts other than the building environment, such as civil engineering applications including railways, bridges, and road projects, or design stage applications.

In addition to the initial search, a more traditional rapid review using the Google Scholar search engine and the Science Direct database was carried out to identify trends in the field of MOM, climate change, and the need for adaptation.

Table 2. Keywords, Boolean operators, and search strings.

WHO	WHAT	HOW
Intervention	*Context*	*Outcomes/Mechanism*
Building	Climate adaptation	Maintenance
Architecture		Facility management
Construction		Operation management
		Upgrade
		Refurbishment

(TITLE-ABS-KEY ("climate adaptation") AND TITLE-ABS-KEY (building OR architecture OR construction) AND TITLE-ABS-KEY (maintenance OR "facility management" OR "operation management" OR upgrade OR refurbishment)).

2.3. Mapping of Research Projects and Initiatives

The main author has previously published a paper involving a literature study and a review of research projects involving Norwegian building owners in which MOM and upgrades were investigated [23]. An internal workshop was held to sort these projects into categories in matrix form according to their principal research theme and the type of decision-maker (main actor) involved.

The project review also involved dialogue with experts and major research project representatives in order to obtain an understanding of the current status of contemporary research. A draft of the project overview was also discussed with three major Norwegian building owners/managers and one of the largest construction consultant companies in Norway.

2.4. Case Studies

Three cases were selected to represent the building owner perspective. Selection was carried out to ensure representative variation in scale in terms of geographical location and climate exposure, building portfolio, and size.

1. Municipality. The municipality selected is one of the most highly populated in Norway. It maintains one million square meters of building stock, but is geographically categorized as a local actor with a relatively uniform level of climate exposure.
2. A large Norwegian actor. This actor is the largest owner of civil buildings in Norway. It owns a highly varied portfolio of more than 2300 buildings covering more than 2.8 million square meters. The buildings are distributed across the entire country and their levels of exposure to climate stresses vary according to location.
3. A medium-sized Norwegian actor. This actor was selected in order to offer insights into a smaller organization with a limited, though geographically widespread, portfolio of buildings. The actor manages 45 airports of varying size.

Three principal methods were adopted for this investigation and were applied according to the guidelines presented by Yin [24] as follows:

(a) Documentation study. An initial examination of specific MOM components was carried out based on a documentation study. This included a review of drawings, operational plans and condition state analyses as set out in guidelines developed by Weber [25]. The number of case-specific documents examined is presented in Figure 2.
(b) On-site inspection. A single on-site inspection for each of the actors was carried out by two of the authors of this paper. During the inspection, they were accompanied by a maintenance officer who commented on maintenance needs and the extent to which different solutions were

working. The main objects of analysis in this case were the building envelope and adjacent terrain, with a specific focus on identifying climate related damage and areas of weakness of the building envelopes.

(c) Semi-structured interviews. Methods (a) and (b) were supplemented with semi-structured interviews (Yin [24]) with representatives from the municipality (operations officer, project manager, and maintenance and sustainability advisor), the large Norwegian actor (senior project managers, innovation officers, building workers, and administrative personnel) and the medium-sized Norwegian actor (maintenance officer). The interviews were also used to extend the results from the on-site inspection of one building to a more general perspective in order to provide a more generic representation of the MOM procedures of each actor.

	Case studies		
	Municipality	Large national actor	Mid-size national actor
Documentation study	21	25	11
Onsite Inspection	1	1	1
Semi-structured Interview	3	10	1

Figure 2. Numbers of case-specific documents examined as part of the case studies.

2.5. Joint Workshop Series

A series of joint workshops involving experts in the fields of building engineering and climate adaptation were carried out in order to develop a climate-adaptive MOM framework. The knowledge gap that emerged as a result of the literature review, combined with an identified need to incorporate climate adaptation in MOM practices based on examination of the case studies, served as the starting point for preparation of the proposed framework. The workshop also established a set of four requirements that the framework would have to meet:

- Compliance with the Norwegian standard EN 15331—"Criteria for design, management, and control of maintenance services for buildings" [26].
- Compliance with the ISO 9001 standard "Quality Management Systems—Requirements" [27], which states that "Management of the processes and the system as a whole can be achieved using the PDCA (PDCA stands for Plan-Do-Check-Act) cycle with an overall focus on risk-based thinking aimed at taking advantage of opportunities and preventing undesirable results". Concepts set out by the IPCC [1] regarding risk assessment are also adopted in this context.
- The framework should be generic and thus applicable at all scales and for all actors carrying out maintenance and operation management on buildings and other facilities.
- The framework should be specifically applicable in a Norwegian context. The findings from various projects linked to the Klima 2050 research centre [9], as well as the report "Climate in Norway 2100" [28], were taken into account.

3. Review of Literature and Initiatives

3.1. Scientific Literature Review

Climate change is forcing society to address factors affecting building maintenance needs from a life cycle perspective. This includes the use of proactive maintenance to extend the operational life of buildings and equipment. A key to this proactive approach is the search for smart and innovative MOM assessment tools.

The ISO standard "15686-8:2012—*Life Cycle Planning: Reference service life and service-life estimation*" [29] defines the so-called "factor method" for the estimation of the expected service life of a component or assembly under the influence of a well-defined set of conditions. The method addresses a number of different factors: (i) material (properties), (ii) design (details), (iii) execution (on-site factors), (iv) climate stresses, such as rain, wind, snow, and chemicals, and (v) maintenance (preventive measures). The first three factors address resilience in the face of deterioration. The fourth influences the speed of deterioration and the fifth addresses measures designed to extend service life. All of these factors are important and should be taken into consideration from the beginning of the design process to project completion, not least as a means of ensuring the incorporation of life cycle planning (LCP). Further development of the standard should include need assessments for the technical and functional upgrade of buildings. New standards are currently in preparation (e.g., CEN TC50, WG8) that address topics within the field of so-called sustainable refurbishment.

Maintenance and operation management takes place during the operational life of a building. Construction projects related to this phase of a building's lifetime have been the subject of increased interest from project management researchers in recent years. In Norway, for example, the research project OSCAR (http://www.oscarvalue.no/) is actively seeking to identify methods for the optimization of building projects, with the primary aim of contributing to value creation and capture in the interests of owners and users during the building life cycle.

An understanding of the role of facility management (FM) and MOM is crucial to ensure the sustainable extension of building lifetimes, and it is essential that proper emphasis is given during the early planning phase. Recent research in Norwegian contexts indicates that qualitative early-phase planning will assist in meeting sustainability requirements and contribute towards ensuring more secure financial conditions for refurbishment projects [30]. Initiatives such as OSCAR rely heavily on the insights of Cooke-Davies [31]. In recent years, such insights have been applied under Norwegian conditions by Hjelmbrekke et al. [32], and studies such as this have helped to illustrate the challenges encountered in the Norwegian construction sector. Hjelmbrekke et al. [32] have proposed a model for the inclusion of strategic perspectives during the operational phase of construction projects. The core insight shared by the foregoing authors is that building-related operations, and maintenance strategies in particular, must be considered during the planning of new projects. In the absence of a proper knowledge of the actual practices that a typical maintenance scheme may involve, any alignment of project execution with the uoperational phase may be highly problematic.

Fregonara et al. [33] proposed a multidisciplinary approach to decision-making in this context taking into account the property market, project economics, architectural technology and building physics, life cycle costing methodologies and energy consumption analyses. Kamari et al. [34] have presented a comprehensive sustainability framework for the development, assessment, and auditing of building renovation performance. Key components in this model include the support given to decision-makers during the project's lifecycle and the need to address the sustainability of entire renovation projects, including the introduction of new categories, criteria, and indicators.

The financial consequences of climate change have been identified by the statistical analysts Finance Norway, showing a 30% increase in insurance claims payments during the last five years [35]. The need for well-functioning strategies and technical systems to protect the building envelope and other components has also been demonstrated in references [36,37]. An examination by the present authors of the building defects archive held by SINTEF Building and Infrastructure reveals that damage to, and defects in, existing Norwegian building stock can be explained as follows:

- 75% of investigated defects are caused by moisture
- 67% of investigated defects are related to building envelopes
- 25% of investigated defects are caused by precipitation
- 33% of investigated defects linked to exterior walls above the terrain surface are caused by moisture
- 50% of investigated defects linked to roofs and terraces are caused by moisture.

The most critical aspects of climate change that are relevant to future building adaptation are increases in annual mean temperatures and levels of precipitation, increasing volumes of winter precipitation falling as rain, and increases in the wind-driven rain component. An increase in the frequency of freeze–thaw cycles (temperature oscillation around 0 °C) is also anticipated in the future [4]. Other factors such as the higher frequency of events such as avalanches, floods, and storm surges, combined with rising sea levels, are not considered relevant to this study, and are not discussed further.

This study has identified four major climate change factors. These are summarized in Table 3, together with a list of the key technical challenges associated with them.

Table 3. Climate change factors and associated technical challenges. This table has previously been presented in Grynning et al. [38].

1. Increase in Annual Temperatures	2. Increase in Precipitation (Rain)	3. Winter Precipitation as Rain	4. Driving Rain
Increased mould growth potential Increased rot decay risk Greater frost-cycle variation Reduced heating demand Increased cooling demand	Increased mould growth potential Increased rot decay risk Longer periods of free-standing water on roofs Stress on the robustness of roof membranes Freeze–thaw cycles Stresses on membrane joints Gaskets and protrusions Drain/gutter capacities Blocking of drains Overflow in drains Standing water due to limited drain capacity	Increased structural loads Stress on roofing robustness Increased water pressure on ground constructions Ice formation on surfaces and in pore structures of materials	Increased mould growth potential Increased rot decay risk Need to upgrade surface treatment of facades Drying out of walls Flashing details Need to identify better window/door mounting solutions

One can see from the challenges listed above that the design of the building envelope is key to the climate adaptation of buildings. Adaptation is a general term that denotes a building's physical properties and its flexibility to adjust to changes in use, function and size [20]. Increases in temperature will promote mould growth within building envelopes. A study carried out by Almås et al. [13] has shown that an estimated 615,000 buildings in Norway are located in areas with a potentially high risk for rot decay. In 2100, this number is anticipated to rise to 2.4 million. Increased temperatures will also result in a general decline in demand for heating and corresponding increases in cooling demand.

3.2. Review of Research Projects and Initiatives

Table 4 provides a summary of research projects identified as being useful to this study, categorized in matrix form according to main actor and research theme. A total of 28 projects are listed, and ten of these address climate adaptation issues. The reviews of existing literature and relevant research projects conducted as part of this study have revealed a number of thematic research needs. One of the general findings is that the research projects relevant to MOM and building upgrades exhibit a notable bias towards energy efficiency in buildings. A more detailed description of these projects is presented in a previous study by the present authors [23].

Table 4. Summary of research projects identified as being useful to this study, categorized in matrix form according to main actor and research theme. Green shading indicates cases where the topic in question is considered highly-researched. Red shading indicates that the topic has been the subject of little or no research.

Level	Main Actor of Interest	Research Theme		
		1. Climate Adaptation	2. Energy Efficiency	3. Economy
A. Law and legislative	Ministry/Directorate	[13,39–42]	[43–47]	[47–49]
B. Legislative/planning/strategy	Municipality/local authority	[13,39–42,50–59]	[43–46,57,58,60,61]	[62]
C. Strategy/system	Managers/MOM operators	[24,51–56,63]	[43–46,60,64–77]	[62,78–80]
D. System/solution	Consultant/contractor	[13,81,82]	[13,43–46,64–66,69, 70,75–77,82–86]	[62,87,88]
E. Solution/component	Product manufacturer		[13,43–46,82]	

3.3. Knowledge Gap

Due to anticipated changes in climate, building envelopes will be subjected to increasing levels of stress in the years to come. Our research review demonstrates that projects addressing the climate adaptation of buildings using MOM focus mostly on the effects of moisture and resilience to potential moisture-related problems. This is mainly due to a broad understanding that future increases in precipitation, combined with annual average increases in temperature, will become the greatest sources of stress on the building envelope in the future. These impacts may reduce the lifetimes of individual building components and increase overall damage risk, and underline the importance of promoting appropriate maintenance strategies and schedules.

Implementation of climate adaptation measures for buildings has been on the agenda in Norway for some years, but the climate adaptation of buildings involving MOM has received very little attention from the research community, and Grynning et al. [23] have identified key knowledge gaps in this field. This study has also been able to identify a need to provide an account of what MOM practices entail and to develop a framework for industry application.

Hauge et al. [89] found that numerous user guides exist to prepare societies for the coming climate challenges, but none of the user guides describes decision processes in depth, and target groups are not specified. Stagrum et al. [90] found little research on the consequences of climate change on buildings in cold regions. The majority of the identified literature concerns climate change impacts on buildings in warm climates, with overheating being seen as the greatest challenge.

Hauge et al. [91] have studied barriers and drivers for climate adaptation in Norway. They conclude that pursuing changes across the practical, political, and personal spheres is essential.

4. Case Studies

4.1. General Remarks

The three organizations introduced in Section 2.4 have all stated that climate adaptation is an explicit ambition as part of their corporate strategies. However, the levels of detailed planning and, perhaps also, of commitment to this ambition vary somewhat. It is clear that there exists wide variation in facility management strategies and implementation within these organizations—ranging from incident-based corrective maintenance programs to long-term preventive plans and strategies.

4.2. The Municipality

The MOM strategies employed by the municipality revolve around a five-year cycle, as shown in Table 5. Every fifth year an extensive condition status analysis is carried out by an external consultant. This analysis forms the basis for longer term MOM plans addressing the needs of the building envelope, technical installations (heating, cooling, ventilation, etc.) and interior MOM and building upgrade needs. On this basis, the municipality prepares annual MOM plans and reporting procedures.

This structured approach to planning ensures adequate levels of supervision of day-to-day maintenance and upgrade needs. It also helps in the facilitation of longer term planning.

Table 5. Stepwise procedure for the management of MOM needs and tasks as exercised by the municipality in this case study.

Step 1—Maintenance Needs	Step 2—Long-Term Planning	Step 3—Annual Plans and Work Orders	Step 4—Revision of Plans
Task:	**Task:**	**Task:**	**Task:**
Assess building condition Define long-term maintenance needs	Prepare 5-year maintenance plans Adjustment of maintenance needs and transfer to budgets	Year on year MOM procedure for building maintenance officers	Revise annual plans and booklets including checklists and tasks
Persons responsible	**Persons responsible**	**Persons responsible**	**Person responsible**
External consultant and internal administrators	Building and head administrators	Administrator and building maintenance officer	Building maintenance officer
Actions:	**Actions:**	**Actions:**	**Actions:**
Condition status analysis of buildings Prepare long-term plans (also for funding purposes)	Coordination of maintenance and potential upgrade needs	Prepare checklists for the maintenance procedure booklets	Submission of plans/booklets to central property department at year end

Dedicated maintenance officers are responsible for the day-to-day maintenance and operation of municipal buildings. As a rule, a maintenance officer may have responsibility for one or several buildings, depending on the size of the structures in question. External contractors are brought in if major MOM or upgrade tasks have to be carried out.

A set of annual MOM procedures is provided to responsible maintenance officers, consisting of a booklet describing all the detailed procedures and mandatory checks that must be carried out during the year in question. An ongoing series of checklists is completed as the year progresses. At the end of the year, each officer submits his or her booklet to the municipality's central property department. Booklets are laid out in nine operational chapters as follows (taken from the 2013 edition): 1. A description of tasks with checklists. 2. A checklist for mandatory tasks. 3. A checklist for critical operational tasks. 4. A checklist for (less critical) operational tasks. 5. A checklist for electrical systems and equipment. 6. A checklist for outdoor areas (e.g., municipal playgrounds). 7. Internal building safety checks. 8. MOM procedures for bomb shelters. 9. Internal checks of operational tasks.

The procedures and systems descriptions in the booklets are for the most part in paper format, making revisions cumbersome. It was also found that much of the knowledge in the organization is tacit and linked to the know-how of experienced individuals, making the organization vulnerable to changes in personnel. Such practices may impair organizational knowledge transfer on a broader scale.

In order to assist the organization in the further development of its systems, we recommend that greater focus should be directed on the digitalization of tools (e.g., using Building information modeling (BIM)), communication, and processes. This may help to promote proactive knowledge sharing at all levels in the organization. Moreover, higher levels of digitalization may promote the transfer of knowledge from individuals into a structured organizational knowledge base.

Table 5 presents a summary of the overall MOM strategy exercised by the organization. The review of documents carried out for this case study indicate that specific challenges exist linked to a lack of climate change planning, and a failure to implement climate adaptation initiatives. For example, the present authors are skeptical of the resilience of some technical systems to increased levels of precipitation, and no strategy exists for the upgrade of these systems.

4.3. The Large Norwegian Actor

Climate adaptation was introduced as an area of focus in this property owner's strategy in 2014 [92]. Currently, the organization's focus is at a more strategic level—*"climate adaptation shall be accounted for in the light of climate changes"*. This statement corresponds with the findings resulting from our research project review that the majority of current research is limited to high-level strategy applications. This actor currently includes no operational plans or specific proposals regarding strategic implementation, which underlines the need to develop a knowledge framework within organizations in general to enable them to implement climate adaptation measures in practice. Of the three case study organizations, this actor possessed the most comprehensive facilities management (FM) system. However, we recognize a need to introduce actions to strengthen the organization's climate adaptation implementation strategy. The existing climate adaptation system for buildings adheres to a structure similar to that described by Cooke-Davies [31], which accentuates the need to establish strong links between the operations and project management levels within the organization in order to meet corporate strategies and goals. The facility management system and tasks are assembled in a database tool and adhere to a stepwise procedure as shown in Table 6. The tool collects all technical and practical information regarding the owners' properties. The property maintenance officers have overall responsibility for entering maintenance needs in the tool. Building administrators register the entries and transfer these to short- and mid-term maintenance plans and budgets. The tool can then be accessed by maintenance officers to obtain activities and work orders. "Long-term" in the context of this organization is a period of five years.

Table 6. Stepwise procedure for the management of MOM needs and tasks as exercised by the large Norwegian actor in this case study.

Step 1—Maintenance Needs	Step 2—Long-Term Plans	Step 3—Annual Plans	Step 4—Activities and Follow-Up
Task:	**Task:**	**Task:**	**Task:**
Definition of maintenance needs	Define 5-year maintenance plans Adjustment of maintenance needs and transfer to budgets On budget approval, needs are transferred to annual plans	Access annual action budgets and preparation of work orders	Implement work orders
Persons responsible	**Persons responsible**	**Persons responsible**	**Persons responsible**
Maintenance officers or administrators	Building and head administrators	Building administrator	Maintenance officer
Actions:	**Actions:**	**Actions:**	**Actions:**
Register needs Transfer to long-term plans	Edit action and budget approval Periodization of actions	Edit action Prepare work orders Closing of actions	Edit work orders Prepare orders for external assistance Closing of work orders

The authors have identified some challenges related to the implementation of the system for this case study. For example, defects in building envelopes that have resulted in ongoing water leakages were revealed during the on-site inspection and in interviews, underlining the importance of strengthening the links between operations and management levels within the organization.

Secondly, the level of detail inherent in the management tool could be improved. Critical building components should be the subject of greater attention from a climate adaptation perspective. Maintenance intervals for such components should be introduced and adapted to local, site-specific, needs.

Finally, building damage and critical events caused by "extreme weather" incidents may be avoided by implementing early warning and weather forecasting systems. The need for such systems was underlined by a flooding incident in a building adjacent to one of the actor's properties. This incident emphasizes the need for greater cross-organizational communication, as well as a need to consolidate internal corporate strategies and plans.

4.4. The Medium-Sized Norwegian Actor

The building considered in this case study is located in an airport for domestic and international flights. This actor's maintenance department employs five maintenance officers and a maintenance manager, who is responsible for day-to-day task planning. The department is responsible for maintenance across the entire airport, including the building considered in this study. The actor uses an interactive management tool for day-to-day building management. The users of the building report their needs to the maintenance department, which in turn plans and coordinates a daily schedule. Recurring maintenance tasks are also plotted in the system, enabling the department to plan and prioritize its assignments according to criticality and urgency. We observed no long-term planning integrated into the MOM system. Instead, it was the task of the maintenance manager to report annual funding needs based on the condition of the building. In contrast to the two previous case studies, no strategic planning structure exists for this actor. We recommend that this situation be remedied.

This actor's management system makes it easy for building managers to keep track of the maintenance needs at any given time. The aim here is to ensure that functional failures can be rectified quickly, enabling building managers and the operations department to prioritize issues based on critical need.

However, we recognize that it would be beneficial if the management system also incorporated a comprehensive and long-term maintenance and upgrade plan based on condition status analyses and the estimated lifetimes of key components and products. The development of a more long-term approach to the management of building stock may make a useful contribution towards major maintenance and upgrade planning, and at the same time facilitate long-term expenditure budgeting. This would provide a detailed overview of future maintenance tasks and enable potential synergies in situations where two or more major improvements could be carried out at the same time with adequate quality and at lower cost.

Periodic tasks such as the supervision of technical installations are guided by procedures. Needs and corresponding tasks are reported on a continuous basis. However, the lack of a structured long-term plan for MOM tasks presents a challenge. In an interview carried out with the operator, it emerged that aspects of the main terminal roof could be used to illustrate this issue. The roof was installed in 1986 and had been subject to considerable wear and tear. However, since no leaks had been detected, no funds had been assigned for a thorough condition status analysis or replacement, even though the roof's expected lifetime was close to expiry. This approach flies in the face common sense that dictates that waiting for a component to fail before initiating a response will lead to additional work as a result of consequential damage.

5. A Climate-Adaptive Maintenance and Operation Management (MOM) Framework

5.1. Definitions

The term climate-adapted buildings and building structures is commonly applied to structures that are planned, designed and built to withstand various types of external climatic stresses including precipitation, snow deposition, wind, temperature, storm water and exposure to the sun. However, adaptation to a changing climate must also incorporate reductions in the risks to wider society. It is not sufficient simply to restrict our focus to historical weather data when designing and maintaining buildings [93].

Climate-adaptive maintenance is hereby defined for the purposes of this study as the combination of technical, administrative, and managerial actions carried out with the intention of maintaining, allocating, and adjusting the condition of an item to a level that fulfils the functional requirements throughout the life-cycle of the item in response to actual or expected climatic stimuli.

Climate-adaptive maintenance and operation management is hereby defined as a combination of all technical, administrative, and managerial actions, including maintenance actions, that result in the item being used and maintained in response to actual or expected climatic stimuli.

5.2. A Climate-Adaptive Maintenance and Operation Management (MOM) Framework Workflow

Figure 3 illustrates the climate-adaptive MOM framework workflow that is proposed in this paper. Climate adaptation exerts most influence during the planning and decision-making stages of the planning phase. For this reason, the workflow focuses primarily on the planning stage of the PDCA cycle since this is intended only to account for the climate change aspect. The components of the workflow are discussed in detail in the following.

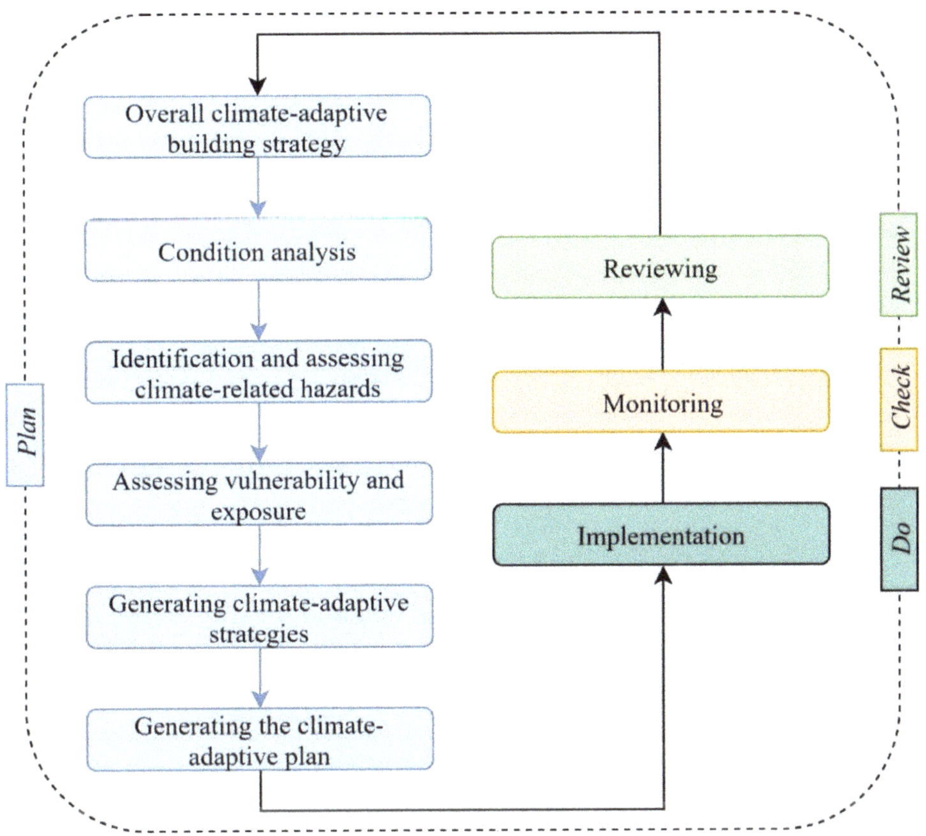

Figure 3. Proposed climate-adaptive maintenance and operation management (MOM) workflow.

Overall climate-adaptive building strategy: This step is designed to determine the applicable climate-related service and performance specification for the building asset in question in compliance with the EN 15331 standard [26]. It will also include a strategy to safeguard property value in the face of exposure to climate change. The step shall be reviewed at specified regular intervals and in situations where a major change in performance requirements is required in response to new knowledge.

Condition analysis: This step is designed to determine the condition of the building asset in relation to a given reference level (performance requirement). The reference level will be set out in a climate-adaptive building strategy. The data required to carry out a risk analysis shall be collected, and the analysis performed according to guidelines set out in the Norwegian standard NS 3424.E:2012 [17].

Identification and assessment of climate-related hazards: This step is designed to identify potential hazards to the building asset in question, and impacts in response to actual or expected climatic stimuli in a Norwegian context. To support this step, Table 7 provides a structured overview of hazards that may be anticipated in a Norwegian context as a result of climate change and their expected impacts on the building asset.

Assessing vulnerability and exposure: This step is designed use the data and knowledge gathered during previous steps to assess the vulnerability and capacity of the exposed building assets in response to identified hazards.

Assessing climate-related risk: This step is designed to identify, assess and prioritize the risks resulting from actual or anticipated climatic stimuli. A graphical risk matrix is constructed illustrating probabilities and consequences. A priority list shall be drawn up to categorize and rank risks based on the level of response required (from immediate to no action).

Generating climate-adaptive strategies: This step is designed to generate and evaluate climate-adaptive strategies in response to identified risks. Strategies may be generated by following the structured matrix proposed in this paper that recommends strategy categorization based on (a) maintenance type (corrective, preventive or predictive) and (b) the risk factor(s) that require mitigation (hazard, exposure, and vulnerability). For example, a hypothetical case study involves a flat roof showing no signs of leakages (see Table 8). The roof is then exposed to levels of precipitation and increased temperatures that were not built in during the design stage. The strategy categorization provides a detailed overview of potential impact in terms of risk reduction and enables identification of the most useful strategy for a given building asset. This approach may prove to be beneficial to actors with responsibility for the management of large building asset portfolios, within which buildings or building stock may require component-specific adaptations to changing climate stimuli and consequent exposure risk.

Generating climate-adaptive plan: This step is designed to facilitate the preparation of an adaptation plan and strategy document that incorporates the main findings of previous steps, and which provides clear direction for implementation and long-term execution.

Implementation: This step involves implementation and execution of the climate-adaptation plan generated in the previous step.

Monitoring: This step is designed to facilitate monitoring and performance measurement of the climate adaptation plan and the effectiveness of its risk mitigation measures. Performance is rated with reference to the objectives, requirements and planned activities identified in the strategy set out for the building asset in question.

Reviewing: This step is designed to review the adopted climate adaptation strategy and plan based on the monitoring and performance indicators emerging from the previous step. Risk assessments and strategies may be upgraded based on the findings of the review process and subsequently implemented in new plans.

Table 7. Overview of potential hazards in response to anticipated climatic stimuli and their implications. Red shading indicates a high degree of influence and white very low or no influence.

		Hazards							
		Increased Precipitation				Increased Temperature	Sea Level Rise	Flooding	Freeze–Thaw Cycles
		Wind-Driven Rain	*Accumulative Precipitation*	*Wet Winter Precipitation*	*Torrential Precipitation*				
Implications	Mould growth								
	Rot decay								
	Biological growth								
	Moisture creep								
	Other fungi								
	Structural overload								
	Heating/ cooling demand								
	Outages/downtime								
	Cracks								
	Spalling								
	Groundwater pressure								
	Corrosion and/or carbonation								
	Blocking of drains								

Table 8. Matrix showing the climate-adaptive strategy proposed in this paper. A (potentially) leaking flat roof is taken as an illustrative example.

Maintenance Type	Hazard	Exposure	Vulnerability
Corrective	Fast cleaning (carry water in buckets)	Change roof covering when damaged	Not applicable
Preventive	Increase dimensions of roof guttering and drainage system	Change roof covering before end of expected service life	Training of MOM personnel
Predictive	Sensor system warning	Change roof covering when embedded sensor alerts that service life is about to end	Direct weather forecast warning to all involved stakeholders

6. Conclusions and Further Work

This study has adopted a multimethod research approach to the analysis of the basic criteria, trends, applications, and developments related to climate adaptation in connection with building maintenance and operation management (MOM) practices in Norway. The study concludes that there exists a significant knowledge gap between current maintenance and operation management practice and the strategies that are required to safeguard adequate adaptation to climate change. The study recommends that future research should focus on resolving this gap by addressing the two factors in combination. A review of practices and research projects has shown that climate adaptation is considered only as a high-level strategic issue in many organizations and that there is a need to incorporate the concept at lower organizational levels. An analysis of the three case studies has served to emphasize the need to adopt a systematic approach to the integration of climate adaptation considerations in current MOM practice.

This study proposes the adoption of a climate-adaptive maintenance and operation management framework. The framework involves a generic and structured process that facilitates the incorporation of climate adaptation in MOM practices on different scales and at different organizational levels. It is anticipated that implementation of this flexible and transferable framework will provide a basis for acquiring more knowledge on the topic of climate adaptation. Further development of the framework should include the introduction into organizations of more tangible and tailored tools and processes, including checklists or scoring systems accompanied by relevant climate adaptation factors and plans.

Author Contributions: Conceptualization—S.G., B.T., and T.K.; Methodology—S.G., J.L., and K.G.; Formal analysis—S.G., K.G., and J.E.G.; Investigation—S.G. and J.E.G.; Writing (original draft)—all authors; Writing—review and editing, S.G., K.G., and J.L.; Visualization—S.G. and K.G.; Supervision—B.T., J.L., and T.K. All authors have read and agreed to the published version of the manuscript.

Funding: This research was funded by the Research Council of Norway and private industry partners under grant number 237859.

Acknowledgments: We would like to express our gratitude to N.N. for contributing towards the funding of this study.

Conflicts of Interest: The authors wish to declare no conflicts of interest.

References

1. IPCC. Climate Change 2013-The Physical Science Basis. In *The fifth Assessment Report of the Intergovernmental Panel on Climate Change*; IPCC: New York, NY, USA, 2013.
2. Hanssen-Bauer, I.; Førland, E.J.; Haddeland, I.; Hisdal, H.; Mayer, S.; Nesje, A.; Nilsen, J.E.Ø.; Sandven, S.; Sandø, A.B.; Sorteberg, A.; et al. *Klima i Norge 2100. Kunnskapsgrunnlag for Klimatilpasning Oppdatert 2015 (in Norwegian)*; NCCS: Oslo, Norway, 2015.
3. RIF-Association of Consulting Engineers, N. *State of the Nation 2015*; RIF-Association of Consulting Engineers, N: Oslo, Norway, 2015.

4. Flyen, C.; Almås, A.-J.; Hygen, H.O. Warm, wet and wild-Climate change vulnerability analysis applied to built environment. In Proceedings of the BEST3 Conference–Building Enclosure Science and Technology, Atlanta, GA, USA, 2–4 April 2012.

5. Barbosa, R.; Vicente, R.; Santos, R. Comfort and buildings: Climate change vulnerability and strategies. *Int. J. Clim. Chang. Strateg. Manag.* **2016**, *8*, 670–688. [CrossRef]

6. Meacham, B.J. Sustainability and resiliency objectives in performance building regulations. *Build. Res. Inf.* **2016**, *44*, 474–489. [CrossRef]

7. Lisø, K.R.; Myhre, L.; Kvande, T.; Thue, J.V.; Nordvik, V. A Norwegian perspective on buildings and climate change. *Build. Res. Inf.* **2007**, *35*, 437–449. [CrossRef]

8. Lisø, K.R.; Kvande, T.; Time, B. Climate adaptation framework for moisture-resilient buildings in Norway, 11th Nordic Symposium on Building Physics. *Energy Procedia* **2017**, *132*, 628–633. [CrossRef]

9. Klima2050. Klima 2050 Webpage. Available online: www.klima2050.no (accessed on 29 May 2020).

10. Grynning, S.; Wærnes, E.G.; Kvande, T.; Time, B. *Klimatilpasning av Bygninger Gjennom Vedlikehold og Oppgradering–Kunnskaps- og Forskningsunderlag*; in Klima 2050 Note; SINTEF: Trondheim, Norway, 2016.

11. Flyen, C.; Mellegård, S.E.; Bøhlerengen, T.; Almås, A.-J.; Groven, K.; Aall, C. *Buildings and Infrastructure–Vulnerability and Adaptive Capacity to Climate Change-Final Report from the BIVUAC Research Project (in Norwegian)*; NCCS: Oslo, Norway, 2014.

12. Norway, S. Yearly National Accounts for 2016. Available online: https://ssb.no/nasjonalregnskap-og-konjunkturer/statistikker/nr/aar/2017-02-09#content (accessed on 1 March 2016).

13. Almås, A.-J.; Lisø, K.R.; Hygen, H.O.; Øyen, C.F.; Thue, J.V. An approach to impact assessments of buildings in a changing climate. *Build. Res. Inf.* **2011**, *39*, 227–238. [CrossRef]

14. Pachauri, R.K.; Meyer, L.A. *IPCC Climate Change 2014: Synthesis Report. Contribution of Working Groups I, II and III to the Fifth Assessment Report of the Intergovernmental Panel on Climate Change-Annex II: Glossary*; IPCC: Geneva, Switzerland, 2014; pp. 117–130.

15. Vennemo, H.; Rasmussen, I. *Samfunnsøkonomiske Virkninger av Klimaendring i Norge (in Norwegian)*; NCCS: Oslo, Norway, 2010.

16. UNFCCC. Glossary of Climate Change Acronyms and Terms. United Nations Framework Convention on Climate Change. Available online: http://unfccc.int/essential_background/glossary/items/3666.php (accessed on 3 October 2019).

17. NS. *NS 3424:2012-Condition Survey of Construction Works-Content and Execution*, 1st ed.; Norsk Standard: Oslo, Norway, 2012; Available online: www.standard.no (accessed on 29 May 2020).

18. NS. *NS 3456:2010-Documentation for Management, Operation, Maintenance and Development of Construction Works*, 1st ed.; Norsk Standard: Oslo, Norway, 2010.

19. NS. *NS-EN 13306:2017, Maintenance-Maintenance Terminology*, 1st ed.; Norsk Standard: Oslo, Norway, 2017.

20. Støre-Valen, M.; Lohne, J. Analysis of assessment methodologies suitable for building performance. *Facilities* **2016**, *344*, 726–747.

21. Andrew Booth, A.S. Diana Papaioan. In *Systematic Approaches to a successful Literature Review*; Sage: Newcastle upon Tyne, UK, 2016.

22. M Petticrew, H.R. *Systematic Reviews in the Social Sciences: A Practical Guide*; Blackwell Publishing: Hoboken, NJ, USA, 2006.

23. Grynning, S.; Wærnes, E.; Kvande, T.; Time, B. Climate adaptation of buildings through MOM- and upgrading-State of the art and research needs. *Energy Procedia* **2017**, *132*, 622–627. [CrossRef]

24. Yin, R.K. *Case Study Research: Design and Methods*, 5th ed.; SAGE: Los Angeles, CA, USA, 2014; ISBN 978-14522425692014.

25. Weber, R.P. *Basic Content Analysis*; Sage Publishing: Newcastle upon Tyne, UK, 1990.

26. EN. *EN 15331 Criteria for Design, Management and Control of Maintenance Services for Buildings*, 1st ed.; Norsk Standard: Oslo, Norway, 2005.

27. ISO. *ISO 9001 Quality Management Systems—Requirements*, 5th ed.; Norsk Standard: Oslo, Norway, 2015.

28. Hanssen-Bauer, I.; Drange, H.; Førland, E.J.; Roald, L.A.; Børsheim, K.Y.; Hisdal, H.; Lawrence, D.; Nesje, A.; Sandven, S.; Sorteberg, A.; et al. *Climate in Norway 2100*; NCCS: Oslo, Norway, 2017.

29. ISO. *15686-8:2012 Life Cycle Planning: Reference Service Life and Service-Life Estimation*, 2nd ed.; Norsk Standard: Oslo, Norway, 2012.

30. Lund, O.B.; Haddadi, A.; Bjørberg, S. Sustainable Planning in Refurbishment Projects–An Early Phase Evaluation, BE16 Tallinn and Helsinki Conference. *Energy Procedia* **2016**, *96*, 425–434. [CrossRef]

31. Cooke-Davies, T. The "real" success factors on projects. *Int. J. Proj. Manag.* **2002**, *20*, 185–190. [CrossRef]

32. Hjelmbrekke, H.; Klakegg, O.J.; Lohne, J. Governing value creation in construction project: A new model. *Int. J. Manag. Proj. Bus.* **2017**, *10*, 60–83. [CrossRef]

33. Fregonara, E.L.V.; Verso, V.R.L.; Lisa, M.; Callegari, G. Retrofit Scenarios and Economic Sustainability. A Case-study in the Italian Context. *Energy Procedia* **2017**, *111*, 245–255. [CrossRef]

34. Kamari, A.; Corrao, R.; Kirkegaard, P.H. Sustainability focused decision-making in building renovation. *Int. J. Sustain. Built Environ.* **2017**, *6*, 330–350. [CrossRef]

35. Norge, F. Water Damage History in Norway. Available online: www.finansnorge.no (accessed on 28 February 2020).

36. Lisø, K.R. Building Envelope Performance Assessments in Harsh Climates: Methods for Geographically Dependent Design. Ph.D. Thesis, NTNU, Trondheim, Norway, 2006.

37. Lisø, K.R.; Kvande, T.; Thue, J.V. Learning from experience-an analysis of process induced building defects in Norway. In Proceedings of the Research in Building Physics and Building Engineering-Proceedings of the 3rd International Building Physics Conference, Montreal, QC, Canada, 27–31 August 2006.

38. Grynning, S.; Gaarder, J.E.; Lohne, J. Climate Adaptation of School Buildings through MOM–A Case Study. *Procedia Eng.* **2017**, *196*, 864–871. [CrossRef]

39. Øyen, C.F.; Almås, A.J.; Hygen, H.O.; Sartori, I. *Klima- og Sårbarhetsanalyse for Bygninger i Norge. Utredning som Grunnlag for NOU Klimatilpasning*; SINTEF Rapport: Oslo, Norway, 2010.

40. NOU. *Tilpassing til Eit Klima i Endring-Samfunnet si Sårbarheit og Behov for Tilpassing til Konsekvensar av klimaendringane*; NOU: Xinbei, Taiwan, 2010.

41. Miljøverndepartementet, D.K. *Stortingsmelding 33-Klimatilpasning i Norge*; Miljødepartementet, K.-O., Ed.; Publisher Miljødepartementet: Oslo, Norway, 2013; Available online: www.regjeringen.no (accessed on 29 May 2020).

42. Aall, C.; Heiberg, E.; Flyen, C.; Martin, M.; Hafskjold, S.; Bruaset, S.; Almås, A.J.; Gjerde, O.I. *Klimaendringenes Konsekvenser for Kommunal og Fylkeskommunal Infrastruktur. Delrapport 3: Egne Analyser av sårbarhet overfor Klimaendringer Belyst Med eksempler fra ulike kommuner*; Vestlandsforsking: Sogndal, Norway, 2011; Available online: www.ks.no (accessed on 29 May 2020).

43. Risholt, B. Zero Energy Renovation of Single Family Houses. Ph.D. Thesis, Norwegian University of Science and Technology, Trondheim, Norway, 2013.

44. Lolli, N. Life Cycle Analyses of CO_2 Emissions of Alternative Retrofitting Measures. Ph.D. Thesis, Norwegian University of Science and Technology, Trondheim, Norway, 2014.

45. Johansson, P. Building retrofit using vacuum insulation panels, Hygrothermal performance and durability. Ph.D. Thesis, Chalmers University of Technology, Göteborg, Sweden, 2014.

46. De De Freitas, V.P.; Delgado, J.M. (Eds.) Advanced materials technologies, Climate-adapted low-energy envelope technologies. In *Hygrothermal Behavior, Building Pathology and Durability*; Springer: Berlin, Germany, 2013; pp. 183–210.

47. GreenConserve. Green Service Innovation Vouchers-Experiences from Testing Voucher Schemes for Sustainable Construction Service Innovators. 2012. Available online: www.greenovate-europe.eu (accessed on 29 May 2020).

48. Holte, K.; Folvik, K.; Kalbakk, T.; Kristjansdottir, T.; Rønning, A.; Lyng, K.; Vold, M.H. KONSENSUS: Metode for Miljøvurdering (LCA) av Bygninger–Dagens Praksis og Anbefalinger: Rapport Delprosjekt 1–Konsensus Arena. In *SINTEF Byggforsk Prosjektrapport*; SINTEF: Oslo, Norway, 2011; Available online: sintefbok.no (accessed on 29 May 2020).

49. Rønning, A.; Lyng, K.; Vold, M.; Holthe, K.; Folvik, K. Konsensus-Modeller for Miljøvurdering av Bygningsdeler-DP2-PCR for Ytterveggskonstruksjon. In *Østfoldforskning Rapport*; Østfoldforskning: Fredrikstad, Norway, 2011.

50. Aall, C.; Øyen, C.; Hafskjold, S.; Almås, A.; Groven, K.; Heiberg, E. Klimaendringenes konsekvenser for kommunal og fylkeskommunal infrastruktur. In *nr. 4/2011 Vestlandsforskingsrapport*; Vestlandsforskning: Sogndal, Norway, 2011; Available online: www.ks.no (accessed on 29 May 2020).

51. Flyen, C.; Mellegård, S.E.; Bøhlerengen, T.; Almås, A.J.; Groven, K.; Aall, C. Bygninger og infrastruktur—Sårbarhet og tilpasningsevne til klimaendringer. In *SINTEF Byggforsk Prosjektrapport*; SINTEF: Oslo, Norway, 2014; Available online: www.sintefbok.no (accessed on 29 May 2020).

52. Nie, L.I.N.M.E.I.; Roald, L.A.; Mellegard, S.; Maksimovic, C. *Flood Risk Management in Cold Climate Experience in Norway*; IAHS-AIHS Publication: Oxfordshire, UK, 2013; Volume 357, pp. 198–207.
53. Campisano, A.; Nie, L.M.; Li, P.J. Retention performance of domestic rain water harvesting tank under climate change conditions. *Appl. Mech. Mater.* **2013**, 438–439. [CrossRef]
54. Nie, L.I.N.M.E.I.; Roald, L.A.; Mellegard, S.; Maksimovic, C. *Flood Risk Management in Cold Climate? Experience in Norway*; IAHS Publication: Wallingford, UK, 2012; p. 325.
55. Øyen, F.C.; Mellegård, S. Legal Steps to Achieve Climate Adaptation in the Norwegian Built Environment. In Proceedings of the CIB/RICS COBRA 2012, Las Vegas, NV, USA, 11–13 September 2012.
56. Nie, L.M.; Øyen, C.F.; Groven, K.; Aall, C. Risk and vulnerability assessment of the built environment. In *Sustainable Building; 8–21 October 2011*; Helsinki Finland, 2011.
57. Flyen, C.; Flyen, A.C.; Hauge, Å.L.; Godbolt, Å.L. Climate for change. Urban regeneration in cultural heritage housing. In *European Network for Housing Research (ENHR)*; ENHR: Lisbon, Portugal, 2015; Volume 6.
58. Flyen, C.; Flyen, A.C.; Hauge, Å.L.; Godbolt, Å.L. The ethos of energy efficiency: Framing consumer considerations in Norway. In *European Network for Housing Research*; ENHR: Lisbon, Portugal, 2015; Volume 6.
59. Oort, B.V.; Hovelsrud, G.K.; Dannevig, H.; Rybråten, S. NORADAPT-Community Adaptation and Vulnerability in Norway. In *CICERO Report*; Cicero: Oslo, Norway, 2012.
60. Rebo. Avslutningskonferansen 2013. 2013 [cited 2015 16.12]. Available online: https://www.sintef.no/projectweb/rebo/avslutningskonferansen-2013/ (accessed on 29 May 2020).
61. Innovation, N. *Sustainable Refurbishment Decision Support Tool and Indicator Requirements*; Standards Norway: Oslo, Norway, 2015.
62. Sødal, A.H. Early Contractor Involvement: Advantages and Disadvantages for the Design Team. In *NTNU, Institutt for Bygg, Anlegg og Transport*; NTNU: Trondheim, Norway, 2014.
63. Board, T.R. *ACRP Synthesis 33, Airport Climate Adaptation and Resilience*; ACRP: Washington, DC, USA, 2012.
64. RetroKit. Public Deliverables. 2015. Available online: http://www.retrokitproject.eu/public-documents/public-deliverables/ (accessed on 29 May 2020).
65. Kleiven, T.; Woods, R.; Risholt, B. Retrofitting multifamily buildings with prefabricated modules-RETROKIT. In *Byggforsk Fag*; Byggforsk, S., Ed.; SINTEF: Oslo, Norway, 2014; Available online: www.sintefbok.no (accessed on 29 May 2020).
66. Heikkinen, P.; Kaufmann, H.; Winter, S.; Larsen, K.E. TES EnergyFaçade–Prefabricated Timber Based Building System for Improving the Energy Efficiency of the Building Envelope. 2010. Available online: https://mediatum.ub.tum.de/doc/1355420/file.pdf (accessed on 3 June 2020).
67. Lattke, F. smartTES Innovation in Timber Construction for the Modernization of the Building Envelope. Ph.D. Thesis, Technische Universität München, München, Germany, 2013.
68. Kolstad, T.S.; Time, B. Retningslinjer for Forankring av Prefabrikkerte Elementer i tre (TES-Elementer) til Fasader ved Oppgradering. In *SINTEF Byggforsk Prosjektrapport*; SINTEF: Oslo, Norway, 2013; Available online: www.sintefbok.no (accessed on 29 May 2020).
69. Skeie, K.S.; Kleiven, T.; Lien, A.G.; Risholt, B.D. Energiplan-tre trinn for tre epoker. In *SINTEF Byggforsk Prosjektrapport*; SINTEF: Oslo, Norway, 2014; Available online: www.sintefbok.no (accessed on 29 May 2020).
70. Thomsen, J.; Hauge, Å.L. Boligeieres beslutningsprosess ved oppgradering. In *SINTEF Byggforsk Prosjektrapport*; SINTEF: Oslo, Norway, 2014; Available online: https://www.sintefbok.no (accessed on 3 June 2020).
71. Blom, P. Fuktsikker innvendig etterisolering av mur- og betongvegger-Ambisiøs energioppgradering når fasaden skal bevares. In *SINTEF Byggforsk Prosjektrapport*; SINTEF: Oslo, Norway, 2014; Available online: www.sintefbok.no (accessed on 29 May 2020).
72. Mellegård, S.; Svensson, A. UPGRADE-Veileder for energiambisiøs oppgradering av yrkesbygg. In *SINTEF Byggforsk Prosjektrapport*; SINTEF: Oslo, Norway, 2014; Available online: www.sintefbok.no (accessed on 29 May 2020).
73. Klinski, M. Ambisiøs energioppgradering med etterisolert fasade. In *SINTEF Byggforsk Prosjektrapport*; SINTEF: Oslo, Norway, 2014.
74. Svensson, A.; Almås, A.J.; Blom, P.; Mysen, M. Syv energiambisiøse oppgraderinger av yrkesbygg. In *SINTEF Byggforsk Prosjektrapport*; SINTEF: Oslo, Norway, 2014; Available online: www.sintefbok.no (accessed on 29 May 2020).
75. Cohereno. Final Conference, Brussels. 2015. Available online: http://www.cohereno.eu/about/final-conference-14102016.html (accessed on 14 October 2015).

76. Cohereno. Outcomes of the Project. 2015. Available online: http://www.cohereno.eu/about/project-outcomes.html (accessed on 29 May 2020).
77. Emrob. Veiledning til Energieffektiv, Miljøvennlig og Robust Oppgradering av Bygninger. 2009. Available online: https://dibk.no/globalassets/energi/verktoy/energieffektiv-miljovennlig-og-robust-oppgradering-av-bygninger.no.pdf (accessed on 7 April 2016).
78. Kristin, H.; Wærp, S. GLITNE-Mer miljøvennlig bygg-Metode og verktøy for økonomisk verdsetting av miljøeffekter. In *SINTEF Byggforsk Prosjektrapport*; SINTEF: Oslo, Norway, 2012; Available online: www.sintefbok.no (accessed on 29 May 2020).
79. Kristin, H.; Wærp, S. Forslag til Byggenæringens Miljøfond og Tiltaksmodellenin. In *SINTEF Byggforsk Prosjektrapport*; SINTEF: Oslo, Norway, 2010.
80. The CILECCTA Partners. *Sustainability within the Construction Sector CILECCTA—Life Cycle Costing and Assessment*; Byggforsk, S., Ed.; SINTEF: Oslo, Norway, 2013; Available online: www.sintefbok.no (accessed on 29 May 2020).
81. Woodwisdom, E.-N. Tall Timber Facades-Cost-effective and Resilient Envelopes for Wood Constructions. 2014–2017. Available online: http://fungi.tallfacades.eu (accessed on 10 January 2017).
82. Haugbølle, K.; Almås, A.J. Nordic solutions for SUstainable REfurbishment (SURE)-Case studies. In *Sustainable Community-Building Smart-SB10*; VTT: Espoo, Finland, 2010.
83. Several, Sustainable refurbishment of exterior walls and building facades Final report, Part A–Methods and recommendations. In *VTT Technology*; VTT: Espoo, Finland, 2012.
84. Several, Sustainable refurbishment of exterior walls and building facades Final report, Part B–General refurbishment concepts. In *VTT Technology*; VTT: Espoo, Finland, 2012.
85. Several, Sustainable refurbishment of exterior walls and building facades Final report, Part C–Specific refurbishment concepts. In *VTT Technology*; VTT: Espoo, Finland, 2012.
86. Effesus. Publications & Dissemination. 2014. Available online: http://www.effesus.eu/dissemination/publications (accessed on 16 December 2015).
87. Gullbrekken, L.; Woods, R.; Kjølle, K.H. *Overhalla Housing: Affordable Retrofitting of Municipal Housing in Partnership with Secondary Schools*; NESS: Trondheim, Norway, 2015.
88. Woods, R.; Gullbrekken, L. User Participation and Upgrading Municipal Housing in Partnership with Secondary Schools. In Proceedings of the Nordic Ruralities 2016, Akyreri, Island, 22–24 May 2016.
89. Hauge, Å.L.; Almås, A.J.; Flyen, C.; Stoknes, P.E.; Lohne, J. User guides for the climate adaptation of buildings and infrastructure in Norway—Characteristics and impact. *Clim. Serv.* **2017**, *6*, 23–33. [CrossRef]
90. Stagrum, A.E.; Andenæs, E.; Kvande, T.; Lohne, J. Climate Change Adaptation Measures for Buildings—A Scoping Review. *Sustainabilirefty* **2020**, *12*, 1721. [CrossRef]
91. Hauge, Å.L.; Flyen, C.; Almås, A.J.; Ebeltoft, M. Klimatilpasning av bygninger og infrastruktur-samfunnsmessige barrierer og drivere. In *Klima 2050 Report*; SINTEF: Oslo, Norway, 2017; Available online: www.klima2050.no (accessed on 29 May 2020).
92. Statsbygg. Miljøstrategi 2015–2018. 2014. Available online: www.statsbygg.no (accessed on 6 October 2018).
93. Kvande, T.; Ime, B.T. Rammeverk for klimatilpassing. In *Byggeindustrien*; Byggeindustrien: Oslo, Norway, 2019; p. 46.

MDPI

St. Alban-Anlage 66

4052 Basel

Switzerland

Tel. +41 61 683 77 34

Fax +41 61 302 89 18

www.mdpi.com

Buildings Editorial Office

E-mail: buildings@mdpi.com

www.mdpi.com/journal/buildings